A Modern Approach to Dynamical Systems

A Modern Approach to Dynamical Systems

Edited by
Hallie McHugh

MURPHY & MOORE
www.murphy-moorepublishing.com

Published by Murphy & Moore Publishing,
1 Rockefeller Plaza,
New York City, NY 10020, USA

ISBN: 978-1-63987-003-5

Cataloging-in-Publication Data

A modern approach to dynamical systems / edited by Hallie McHugh.
p. cm.
Includes bibliographical references and index.
ISBN 978-1-63987-003-5
1. Dynamics. 2. Mathematics. 3. Chaotic behavior in systems. I. McHugh, Hallie.
QA845 .M63 2022
531.11--dc23

For information on all Murphy & Moore Publications
visit our website at www.murphy-moorepublishing.com

MURPHY & MOORE

Contents

Preface

A system in which a function describes the time dependence of a point in a geometrical space is referred to as a dynamical system. The mathematical models that describe the swinging of a clock pendulum, the number of fish present each springtime in a lake and the flow of water in a pipe are some of the examples of dynamical systems. A domain of mathematics that is used to describe the behavior of complex dynamical systems by using differential equations is referred to as dynamical systems theory. The focus of dynamical systems theory is the study of dynamical systems which has applications in a wide variety of fields such as mathematics, physics, chemistry, biology, engineering, economics, history and medicine. Dynamical systems are a crucial part of logistic map dynamics, chaos theory, bifurcation theory, the self-assembly and self-organization processes, and the edge of chaos concept. This book brings forth some of the most innovative concepts and elucidates the unexplored aspects of dynamical systems. Its aim is to present researches that have transformed this discipline and aided its advancement. This book will serve as a reference to a broad spectrum of readers.

The researches compiled throughout the book are authentic and of high quality, combining several disciplines and from very diverse regions from around the world. Drawing on the contributions of many researchers from diverse countries, the book's objective is to provide the readers with the latest achievements in the area of research. This book will surely be a source of knowledge to all interested and researching the field.

In the end, I would like to express my deep sense of gratitude to all the authors for meeting the set deadlines in completing and submitting their research chapters. I would also like to thank the publisher for the support offered to us throughout the course of the book. Finally, I extend my sincere thanks to my family for being a constant source of inspiration and encouragement.

Editor

1

Predictability in Deterministic Dynamical Systems with Application to Weather Forecasting and Climate Modelling

Sergei Soldatenko and Rafael Yusupov

Abstract

Climate system consisting of the atmosphere, ocean, cryosphere, land and biota is considered as a complex adaptive dynamical system along with its essential physical properties. Since climate system is a nonlinear dissipative dynamical system that possesses a global attractor and its dynamics on the attractor are chaotic, the prediction of weather and climate change has a finite time horizon. There are two kinds of predictability of climate system: one is generated by uncertainties in the initial conditions (predictability of the first kind) and another is produced by uncertainties in parameters that describe the external forcing (predictability of the second kind). Using the concept of the 'perfect' climate model, two kinds of predictability are considered from the standpoint of the mathematical theory of climate.

Keywords: climate system, deterministic chaos, predictability, stability

1. Introduction

High-complexity computational models that simulate earth's climate system (ECS) have earned well-deserved recognition as the indispensable and primary instrument for numerical weather prediction (NWP) as well as for the study of climate change and variability caused by both natural processes and human activities [1–4]. In spite of dramatic progress achieved over the past few decades in weather forecasting and climate simulation thanks to the advances in computing hardware and algorithms and to a substantial increase in the volume of climatological data, contemporary computational climate models can reconstruct the real world only with a certain degree of validity [3]. There are several major sources of discrepancy between climate model simulation results and reality. First of all, climate models remain an ideal mathematical abstraction of a real physical system, namely the ECS. These models ignore some physical, dynamical and chemical processes or, at least, represent them in a simplified fashion. As a result, various physical simplifications in the formulation of climate models substantially influence their adequacy [5]. Second, the NWP and climate simulation are mathematically an initial-value (Cauchy) and/or a boundary-value (Dirichlet or von Neumann) problem, which is solved numerically using finite-difference, spectral or another appropriate method. Consequently, uncertainties emerging in the initial and boundary conditions as well as in the climate model parameters and external forcing, approximation, truncation and round-off errors lead to distinctions between the model output and the observed real state of the ECS. Third, let us suppose that we have the 'perfect' model of the ECS. It means that exact governing equations are known exactly and can be solved. However, even in this, hypothetically ideal, case the ability of climate models to predict the future remains limited. This can be explained by the fact that the atmosphere, which is the most rapidly changing component of the ECS, is strongly nonlinear and exhibits irregular (chaotic) spatial-temporal oscillations on all scales ranging from millimetre seconds (turbulent fluctuations) to thousands of kilometres and several years (climate variability). This phenomenon known as deterministic chaos was first discovered by Lorenz [6]. The chaotic nature of the atmosphere significantly limits our ability to successfully predict the weather and climate since the predicted trajectory of the ECS is unstable with respect to both the infinitesimal errors in initial conditions and external forcing [7]. Even with a perfect atmospheric model and accurate initial condition, we cannot predict the weather beyond approximately two weeks.

For further discussion, we need to clarify that terms 'weather' and 'climate' have different meanings. Weather is defined as the daily conditions of the atmosphere in terms of such atmospheric variables as temperature, humidity, wind direction and velocity, surface pressure, cloud cover and precipitation. In turn, the climate represents an ensemble of states traversed by climate system over a sufficiently long temporal interval (about 30 years, according to the World Meteorological Organization). Here, the ensemble includes not only a set of system states but also the probability measure defined on this set. Therefore, climate, roughly speaking, can be considered as the 'average' weather, in terms of mean and variance, in a certain geographical location over many years.

Time horizon of a forecast's usefulness and validity can be characterized by the specific measure known as predictability. Predictability is commonly understood as the degree to which it is possible to make an accurate qualitative or quantitative forecast of the future system's state. The study of atmospheric predictability was initiated by Thompson [8] and Lorenz [6, 9] more than 50 years ago and was extensively explored theoretically using various numerical and statistical models since then (e.g. [10–17]). One of the obvious measures of predictability that can be used to verify a weather forecast is the mean-squared error (the average of the squared differences between forecasts and observations). This measure increases over time and asymptotically approaches some finite value known as the saturation value. Therefore, predictability is lost when the forecast errors become comparable to the saturation value in magnitude. If this happens, the forecast result is not better than any randomly selected trajectory of the system. However, for a number of reasons, mean-squared error and other weather forecast verification metrics (e.g. mean absolute error and mean error) are rarely used to estimate the climate system predictability in practice.

Predictability characterizes both the physical system itself and the model of this system that is used to make a forecast. However, in atmospheric and climate studies we are interested in the predictability of real dynamical processes rather than the predictability of the model used in simulations.

According to Lorenz [19], in weather and climate modelling we are facing the predictability of two kinds reflecting the internal and external variability of the climate system, respectively. The predictability of the first kind relates to the Cauchy (initial value) problem, namely the prediction of sequential states of the ECS for constant values of external parameters and given variations in the initial conditions. In contrast, the predictability of the second kind refers to a boundary-value problem, specifically to the prediction of response of the climate system in asymptotical equilibrium to perturbations in external parameters (forcing).

This chapter considers both the predictability of atmospheric and climate processes with respect to the initial data errors (predictability of the first kind) as well as the predictability with respect to external perturbations (predictability of the second kind). The stability of dynamical system is also discussed since stability is a key problem related to predictability in dynamical systems.

2. Climate system as a complex adaptive dynamical system

Let us begin with some preliminary notes and definitions which will be used in this chapter.

The term 'system' generally refers to a goal-oriented set of interconnected and interdependent elements that operate together to achieve some objectives [20]. The system is called complex if it possesses such characteristics as emergent behaviour, nonlinearity and high sensitivity to initial conditions and/or to perturbations, self-organization, chaotic behaviour, feedback loop, spontaneous order, robustness and hierarchical structure. Complexity in systems arises from nonlinear spatio-temporal interactions between their components. These nonlinear interactions lead to the appearance of new dynamical properties (for example, synchronous oscillations and other structural changes) that cannot be observed by exploring constituent elements individually.

Complex systems include a special class of systems that have the capacity to adapt to system's environment. These systems are known as complex adaptive systems. In a complex adaptive system, parts are linked together in such a way that the entire system as a whole has the capacity to transform fundamentally the interrelations and interdependences between its components, the collective behaviour of a system and also the behaviour of individual components due to the external forcing. Complex adaptive systems are dynamical systems since they evolve and change over time. These systems have a number of properties that include the following [21, 22]: co-evolution, connectivity, sub-optimality, requisite variety and iteration, edge of chaos and, certainly, emergence and self-organization.

The ECS (S) is understood as a complex, large-scale physical system that consists of the following five basic and interacting constituent subsystems [23]:

1. Atmosphere (A), the gaseous and aerosol envelope of the earth that propagates from the land, water bodies and ice-covered surface outwards to space.
2. Hydrosphere (H), the ocean and other water bodies on the surface of our planet, and water that is underground and in the atmosphere.
3. Cryosphere (C), the sea ice, freshwater ice, snow cover, glaciers, ice caps and ice sheets and permafrost.
4. Lithosphere (L), the solid, external part of the earth.
5. Biosphere (B), the part of our planet where life exists, i.e.

$$S = A \cup \mathrm{H} \cup \mathrm{C} \cup \mathrm{L} \cup \mathrm{B}$$

The ECS components are characterized by a finite set of distributed variables whose values at a given time determine their state. The most unstable and rapidly oscillating component of the ECS is the atmosphere.

The ECS is a large-scale and unique physical system that possesses a number of specific properties (e.g. [24–29]) making the exploration of this system a high complexity problem. In contradistinction to many problems in physics, the study of the climate system, its change and variability cannot be implemented by a direct physical experiment due to climate system's essential features as a large-scale physical system. Laboratory experiments and analytical approaches have a very limited applicability to climate exploration by virtue of extreme complexity of the ECS. As a result, in climate studies the computational simulation represents the primary instrument and as such requires the development of appropriate mathematical models and numerical algorithms.

The utilization of mathematical models in climate research involves the development of a specific mathematical theory that allows one to explore the climate system along with its mathematical models. The contemporary mathematical theory of climate is based on methods of the qualitative theory of differential equations that enables us to explore the behaviour of climate system in its phase space [30]. In other words, the dynamical system theory is currently the theoretical foundation of mathematical climate theory. In this context, the ECS can be viewed as a complex adaptive dynamical system [21, 22].

The ECS belongs to the class of complex adaptive systems due to the following factors:

1. The ECS is a complex large-scale physical system combining the atmosphere, hydrosphere, cryosphere, land and biota together with global biochemical cycles (such as cycles of CO_2, N_2O and CH_4) and aerosols. Components of the climate system are heterogeneous thermo-dynamical subsystems characterized by specific variables that determine their states. Elements of the ECS have strong differences in their structure, dynamics, physics and chemistry. They cover processes with different temporal and spatial scales, and link together via numerous physical coupling mechanisms, which can be either weak or strong. Each subsystem of the ECS can in turn be viewed as being composed of subsystems, which are themselves composed of subsystems. For example, the atmosphere can be divided into several layers based on its vertical temperature distribution. These

layers are respectively the troposphere, stratosphere, mesosphere and thermosphere. The atmosphere can also be divided into surface layer, boundary layer and free atmosphere based on the influence of surface friction.

2. Each component of the ECS is characterized by a specific response time. This fact is very important to building the ECS' models. The relation of a certain component to some ECS' model is determined by the ratio between the temporal scale of processes under consideration and its response time. For example, the atmosphere, which has a response time of about one month in the troposphere, can be considered a sole component of the ECS' model for processes with temporal scales of days to weeks. In this case, oceans, land surface and ice cover are considered as the boundary conditions and/or external forcing. If we study processes which have temporal scales of months to years, the atmosphere and ocean must be included in the ECS' model together with sea ice. Thus, computational models of the ECS are built up from hierarchy of models, forming finally a complex integrated model.

3. The ECS has a large number of positive and negative feedback mechanisms which control the behaviour of the ECS. Some examples of these mechanisms are ice-albedo feedback (positive feedback), water vapour feedback (positive feedback), cloud feedback (both positive and negative feedbacks), carbon cycle feedback (negative feedback), feedback due to Arctic methane release (positive feedback) and many others.

4. Physical and dynamical processes in the ECS cover a broad spectrum of temporal and spatial scales. Time scales are varied from seconds to decades, and spatial spectrum of dynamical processes covers molecular to planetary scales. Dynamical processes in the ECS and its components are nonlinear. Subsystems of the ECS interact with one another nonlinearly producing, under certain conditions, a chaotic behaviour of subsystems and the overall climate system.

5. The ECS and its components inherently have emergent properties. Examples of atmospheric emergent phenomena include but are not limited to clouds, large-scale eddies (cyclones and anticyclones) and small-scale vortices such as tornados. Examples of climate emergent phenomena are the El Niño–Southern Oscillation, which is a quasi-periodical irregular variation in the ocean surface temperature over the Pacific in tropics that strongly influences global climate, ocean circulation patterns and glacial-interglacial cycles. Natural emergent phenomena appear spontaneously under certain favourable conditions.

6. The ECS is a thermodynamically open and non-isolated system because it exchanges energy with its surroundings. However, the ECS is a closed system with respect to the exchange of matter with its surroundings. The energy that drives the ECS is mainly solar energy. The ECS is affected by changes in external driving forces, which imply natural causes such as solar activity variations and volcanic activities, as well as man-made changes in chemical composition of the atmosphere. However, the impact of the ECS on the outer space is insignificant. Currently, changes in climate are mostly affected by variations in the atmospheric composition of particles and gases. In the Arctic, the role of

changes in albedo (reflection coefficient) is also tangible. The most influential gas component to affect the climate is CO_2, which comprises about 70% points of the global warming potential.

7. The components of the ECS are also non-isolated systems. They act as cascading systems and interact with each other in various ways including through the transfer of momentum, sensible and latent heat, gases and particles. All together they compose the climate system, which is a unique large-scale natural system.

8. Dynamical processes in the ECS fluctuate due to both internal factors (natural oscillations) and external forcing (forced oscillations). Natural fluctuations are caused by internal instability (for example, hydrodynamic instability such as barotropic and baroclinic) with respect to stochastic perturbations. Human impacts, both intentional and unintentional, belong to the category of external forcing.

Undoubtedly, there are other specific properties of the ECS that should be taken into account while studying climate as a complex adaptive system and building models of the ECS.

To simulate the ECS, we should assign some mathematical object that is an abstract representation of the real climate system taking into account its essential features mentioned above. This object is known as a perfect model of the ECS. It is usually assumed that a perfect model is deterministic semi-dynamical system that is dissipative, ergodic and possesses a global attractor. It is also assumed that any trajectory generated by the model is unstable [30].

Formally, an abstract climate system model represents a set of multi-dimensional nonlinear differential equations in partial derivatives, which generates finite dimensional deterministic semi-dynamical system of the form [24, 30]

$$dx/dt = F(x,p,f), \quad x \in \mathbf{R}^n, x|_{t=0} = x_0, t \geq 0, \tag{1}$$

where x is the state vector, the components of which characterize the state of a system at a given time t, x_0 is a given initial state of a system, n is the dimension of dynamical system, $p \in \mathbf{R}^p$ is the vector of model parameters and f is the external forcing. The solution to climate model equations (1) cannot be found analytically and one needs to employ available numerical methods. For that reason, in order to obtain numerical solution, the original set of partial differential equations is replaced with discrete spatio-temporal approximations using any appropriate technique (e.g. finite-difference method, Galerkin approach, etc.). Thus, in weather and climate simulation we mainly deal with discrete dynamical systems.

Suppose that the set of n real variables $x_1, x_2, \ldots, x_n$ defines the current state of discrete-time dynamical system representing the ECS. A certain particular state $x = (x_1, x_2, \ldots, x_n)$ corresponds to a point in an n-dimensional space $Q \subseteq \mathbf{R}^n$, known as the system phase space. Let $t_m \in \mathbf{Z}_+$ $(m = 0, 1, 2, \ldots)$ be the discrete time, and let $f = (f_1, f_2, \ldots, f_n)$ be a smooth vector-valued function defined in the domain $Q \subseteq \mathbf{R}^n$. This function describes the evolution of the system state from one moment to another. Then, a deterministic discrete-time semi-dynamical system that approximates the continuous time dynamical system (1) can be specified by the following equation:

$$x(t_{m+1}) = f\left(x(t_m)\right), \quad x(t_0) = x_0, \quad m = 0, 1, 2, \ldots, . \tag{2}$$

It is obvious that a family of operators forms a semi-group:

$$f_{s+p} = f_s \circ f_p, \quad f_0 = I, \forall s, p \in \mathbf{Z}_+, \tag{3}$$

where I is the identity operator. Therefore, the system state $x(t_m)$ at time t_m can be explicitly expressed via the initial condition x_0:

$$x(t_m) = f^m(x_0), \tag{4}$$

where f^m denotes an m-folding application of f to x_0. The sequence $\{x(t_m)\}_{m=0}^{\infty}$ is a trajectory of system (2) in its phase space, which is uniquely defined by the initial values of state variables x_0.

For reference, let us reproduce a couple of definitions [30].

Definition 1. The solution $x(t)$ to system (1) is Lyapunov stable if $\forall \varepsilon > 0, \ \exists \delta(\varepsilon) > 0$ such that

$$\| x_0 - x_0^* \| < \delta(\varepsilon) \Rightarrow \| x(t) - x^*(t) \| < \varepsilon, \forall t \geq 0, \tag{5}$$

where $x^*(t)$ is the solution to the system

$$dx^*/dt = F(x^*, p, f), \quad x^*|_{t=0} = x_0^*. \tag{6}$$

Definition 2. The solution $x(t)$ to system (1) is stable with respect to the continuous perturbation δF if $\forall \varepsilon > 0, \ \exists \delta(\varepsilon) > 0$ such that

$$\| \delta F \| < \delta(\varepsilon) \Rightarrow \| x(t) - x^*(t) \| < \varepsilon, \forall t \geq 0, \tag{7}$$

where $x(t)$ is the solution to the following perturbed equation:

$$dx^*/dt = F(x^*, p, f) + \delta F, x^*|_{t=0} = x_0^*. \tag{8}$$

These definitions are important when considering both kinds of predictability.

The key point for further consideration is the assumption that climate system model described by the set of nonlinear partial differential equations (1) is 'perfect'. We suppose that system (1) is nonlinear dissipative semi-dynamical system ($t \geq 0$) that has an absorbing set in the phase space and its solution exists and is unique for any $t \geq 0$. Next, we assume that the system (1) possesses a global attractor of finite dimension that is a certain set in the system's phase space towards which a system tends to evolve for a wide variety of initial conditions of the system. Global attractor is characterized by the attraction property and invariance [30]. So, the dynamics of system (1) can be formally divided into to two phases: (1) movement towards the attractor and (2) motion on the attractor. When studying the climate system stability and predictability we assume that the system trajectory is on the attractor and its dynamics are chaotic. We also assume that system (1) possesses the property of ergodicity. Thus, statistical characteristics of the climate system (e.g. the first $\overline{x} = \langle x \rangle$ and second $var(x) = \langle x^2 \rangle - \overline{x}^2$ moments) can be calculated by averaging along a certain trajectory.

Structurally, any climate system model represents a set of interacting and interdependent models of lower level (i.e. atmospheric model, model of the ocean, etc.). The number of these lower level models is determined by the objectives of a problem under consideration. For example, to study the large-scale climate variability the model can include the following major components: tropical, mid-latitude and polar troposphere, stratosphere, ocean, land ice, ocean and sea ice, surface and boundary layers, hydrological cycle, clouds (e.g. convective and stratiform), precipitation, aerosols, CO_2 and CH_4 cycles, solar radiation, terrestrial emission, etc. Other subsystems of the ECS (e.g. vegetation, land surface and biota) can be considered as the boundary conditions and external forcing. In numerical weather prediction problem, some atmospheric model (either global, regional or local) is the main component of the forecasting system, while ocean, sea ice, land surface are used only to impose boundary conditions. Note that models of general circulation of the atmosphere and the ocean represent main computational instruments for simulating the ECS.

3. Climate model governing equations

The main energy source of the ECS is the Sun. Spatial inhomogeneity and temporal changes of the heat energy that the earth's surface receives from the Sun are the main cause of motions in the atmosphere and ocean. Equations that govern the atmospheric and oceanic circulation represent the mathematical expressions of fundamental laws of physics: conservation of momentum, conservation of mass, conservation of water and conservation of energy (the first law of thermodynamics). Some diagnostic relationships between variables are also used (i.e. the equation of state). Almost every model uses a slightly different set of equations tailored to a specific problem. However, all climate models include the following basic equations: two equations for horizontal motions (or equation for the vorticity and divergence), equation for the vertical velocity (or hydrostatic equation), continuity equation, as well as thermodynamic and moisture equations. Equations of motion are derived from the law of conservation of momentum applicable to a rotating system. These equations describe all types and scales of atmospheric motions that are important for the formation of weather and climate (i.e. large-scale Rossby waves, planetary waves and gravity waves). Conservation of mass is mathematically expressed in the form of continuity equation, equation for conservation of moisture and equations for conservation of other substances taken into account in a particular climate model.

The set of equations that describes the general circulation of the atmosphere can be written in the spherical co-ordinate system (λ, φ) defined by longitude λ and latitude φ, with normalized pressure as a vertical coordinate $\sigma = p/p_s$, where p is pressure and p_s is the surface pressure [1, 31]. The set of the model equations includes *two momentum equations*:

$$\frac{\partial u}{\partial t} = \eta v - \frac{1}{a\cos\varphi}\frac{\partial}{\partial\lambda}(\Phi + K) - \frac{RT_v}{a\cos\varphi}\frac{\partial \ln p_s}{\partial\lambda} - \dot{\sigma}\frac{\partial u}{\partial\sigma} = F_{uV} + F_{uH}, \tag{9}$$

$$\frac{\partial v}{\partial t} = -\eta u - \frac{1}{a}\frac{\partial}{\partial\varphi}(\Phi + K) - \frac{RT_v}{a}\frac{\partial \ln p_s}{\partial\varphi} - \dot{\sigma}\frac{\partial v}{\partial\sigma} = F_{vV} + F_{vH}, \tag{10}$$

where u and v are zonal and meridional velocities, a is the earth's average radius, $\sigma = d\sigma/dt$ is the vertical velocity in the σ co-ordinate system, Φ is geopotential, T is temperature, R is the gas constant for dry air, $K = (u^2 + v^2)/2$ is the kinetic energy, $\eta = \varsigma + f$ is the absolute vorticity, f is the Coriolis parameter and ς is the relative vorticity that is given by

$$\varsigma = \frac{1}{a\cos\varphi}\left[\frac{\partial v}{\partial\lambda} - \frac{\partial}{\partial\varphi}(u\cos\varphi)\right]. \tag{11}$$

The virtual temperature T_v is represented as

$$T_v = T\left[1 + \left(\frac{R_v}{R} - 1\right)q\right], \tag{12}$$

where T is the temperature, q is the specific humidity and R_v is the gas constant for water vapour. The terms F_{uV} and F_{vV} describe the vertical friction and terms F_{uH} and F_{vH} the horizontal diffusion. Generally, however, the momentum equations are transformed into the equations for the absolute vorticity η and the divergence D using new independent variable $\mu = \sin\varphi$:

$$\frac{\partial\eta}{\partial t} = \frac{1}{a(1-\mu^2)}\frac{\partial}{\partial\lambda}(N_v + \cos\varphi F_{vV}) - \frac{1}{a}\frac{\partial}{\partial\mu}(N_u + \cos\varphi F_{uV}) + F_{\eta H}, \tag{13}$$

$$\begin{aligned}\frac{\partial D}{\partial t} = {} & \frac{1}{a(1-\mu^2)}\frac{\partial}{\partial\lambda}(N_u + \cos\varphi F_{uV}) + \frac{1}{a}\frac{\partial}{\nabla\mu}(N_v + \cos\varphi F_{vV}) + F_{DH} \\ & -\nabla^2(\Phi + K + RT_0 \ln p_s),\end{aligned} \tag{14}$$

where the horizontal divergence is given by

$$D = \frac{1}{a\cos\varphi}\left[\frac{\partial u}{\partial\lambda} + \frac{\partial}{\partial\varphi}(v\cos\varphi)\right]. \tag{15}$$

The spherical horizontal Laplacian can be written as

$$\nabla^2 = \frac{1}{a^2(1-\mu^2)}\frac{\partial^2}{\partial\lambda^2} + \frac{1}{a^2}\frac{\partial}{\partial\mu}\left[(1-\mu^2)\frac{\partial}{\partial\mu}\right]. \tag{16}$$

To provide the computational effectiveness of numerical integration scheme, the virtual temperature is partitioned into two parts, one of which T_0 is a function of the vertical coordinate only, i.e. $T_v(\lambda, \mu, \sigma, t) = T_0(\sigma) + T_v'(\lambda, \mu, \sigma, t)$. Then, the nonlinear dynamical terms N_u and N_v can be represented in the following form:

$$N_u = \eta V - RT_v' \frac{1}{a}\frac{\partial \ln p_s}{\partial \lambda} - \dot{\sigma}\frac{\partial U}{\partial \sigma}, \tag{17}$$

$$N_v = -\eta U - RT_v' \frac{(1-\mu^2)}{a}\frac{\partial \ln p_s}{\partial \mu} - \dot{\sigma}\frac{\partial V}{\partial \sigma}, \tag{18}$$

where $U = u\cos\varphi$ and $V = v\cos\varphi$.

The thermodynamic equation, which represents the mathematical expression of the first law of thermodynamic, is written for a perturbation in temperature T' calculated with respect to the mean $T_0(\sigma)$ mentioned above:

$$\begin{aligned}\frac{\partial T'}{\partial t} = -\frac{1}{a(1-\mu^2)}\frac{\partial}{\partial \lambda}(UT') - \frac{1}{a}\frac{\partial}{\partial \mu}(VT') + T'D - \dot{\sigma}\frac{\partial T'}{\partial \sigma} + \frac{RT_v}{c_p^*}\frac{\omega}{p} \\ +Q + F_{TV} + F_{TH} - \frac{1}{c_p^*}[u(F_{uV} + F_{uH}) + v(F_{vV} + F_{vH})],\end{aligned} \tag{19}$$

where Q is the diabatic heating rate, ω is the pressure vertical velocity and c_p^* is given by

$$c_p^* = c_p\left[1 + \left(\frac{c_v}{c_p} - 1\right)\right]. \tag{20}$$

Here, c_p is the specific heat of dry air at a constant pressure and c_v is the specific heat of water vapour at a constant pressure.

The equation for specific humidity is used to describe the hydrologic cycle in the atmosphere:

$$\frac{\partial q}{\partial t} = -\frac{1}{a(1-\mu^2)}\frac{\partial}{\partial \lambda}(Uq) - \frac{1}{a}\frac{\partial}{\partial \mu}(Vq) + qD - \dot{\sigma}\frac{\partial q}{\partial \sigma} + S + F_{qV} + F_{qH}, \tag{21}$$

where the term S describes the source/sink processes for water vapour, and F_{qV} and F_{qH} are the vertical and horizontal water vapour diffusion.

Let us consider now the continuity equation that represents the conservation of mass law:

$$\frac{\partial \ln p_s}{\partial t} = -\frac{U}{a(1-\mu^2)}\frac{\partial \ln p_s}{\partial \lambda} - \frac{V}{a}\frac{\partial \ln p_s}{\partial \mu} - D - \frac{\partial \dot{\sigma}}{\partial \sigma}. \tag{22}$$

Integrating this equation from the top ($\sigma = 0$) to the bottom ($\sigma = 1$), with the vertical boundary conditions $\dot{\sigma} = 0$ at $\sigma = 1$ and $\sigma = 0$, one can obtain *the equation for surface pressure* prediction:

$$\frac{\partial \ln p_s}{\partial t} = \int_0^1 \left[D + \frac{U}{a(1-\mu^2)}\frac{\partial \ln p_s}{\partial \lambda} + \frac{V}{a}\frac{\partial \ln p_s}{\partial \mu}\right] d\sigma. \tag{23}$$

Combining the continuity equation and the equation for the surface pressure, one can derive the diagnostic equation for the vertical velocity $\dot{\sigma}$:

$$\dot{\sigma} = \sigma \int_0^1 \left[D + \frac{U}{a(1-\mu^2)} \frac{\partial \ln p_s}{\partial \lambda} + \frac{V}{a} \frac{\partial \ln p_s}{\partial \mu} \right] d\sigma - \int_0^{\sigma} \left[D + \frac{U}{a(1-\mu^2)} \frac{\partial \ln p_s}{\partial \lambda} + \frac{V}{a} \frac{\partial \ln p_s}{\partial \mu} \right] d\sigma. \quad (24)$$

Two diagnostic equations, the hydrostatic equation and the equation of state, are also components of a set of equations that are used to simulate the atmospheric general circulation. *The hydrostatic equation* is

$$\partial \Phi / \partial \ln \sigma = -RT_v. \quad (25)$$

In the integral form, this equation can be written as

$$\Phi = \Phi_s - \int_1^{\sigma} RT_v d\ln \sigma, \quad (26)$$

where Φ_s is the geopotential at the earth's surface. *The equation of state* is written as

$$p = \rho RT_v, \quad (27)$$

where ρ is the air density.

Boundary conditions in the longitudinal direction are periodic, and the solution to the atmospheric model equations is bounded at the north and south poles. Vertical boundary conditions represent the vanishing of vertical velocity both at the bottom and at the top of the atmosphere: $\dot{\sigma} = 0$ at $\sigma = 1$ and $\sigma = 0$.

Equations used in the ocean model are written in the Boussinesq hydrostatic approximation with a rigid lid in the spherical coordinate system, with depth z as a vertical coordinate defined as negative downwards from $z = 0$, which denotes the ocean surface [1, 31]. The set of model equations include the following:

1. The horizontal equations of motion:

$$\frac{\partial u}{\partial t} + L(u) - \left(f + \frac{u}{a} \tan \varphi \right) v + \frac{1}{a \rho_o \cos \varphi} \frac{\partial p}{\partial \lambda} = k_V \frac{\partial^2 u}{\partial z^2} + F_u, \quad (28)$$

$$\frac{\partial v}{\partial t} + L(v) + \left(f + \frac{u}{a} \tan \varphi \right) u + \frac{1}{a \rho_o} \frac{\partial p}{\partial \varphi} = k_V \frac{\partial^2 v}{\partial z^2} + F_v, \quad (29)$$

where k_V is the vertical eddy viscosity coefficient, ρ_0 is the density of sea water and the advection operator, $L(\alpha)$, is given by

$$L(\alpha) = \frac{1}{u \cos \varphi} \left(\frac{\partial u \alpha}{\partial \lambda} + \frac{\partial v \alpha \cos \varphi}{\partial \varphi} \right) + \frac{\partial w \alpha}{\partial z}. \quad (30)$$

The horizontal viscosity terms, F_u and F_v, are defined as

$$F_u = k_H \left[\nabla^2 u + \frac{(1-\tan^2\varphi)u}{a^2} - \frac{2\sin\varphi}{a^2\cos^2\varphi} \frac{\partial v}{\partial \lambda} \right], \quad (31)$$

$$F_v = k_H \left[\nabla^2 v + \frac{(1-\tan^2\varphi)v}{a^2} + \frac{2\sin\varphi}{a^2\cos^2\varphi} \frac{\partial u}{\partial \lambda} \right], \quad (32)$$

where k_H is the horizontal eddy viscosity coefficient. The given form of the diffusion terms, F_u and F_v, is required for conserving angular momentum property.

2. The hydrostatic equation:

$$\partial p / \partial z = -g\rho. \quad (33)$$

3. The thermodynamic equation:

$$\frac{\partial T}{\partial t} + L(T) = \kappa_V \frac{\partial^2 T}{\partial z^2} + \kappa_H \nabla^2 T, \quad (34)$$

where κ_V and κ_H are, respectively, the vertical and horizontal eddy diffusivity coefficients.

4. The equation for the mass continuity of salinity:

$$\frac{\partial S}{\partial t} + L(S) = \kappa_V \frac{\partial^2 S}{\partial z^2} + \kappa_H \nabla^2 S. \quad (35)$$

5. The equation of continuity:

$$\frac{\partial w}{\partial z} = -\frac{1}{a\cos\varphi} \frac{\partial u}{\partial \lambda} - \frac{1}{a\cos\varphi} \frac{\partial v \cos\varphi}{\partial \varphi}. \quad (36)$$

6. The equation of state:

$$\rho = \rho(T, S.p). \quad (37)$$

Due to their extreme complexity, weather and climate models can be implemented on computers only using numerical techniques. Since models are based on partial differential equations, it is necessary, first, to ensure that the problem under consideration is well posed, i.e. it has a unique solution that depends on the boundary and initial conditions. Thus, both the initial and boundary conditions must be properly specified. Next, weather and climate mathematical models should be transformed into numerical models that can be implemented on computers. The most widely used technique for solving differential equations of weather and climate models is the finite-difference method according to which the derivatives in the partial differential equations are approximated on a certain temporal-spatial grid. Thus, instead of continuous functions, which describe the state of climate system and its components, we deal with discrete functions defined only for specific times separated by the time step Δt and specific space locations separated by spatial (horizontal Δs and vertical Δh) steps. As a result, instead of partial differential equation we obtain finite-difference equations (numerical model). It is very important that numerical schemes used for the discretization of model differential equations must satisfy several fundamental requirements: finite-difference equations must be consistent with model differential equations, the solution of finite-difference equations must converge to the solution of differential equations and numerical schemes must be computationally stable. In practice, finite difference is not the only method used to solve weather and climate problems. The most popular among other methods are the family of Galerkin techniques, spectral, finite-volume and finite element approaches.

In contemporary climate models, due to their discrete spatial and temporal structure, a large number of physical processes and cycles cannot be clearly represented and formulated by model equations. Climate models are theoretically incapable of simulating processes on spatial scales of the order of magnitude that is twice the model grid length [32]. Such thermo-dynamical, physical and chemical processes and cycles are parameterized, i. e. expressed parametrically using simplified description. Study of the climate system by computer simulation requires extensive computational resources. As a result, the predictability problem is usually studied either on the basis of low-order models, which possess the main properties of the climate system (nonlinearity, chaos, dissipative, etc.), or on the basis of complex climate models using the ensemble approach or the Monte Carlo method.

4. Predictability of climate system

4.1. Predictability of the first kind

The first kind predictability of climate processes (predictability of climate processes with respect to the initial conditions) will be considered under the assumption that the climate system (1) evolves on its attractor. Since system (1) is a nonlinear dissipative dynamical system, its attractor, known as a strange attractor, has an extremely complex fractal structure and can be characterized by such parameters as dimension, characteristic Lyapunov exponents, invariant measure and asymptotically steady solution and others. If some trajectory of system (1) is enclosed in a bounded phase volume (attractor), then the system's dynamics demonstrate deterministic chaos: the behaviour of simulated system resembles a stochastic process despite the fact that the system is described by deterministic laws and its evolution is governed by deterministic differential equations. So, all orbits of a system that start close enough will diverge from one another, however, will never depart from the attractor. The rate of separation of infinitesimally close orbits is characterized by positive Lyapunov exponents. The number of directions along which the orbit is unstable is equal to the number of positive Lyapunov exponents n_λ (note that $n_\lambda < n$, where n is a system's dimension). Thus, trajectories of climate dynamical systems are Lyapunov unstable.

To consider the initial growth rates of errors in the initial conditions let us linearize Eq. (1) around some trajectory to obtain the equation in variations:

$$dx^{'}/dt = M_t x_0^{'}, \tag{38}$$

where $M_t = \partial F / \partial x$ is the tangent propagator along the trajectory between the initial state x_0' and the forecast state x' at a certain time t (actually M_t is a Jacobian matrix). Obviously, one can obtain

$$\| x^{'}(t) \|^2 = (M_t x_0^{'}, M_t x_0^{'}) = (M_t^* M_t x_0^{'}, x_0^{'}), \tag{39}$$

where $(\cdot,\cdot)$ is the inner product in $\mathbf{R}^n$ and M^* is the transpose of M. Since the operator $M_t^* M_t$ is self-adjoint, then for any t one can consider the following eigenvalue problem:

$$M_t^* M_t \psi_i = \sigma_i \psi_i, \tag{40}$$

where σ_i is the ith eigenvalue of the matrix $M_t^* M_t$ and ψ_i is the corresponding eigenvector. Representing x_0' in the form of series as $x_0' = \sum_i \alpha_i \psi_i$, one can get $\| x^{'}(t) \|^2 = \sum_i \sigma_i {\alpha_i}^2$. So, the forecast error on temporal interval $[0, t]$ depends on errors in the initial distribution of eigenvectors ψ_i and singular values of the tangent linear propagator M_t. Since system (1) is ergodic, we can also calculate the Lyapunov exponents λ_i in accordance with the multiplicative theorem [33]:

$$\lambda_i = \lim_{t \to \infty} \frac{1}{t} \ln \sigma_i (M_t^* M_t), \quad i = 1, \ldots, n. \tag{41}$$

The Lyapunov exponents define the exponential growth (decay) of linear independent components of x' at $x^{'} \to 0$. The knowledge of the Lyapunov exponent spectrum of a dynamical system allows one to estimate the attractor fractal dimension, the rate of Kolmogorov-Sinai entropy production and the characteristic e-folding time. Knowledge of these parameters is very important for the stability and predictability analysis of dynamical systems. The fractal dimension of attractors of dissipative dynamical systems can be determined by applying the Kaplan-Yorke conjecture [34]:

$$D_{KY} = J + \sum_{i=1}^{J} \lambda_i / |\lambda_{i+1}|, \tag{42}$$

where J is the maximum integer such that the sum of the J largest exponents is still non-negative, i.e. $\sum_{i=1}^{J} \lambda_i > 0$. The sum of all positive Lyapunov exponents, according to theorem [35], gives an estimate of the Kolmogorov-Sinai entropy, i.e. the value showing mean divergence of the trajectories on attractors. The arrangement of the Lyapunov exponents in (42) is as follows: $\lambda_1 \geq \lambda_2 \geq \ldots \geq \lambda_{n_d}$. The multiplicative inverse (reciprocal) of the largest Lyapunov exponent is referred to as the characteristic e-folding time.

Let δ_0 be the initial perturbation of x' used to integrate equation (8). Since the system is Lyapunov unstable, after some sufficiently large temporal interval of integration the distance between two hyper-points in the phase space reaches the value of δ_t. Let $\overline{\delta}_t$ be the accepted error tolerance, then the predictability time of a system can be roughly estimated as

$$T_p \approx \lambda_{\max}^{-1} \ln(\overline{\delta}_t/\delta_0), \tag{43}$$

where $\lambda_{\max}$ is the leading Lyapunov exponent. The error doubling time can be calculated as $t = \ln 2/\lambda_{\max}$. However, Lyapunov exponents are very useful instrument to estimate the predictability of low-order dynamical systems [36].

Climate data observations are subject to measurement errors. The simplest way to represent the resulting uncertainty is to define the probability density function (PDF) $\rho(x, t_0)$ or, generally, the set of a finite measure μ_0 on which the initial state x_0 is concentrated. The time evolution of a system leads to a divergence and mixing of points of this set. Since the initial state x_0 is concentrated on a set having the measure μ_0, then after some period of time the measure will become μ_t. Let $\overline{\mu}$ be the invariant ergodic measure. Suppose the convergence theorem $\mu \rightarrow \overline{\mu}$ does exist. Hence, at a certain time $t \rightarrow t_\varepsilon$ the measure μ_t falls into the ε-neighbourhood of $\overline{\mu}$. Consequently, the initial data information characterized by μ_0 will be completely lost. So, one can say that the time t_ε defines the potential predictability of a system under consideration [16]. Thus, a focal point of the predictability problem is to prove the existence of ergodic measure and the existence of convergence theorem. This problem, however, is extremely difficult to solve because the structure of the invariant measure generated on the system attractor is sophisticated and non-smooth. To avoid this problem, the stochastic regularization can be applied [37]. So, in lieu of system (1), the following stochastic dynamical system will be considered [16]:

$$dx/dt = F(x) + \eta(t), \tag{44}$$

where η is a Gaussian stochastic process: $\langle \eta_i(t)\eta_j(t')\rangle = 2d_{ij}\delta(t-t'), d_{ij} \geq 0$. This procedure is correct since our knowledge about the model parameters is always limited, thus real climate models have random errors, which are represented by the term η. Under the assumption that $d_{ij} = d$, one can write the Fokker-Plank equation with respect to PDF $\rho(x, t)$, which describes the evolution of ρ [30]:

$$\partial\rho/\partial t + div\Big(F(x)\rho\Big) = d\Delta\rho, \rho \geq 0, \int \rho dx = 1. \tag{45}$$

Let $\overline{\rho}$ be a stationary solution to Eq. (45), i.e. $\text{div}\Big(F(x)\rho\Big) = d\Delta\rho$. If x belongs to the compact manifold without boundary, then $\overline{\rho}$ is asymptotically stable [37]. The existence of a stationary solution (i.e. attractor) at infinity has been proved for finite-dimensional dynamical systems [38].

Suppose that the initial condition x_0 is specified then the condition $\rho|_{t=0} = \delta(x-x_0)$ is also specified and enable us to solve Eq. (45). The numerical integration of Eq. (45) transforms the PDF $\rho(x, t)$, which asymptotically evolves to the stationary solution $\overline{\rho}$: $\rho \rightarrow \overline{\rho}$ at $t \rightarrow t_\varepsilon$. Thus, at sufficiently large time t_ε predictability is finally lost. There is a question: how can we estimate the time t_ε? Let us consider the following one-variable stochastic dynamical equation [16].

$$dx/dt = -\gamma x + \eta, \tag{46}$$

$$x|_{t=0} = x_0, \langle \eta(t)\eta(t')\rangle = 2\eta^2\delta(t-t'), \langle \eta \rangle = 0, \tag{47}$$

where x_0 is the known initial condition and η is the Gaussian δ-correlated process. If we average Eq. (46) we obtain

$$d\langle x\rangle/dt = -\gamma\langle x\rangle, \langle x\rangle|_{t=0} = x_0, \tag{48}$$

thus $\langle x\rangle = x_0 e^{-\gamma t}$. For the newly introduced variable $\theta(t) = \langle x^2\rangle$, we can obtain the following equation:

$$d\theta/dt = -2\gamma\theta + 2\langle \eta \cdot x\rangle. \tag{49}$$

Since $x(t) = x_0 e^{-\gamma t} + \int_0^t e^{\gamma(t-\tau)}\ \eta(\tau)d\tau$, then

$$d\theta/dt = -2\gamma\theta + 4\eta^2. \tag{50}$$

The solution to this equation is

$$\theta(t) = \frac{2\eta^2}{\gamma}(1-e^{-2\gamma t}). \tag{51}$$

Equation for the PDF ρ has the following form:

$$\partial\rho/\partial t = \partial(\rho x\gamma)/\partial x + \eta^2\partial^2\rho/\partial x^2. \tag{52}$$

The stationary solution to Eq. (52) can be found if we suppose that the left-hand side is equal to zero. Then, we have $\overline{\rho} = \left(1/\sqrt{\pi\overline{\theta}}\right)e^{-x^2/\overline{\theta}}$, where $\overline{\theta} = 2\eta^2/\gamma$. We assume that the solution to Eq. (48) is of the form

$$\rho(t) = \frac{1}{\sqrt{\pi\theta(t)}} e^{-\left(x-\langle x(t)\rangle\right)^2/\theta(t)}. \tag{53}$$

By substituting (53) into (52) one can be convinced that if $\theta(t)$ and $\langle x(t)\rangle$ satisfy Eq. (48) and Eq. (50), respectively, then Eq. (53) is the solution to the Fokker-Planck equation (52). As a result, any initial data that is normally distributed will be attracted to the steady solution of Eq. (52), which is also normally distributed. The dissipation parameter γ determines the rate at which PDF ρ approaches $\overline{\rho}$. The auto-correlation function for the stationary stochastic process (46) can be written as

$$C(\tau) = \frac{2\eta^2}{\gamma} e^{-\gamma\tau} \equiv \overline{\theta} e^{-\gamma t}. \tag{54}$$

Thus, the potential predictability of system (46) can be characterized by the auto-correlation function of the process $x(t)$ and, therefore, the convergence of $\rho(t)$ to $\overline{\rho}$ can be explored based only on function $C(\tau)$ with time lag τ. This conclusion is valid for the set of multi-dimensional differential equations [16]. In this case, however, the covariance matrix is used instead of the auto-correlation function. It is very important that for climate models the convergence of the covariance matrix $C(t)$ to the covariance matrix of stationary process $\overline{C}$ is defined only by climatological values of climate model variables. As a result, potential predictability is also determined by climatological data.

Generally, the potential predictability can be defined as the convergence time of initial distribution to the equilibrium one. To quantify the rate of convergence of one-dimensional distributions to the equilibrium ones, the concept of entropy can be used. If the information entropy $S = \int \rho \ln \rho d\alpha$ is taken as a measure of predictability, then for the Gaussian distribution $\rho = (1/\sqrt{2\pi\sigma^2})e^{-(\alpha-\overline{\alpha})^2/(2\sigma^2)}$ information entropy can be expressed as $S = \ln\sigma^2 + C$. It can be shown that the variance and, therefore, the entropy are directly dependent on the Lyapunov exponents [39]. To study the predictability of climate system, the relative entropy $S_r = \int \rho \ln(\rho/\overline{\rho}) d\alpha$, where $\overline{\rho}$ is an equilibrium PDF, is a more suitable measure [40]. Relative entropy is invariant with respect to nonlinear transformations of α and $\rho \to \overline{\rho}$ at $t \to \infty$.

4.2. Predictability of the second kind

Predictability of the second kind relates to the predictability of changes in climate system caused by infinitesimal perturbations in the parameters that describe the external forcing. Climate prediction does not involve forecasting weather conditions at either a certain geographical region or globally. On the contrary, climate prediction aims to forecast statistics of the climate system averaged over sufficiently long period of time. So, we are interested in how external perturbations affect certain aspects of climate statistic, such as the first $\overline{x}$ (mean) and/or second σ_x^2 (variance) moments. One of the most important problems in the exploration of predictability of the second kind is to distinguish the response signal of the climate system to perturbed external forcing from the noise in the model output results. The signal-to-noise ratio can be used to make the conclusion with respect to the usefulness of the obtained climate system response. Thus, the predictability of the second kind is mathematically reduced to finding the response function of the climate system model [39].

Consider the following finite-dimensional dynamical system that is controlled by some external forcing f (e.g. the concentration of carbon dioxide in the atmosphere):

$$dx/dt = F(x) + f, x|_{t=0} = x_0, \tag{55}$$

Suppose that system (55) possesses the attractor A and let μ be its invariant measure. The behaviour of this system will be explored on the attractor A. Since system (55) *a priori* possesses the property of ergodicity, its statistical characteristics are calculated by averaging along a single, sufficiently long, random trajectory. Thus, the average state $\langle x \rangle$ and variance $\langle \sigma_x^2 \rangle$ of system (55) are defined, respectively, as

$$\langle x \rangle = \lim_{T\to\infty} \frac{1}{T} \int_0^T x(t)dt = \int_A x d\mu, \langle \sigma_x^2 \rangle = \int_A (x - \langle x \rangle)^2 d\mu. \tag{56}$$

Let system (55) be perturbed by an infinitesimal disturbance in the external forcing δf such that $\delta f \ll f$:

$$dx^*/dt = F(x^*) + f + \delta f. \tag{57}$$

For this system $\langle x^* \rangle = \int_A x^* d\mu^*$ and $\langle \sigma_x^2 \rangle = \int_A (x^* - \langle x^* \rangle)^2 d\mu^*$. Let us introduce the new variable $x'(t) = x(t) - x^*(t)$. Assuming that $\|x'\|$ is rather small then, combining (55) and (56), one can obtain the following linear equation for variable x':

$$dx'/dt = J(x)x' + \delta f. \tag{58}$$

where $J(x) = \partial F/\partial x$ is the Jacobian. Let δf be a staircase function that is activated at $t = 0$ then the solution to Eq. (58) can be written in terms of the Green's function:

$$x'(t) = \int_0^t G(t, t')\delta f(t')dt'. \tag{59}$$

The operator $\mathrm{R} = \int_0^t G(t, t')dt'$ is a sought-for response function (operator). If at $t = 0$ the distribution of initial states is identical for both unperturbed (55) and perturbed (57) systems, then one can calculate the average response operator:

$$\langle \mathrm{R} \rangle = \int_0^t \langle G(t, t') \rangle dt' = \int_0^t G(t - t')d(t - t'). \tag{60}$$

By averaging both sides of Eq. (59), one can get the following linear equation to calculate the system's response to the external forcing:

$$\langle x' \rangle = \langle \mathrm{R} \rangle \delta f. \tag{61}$$

Suppose that system (55) is regular, i.e. for this system the quadratic conservation law is valid and system itself satisfies the Liouville equation for incompressibility in the phase space. Assume also that the system is in equilibrium. Taking into consideration the fluctuation dissipation theorem [41], the average impulse response operator of the regular system in equilibrium is expressed via system's statistics:

$$\langle G(t,t') \rangle = G(t\text{–}t') = C(t-t')C^{-1}(0), \tag{62}$$

where $C(t\text{–}t') = \langle x(t)x^{T}(t') \rangle$ is the system's auto-correlation matrix with time lag $\tau = t\text{–}t'$. Now we can combine (60) and (62) to get the following well-known formula [42]:

$$\langle x' \rangle = \int_{0}^{\infty} C(t)C^{-1}(0)dt \cdot \delta f. \tag{63}$$

Thus, the mean response of climate system to external forcing is determined by observations of unperturbed climate oscillation.

5. Concluding remarks

The prediction of climate change caused by natural processes and human-induced drivers is one of the most critical scientific issues facing the mankind in the 21st century. Computer-simulated climate models represent a very powerful and, perhaps, the only research instrument for studying climate and its dynamics. One of the key components of climate models, namely the model of the atmospheric general circulation, currently also serves as a primary tool for the numerical weather prediction all around the globe. However, the climate (atmospheric) system's trajectory calculated via numerical integration of multi-dimensional partial differential equations that describe the climate (atmospheric) system evolution is unstable with respect to both perturbations (errors) in the initial conditions and infinitesimal external forcing expressed by some model parameters and/or boundary conditions. This instability limits the time horizon of the validity of the climate (weather) forecast and leads to predictability problem.

In this chapter, the climate system is considered as a complex adaptive dynamical system that possesses a number of specific properties such as, for example, dissipativity, nonlinearity and chaoticity. From this perspective, the climate predictability problem is best discussed and analysed by formally examine two kinds of predictability. The first kind of predictability refers to the initial value problem (estimating the impact of perturbations in the initial conditions on the forecast skill), while the second kind of predictability relates to the boundary value problem (estimating the impact of external forcing on the system's behaviour).

Author details

Sergei Soldatenko* and Rafael Yusupov

*Address all correspondence to: s.soldatenko@bom.gov.au

St. Petersburg Institute for Informatics and Automation of the Russian Academy of Sciences, St. Petersburg, Russia

References

[1] Washington W.M., Parkinson C.L. Introduction to Three-Dimensional Climate Modelling. 2nd ed. Sausalito, California: University Science Book; 2005. 368 pp.

[2] McGuffie K., Henderson-Sellers A. The Climate Modelling Primer. 4th ed. New York: J. Wiley & Sons; 2014. 456 pp.

[3] Randall D.A., Wood R.A., Bony S., et al. Climate models and their evaluation. In: Solomon S., Qin D., Manning M., et al., editors. Climate Change 2007: The Physical Science Basis. Contribution of Working Group I to the Fourth Assessment Report of the Intergovernmental Panel on Climate Change. Cambridge and New York: Cambridge University Press; 2007. 74 pp.

[4] Coiffier J. Fundamentals of Numerical Weather Prediction. Cambridge: Cambridge University Press; 2012. 368 pp.

[5] Parker W.S. Confirmation and adequacy-for-purpose in climate modelling. Proc. Aristotel. Soc. Suppl. Vol. 2009; **83**: 233–249.

[6] Lorenz E.N. Deterministic nonperiodic flow. J. Atmos. Sci. 1963; **20**: 130–141.

[7] Selvam A. Chaotic Climate Dynamics. Bristol, UK: Luniver Press; 2007. 156 pp.

[8] Thompson P.D. Uncertainty of initial state as a factor in the predictability of large-scale atmospheric flow patterns. Tellus. 1957; **9**: 275–295.

[9] Lorenz E.N. A study of the predictability of a 28-variable atmospheric model. Tellus. 1965; **17**: 321–333.

[10] Smagorinsky J. Problems and promises of deterministic extended range forecasting. Bull. Amer. Meteor. Soc. 1969; **50**: 286–312.

[11] Leith C.E. Predictability in theory and practice. In: Hoskins B.J., Pearce R.P., editors. Large-Scale Dynamical Processes in the Atmosphere. New York: Academic Press; 1983. pp. 365–383.

[12] Fraedrich K. Estimating weather and climate predictability on attractors. J. Atmos. Sci. 1987; **44**: 722–728.

[13] Dalcher A., Kalnay E. Error growth and predictability in operational ECMWF forecasts. Tellus. 1987; **39A**: 474–491.

[14] Chou J.F. Predictability of the atmosphere. Adv. Atmos. Sci. 1989; **6**: 335–346.

[15] Farrell B.F. Small error dynamics and the predictability of atmospheric flows. J. Atmos. Sci. 1990; **47**: 2409–2416.

[16] Dymnikov V.P. Potential predictability of large-scale atmospheric processes. Atmos. Oceanic Phys. 2004; **40**: 579–585.

[17] Palmer T.N. Predictability of weather and climate: From theory to practice. In: Palmer T., Hagedorn R., editors. Predictability of Weather and Climate. New York: Cambridge University Press; 2006. pp. 1–10.

[18] DelSole T. Predictability and information theory. Part I: Measures of predictability. J. Atmos. Sci. 2004; **61**: 2425–2440.

[19] Lorenz E.N. Climate Predictability: The Physical Basis of Climate Modelling. Geneva: World Meteorological Organization; GARP Publication Series; 1975. Vol. 16, pp. 132–136.

[20] Meadows D., Write D. Thinking in Systems: A Primer. Vermont: Chelsea Green Publishing; 2008.

[21] Waldrop M.M. Complexity: The Emerging Science at the Edge of Order and Chaos. New York: Simon & Schuster; 1993. 380 pp.

[22] Gros C. Complex and Adaptive Dynamical Systems: A Primer. Berlin: Springer-Verlag; 2010. 326 pp.

[23] Daede A.P.M., Ahlonsou R., Ding Y., Schimel D. The climate system: An overview. In: Houghton J.T., Ding Y., Grogs D.J., et al., editors. IPCC, 2001: Climate Change 2001: The Scientific Basis. Contribution of Working Group I to the Third Assessment Report of the Intergovernmental Panel on Climate Change. Cambridge: Cambridge University Press; 2001. pp. 85–98.

[24] Dijkstra H.A. Nonlinear Climate Dynamics. New York: Cambridge University Press; 2013. 367 pp.

[25] Trenberth K.E. Climate System Modelling. Cambridge: Cambridge University Press; 2010. 820 pp.

[26] Barry R.G., Hall-McKim E.A. Essentials of the Earth's Climate System. Cambridge: Cambridge University Press; 2014. 271 pp.

[27] Marshall J., Plumb R.A. Atmosphere, Ocean and Climate Dynamics. San Diego: Academic Press; 2016. 489 pp.

[28] Holton J.R. An Introduction to Dynamical Meteorology. New York: Academic Press; 2004. 535 pp.

[29] Pedlosky J. Geophysical Fluid Dynamics. New York: Springer-Verlag; 1992. 710 pp.

[30] Dymnikov V.P., Filatov A.N. Mathematics of Climate Modelling. Boston: Birkhäuser; 1997. 264 pp.

[31] Washington W.M., VerPlamk L. A Description of Coupled General Circulation Models of the Atmosphere and Oceans Used for Carbon Dioxe Studies. Boulder, Colorado: NCAR (National Centre for Atmospheric Research)/TN-271; 1986. 34 pp.

[32] Mezinger F., Arakawa A. Numerical Methods Used in Atmospheric Models. Geneva: World Meteorological Organization; GARP Publication Series; 1979. Vol. 17, 64 pp.

[33] Oseledets V.I. Multiplicative ergodic theorem: Characteristic Lyapunov exponents of dynamical systems. Trans. Moscow Math. Soc. 1968; **19**: 179–210.

[34] Kaplan J.L., Yorke A.J. Chaotic behaviour in multidimensional difference equations. In: Peitgen H.-O., Walter H.-O., editors. Functional Differential Equations and Approximations of Fixed Points. Lecture Notes in Mathematics. Berlin: Springer-Verlag; 1979. pp. 228–237.

[35] Pesin B.J. Characteristic Lyapunov exponents and smooth ergodic theory. Russian Math. Surveys. 1977; **32**: 55–114.

[36] Palmer T.N. Predicting Uncertainty in Forecasts of Weather and Climate. ECMWF Technical Memorandum No. 294. ECMWF Shinfield Park: ECMWF (European Centre for Medium Range Weather Forecasting), Reading; 1999. 64 pp.

[37] Zeeman E.C. Stability of dynamical systems. Nonlinearity. 1988; **1**: 115–155.

[38] Noarov A.I. Sufficient condition for the existence of a stationary solution to the Fokker-Planck equation. J. Comput. Math. Physics. 1997; **5**: 587–598.

[39] Dymnikov V.P. Stability and Predictability of Large Scale Atmospheric Processes. Moscow: INM RAS; 2007. 283 pp.

[40] Kleeman R. Measuring dynamical prediction utility using relative entropy. J. Atmos. Sci. 2002; **59**: 2057–2972.

[41] Kraichnan R.H. Classical fluctuation-relaxation theorem. Phys. Rev. 1959; **113**: 1181–1182.

[42] Leith C.E. Climate response and fluctuation dissipation. J. Atmos. Sci. 1975; **32**: 2022–2026.

2

Preservation of Synchronization Using a Tracy-Singh Product in the Transformation on their Linear Matrix

Guillermo Fernadez-Anaya,
Luis Alberto Quezada-Téllez,
Jorge Antonio López-Rentería,
Oscar A. Rosas-Jaimes, Rodrigo Muñoz-Vega,
Guillermo Manuel Mallen-Fullerton and
José Job Flores-Godoy

Abstract

Preservation is related to local asymptotic stability in nonlinear systems by using dynamical systems tools. It is known that a system, which is stable, asymptotically stable, or unstable at origin, through a transformation can remain stable, asymptotically stable, or unstable. Some systems permit partition of its nonlinear equation in a linear and nonlinear part. Some authors have stated that such systems preserve their local asymptotic stability through the transformations on their linear part. The preservation of synchronization is a typical application of these types of tools and it is considered an interesting topic by scientific community. This chapter is devoted to extend the methodology of the dynamical systems through a partition in the linear part and the nonlinear part, transforming the linear part using the Tracy-Singh product in the Jacobian matrix. This methodology preserves the structure of signs through the real part of eigenvalues of the Jacobian matrix of the dynamical systems in their equilibrium points. The principal part of this methodology is that it permits to extend the fundamental theorems of the dynamical systems, given a linear transformation. The results allow us to infer the hyperbolicity, the stability and the synchronization of transformed systems of higher dimension.

Keywords: preservation, synchronization, Tracy-Singh product, chaotic dynamical system

1. Introduction

In nonlinear autonomous dynamical systems, the study of synchronization is not new. We can see several papers about these themes from different approaches. Some examples show the use of change of variables, that is, through a diffeomorphism of the origin. From this, it is possible to say if a system is stable, asymptotically stable, or unstable. Some results are also obtained by the product in a vector field in the nonlinear dynamical system by a continuously differentiable function at the origin [1]. On one hand, there are studies showing the use of statistical properties to characterize the synchronization [2]. The eigenvalues of a system determine a system dynamics, but they are not derivable from the statistical features of such a system. One way to observe the stability is through a linear part of a dynamical system. But the problem to preserve stability by the transformation of its linear part in a nonlinear autonomous system has just been analyzed recently.

In [3], it is presented a methodology under which stability and synchronization of a dynamical master-slave system configuration are preserved under a modification through matrix multiplication. The conservation of stability is important for chaos control. A generalized synchronization can also be derived for different systems by finding a diffeomorphic transformation such as the slave system written as a function of the master system. One example of preservation for asymptotic stability is the use of transformations on rational functions in the frequency domain [4, 5].

This class of transformation can be interpreted as noise in the system or as a simple disturbance on the value of the physical parameters of the model. The chaotic synchronization problem studied in [6] is mainly related to preservation of the stability of the master-slave system presented in it. Results included therein show that stability is preserved by transforming the linear part of system. The same results can also be used in the chaos suppression problem. In [7], the authors show the viability of preserving the hyperbolicity of a master-slave pair of chaotic systems under different types of nonlinear modifications to its Jacobian matrix.

In [8], the developed methodology is used to study the problem of preservation of synchronization in chaotic dynamical systems, in particular the case of dynamical networks. Given a chaotic system, its transformed version is also a chaotic system. By means of a master-slave scheme obtained a controller for the system using a linear-quadratic regulator, preserving the stability even after the master-slave controller is transformed. This chapter is inspired by the same objective, that is, to preserve the stability in a master-slave system even through a transformation is performed over it. One way to achieve it is by extending some of the results in [8], particularly those of the local stable-unstable manifold theorem and extension of the center manifold theorem based in the preservation of the linear part of the vector field in nonlinear dynamical systems. As we will see, these results depart from the hypothesis of the existence of a constant state feedback as anominal synchronization force. In this work, we elaborate another approach to the problem of preservation of synchronization. We focus particularly on autonomous nonlinear dynamical systems, extending the previous results already mentioned.

This chapter is organized as follows: First, in Section , we will give basic concepts of dynamical systems. The fundamental theorem for linear systems, the local stable-unstable manifold theorem, the center manifold theorem, the Hartman-Grobman theorem and the concept of group action are introduced. In Section, we present some definitions about matrices and Tracy-Singh product of matrices. Also in this section, the main result is presented as a generalization of Proposition 4 in [6]. In Section, we will show that it is possible to preserve synchronization under a class of transformations defined under a certain method. Numerical experiments on the stability preservation for chaotic synchronization are shown in Section. Finally, a set of concluding remarks is given in Section.

2. Classical concepts of dynamical systems

We introduce theorems and classical definitions on properties of dynamical systems in this section. The fundamental theorem for linear systems, the local stable-unstable manifold theorem and the center manifold theorem are those important propositions mainly needed to develop analyses in this chapter. We will combine them with the Hartman-Grobman theorem in order to achieve a necessary generalization for those particular results of this chapter.

Theorem 2.1. (The local stable-unstable manifold theorem [9]). *Let E be an open subset of* $\mathbb{R}^n$ *containing the origin. Let* $f \in C^1(E)$ *and* ϕ_t *be the flow of the nonlinear system of the form* $\dot{x} = f(x)$. *Suppose that* $f(0) = 0$ *and that* $Df(0)$ *are the Jacobian matrix, which has k eigenvalues with negative real part and n−k eigenvalues with positive real part.*

1. *(Stable manifold) Then, there exists a k-dimensional differentiable manifold S tangent to the stable subspace* E^S *of the linear system* $\dot{x} = A(x)$ *at* x_0 *such that for all* $t \geq 0$, $\phi_t(S) \subset S$ *and for all* $x_0 \in S$, $\lim_{t\to\infty} \phi_t(x_0) = 0$.
2. *(Unstable manifold) Also there exists an n−k dimensional differentiable manifold W tangent to the unstable subspace* E^W *of* $\dot{x} = A(x)$ *at* x_0 *such that for all* $t \leq 0$, $\phi_t(W) \subset W$ *and for all* $x_0 \in W$, $\lim_{t\to-\infty} \phi_t(x_0) = 0$.

It should be noted that the manifolds *S* and *W* mentioned in Theorem 2.1 are unique. We define now the central manifold theorem in the following.

Theorem 2.2. (The center manifold theorem [9]). *Let E be an open subset of* $\mathbb{R}^n$ *containing the origin and* $r \geq 1$. *Let* $f \in C^r(E)$, *that is, f is a continuously differentiable function on E of order r. Now we suppose that* $f(0) = 0$ *and that* $Df(0)$ *have k eigenvalues with negative real part, j eigenvalues with positive real part and* $l = n-k-j$ *eigenvalues with zero real part. Therefore, there exists an l*

-dimensional center manifold $W^C(0)$ *of class* C^r *tangent to the center subspace* E^C *of* $\dot{x} = A(x)$ *at 0 which is invariant under the flow* ϕ_t *of* $\dot{x} = f(x)$.

By what it is established in Theorem 2.2, the center manifold $W^C(0)$ is not unique, which is an important difference for the stable character of the systems to be studied.

Theorem 2.3. (The Hartman-Grobman theorem [9]). *Let E be an open subset of* $\mathbb{R}^n$ *containing the origin, let* ϕ_t *be the flow of the nonlinear system* $\dot{x} = f(x)$. *Now, we assume that* $f(0) = 0$, *that is, the origin is an equilibrium point of the dynamical system; also the Jacobian matrix evaluated at the origin,* $A = Df(0)$. *If H is an homeomorphism of an open set W onto an open set V such that for each* $x_0 \in W$, *it exists an open interval* $I_0 \subset \mathbb{R}$ *such that for all* $x_0 \in W$ *and* $t \in I_0$

$$H \circ \phi_t(x_0) = e^{At} H(x_0); \tag{1}$$

that is, H maps trajectories of the nonlinear system $\dot{x} = f(x)$ *near the origin onto trajectories of* $\dot{x} = Ax$ *near the origin and preserves the parametrization.*

From the following argument, it is show that for any matrix $A = U^T T_A U$, there exists an homeomorphism $\hat{H} = UH$ such that for an open set W containing the origin onto an open set V also containing the origin such that for each $x_0 \in W$ and there is an open interval $I_0 \subset \mathbb{R}$ containing zero such that for all $x_0 \in W$ and $t \in I_0$

$$\hat{H} \circ \phi_t(x_0) = e^{T_A t} \hat{H}(x_0); \tag{2}$$

This last equality is a consequence of the Hartman-Grobman theorem and of the fact of $Ue^{At} = e^{T_A t} U$, that is, $\hat{H}$ maps trajectories of the nonlinear system $\dot{x} = f(x)$ near the origin onto trajectories of $\dot{x} = T_A x$ near the origin and preserves the parametrization.

On the other hand, some classical definitions are now included. A linear system of the form $\dot{x} = Ax$ where $x \in \mathbb{R}^n$, A is a $n \times n$ matrix and $\dot{x} = \frac{dx}{dt}$. It is shown that the solution of the linear system together with the initial condition $x(0) = x_0$ is given by $x(t) = e^{At} x_0$. The mapping $e^{At} : \mathbb{R}^n \to \mathbb{R}^n$ is called *the flow* of the linear system.

Definition 2.1. *For all eigenvalues of a matrix* $A(n \times n)$ *have nonzero real part, then the flow* e^{At} *is called a hyperbolic flow and therefore,* $\dot{x} = Ax$ *is called a hyperbolic linear system* [9].

Definition 2.2. *A subspace* $E \subset \mathbb{R}^n$ *is said to be invariant with respect to the flow* $e^{At} : \mathbb{R}^n \to \mathbb{R}^n$ *if* $e^{At} \subset E$ *for all* $t \in \mathbb{R}$ [9].

Lemma 2.1. Let $\mathrm{A} \in \mathbb{R}^{\mathrm{n} \times \mathrm{n}}$. If $\mathbb{R}^{\mathrm{n}} = \mathrm{E^s} \oplus \mathrm{E^u} \oplus \mathrm{E^c}$ where $\mathrm{E^s}, \mathrm{E^u}$ and $\mathrm{E^c}$ are the stable, unstable and center subspaces of the linear system $\dot{\mathrm{x}} = \mathrm{Ax}$. By the above, $\mathrm{E^s}, \mathrm{E^u}$ and $\mathrm{E^c}$ are invariant with respect to the flow $\mathrm{e^{At}}$, respectively [9].

Definition 2.3. *Let E be an open subset of* $\mathbb{R}^n$ *and let* $f \in C^1(E)$, *that is, f is a continuous differentiable function defined on E. For* $x_0 \in E$, *let* $\phi(t, x_0)$ *be the solution of the initial value problem* $\dot{x} = f(x), x(0) = x_0$ *defined on its maximal interval of existence* $I(x_0)$. *Then for* $t \in I(x_0)$, *the mapping* $\phi_t : E \to E$ *defined by* $\phi_t(x_0) = \phi_t(t, x_0)$ *is called the flow of the differential equation* [9].

Definition 2.4. *For any* $x_0 \in \mathbb{R}^n$, *let* $\phi_t(x_0)$ *be the flow of the differential equation through* x_0. *(i) The local stable set S corresponding to a neighborhood V of* x_0 *is defined by* $S = S(0) = \{x_0 \in \mathbb{R}^n : \phi_t(x_0) \in V, t \geq 0$ *and* $\phi_t(x_0) \to 0$ *as* $t \to \infty\}$. *(ii) The local unstable set W of* x_0 *corresponding to a neighborhood V of* x_0 *is defined by* $W = W(0) = \{x_0 \in \mathbb{R}^n : \phi_t(x_0) \in V, t \leq 0$ *and* $\phi_t(x_0) \to 0$ *as* $t \to \infty\}$.

Then, these stable and unstable local sets are submanifolds of $\mathbb{R}^n$ in a sufficiently small neighborhood V of x_0[9].

Definition 2.5. *If G is a group and X is a set, then a (left) group action of G on X is a binary function $G\times X \to X$, denoted by* [9]

$$(g,x)\mapsto g\cdot x \tag{3}$$

which satisfies the following two axioms:

1. $(gh)\cdot x = g\cdot(h\cdot x)$ *for all* $g,h \in G$ *and* $x \in X$;
2. $e\cdot x = x$ *for every* $x \in X$ *(where e denotes the identity element of G).*

The action is faithful (or effective) if for any two different $g,h \in G$, there exists an $x \in X$ such that $g\cdot x \neq h\cdot x$; or equivalently, if for any $g \neq e$ in G, there exists an $x \in X$ such that $g\cdot x \neq x$.

The action is free or semiregular if for any two different $g,h \in G$ and all $x \in X$, we have $g\cdot x \neq h\cdot x$; or equivalently, if $g\cdot x = x$ for some x implies $g = e$.

For every $x \in X$, we define the stabilizer subgroup of x (also called the isotropy group or little group) as the set of all elements in G that fix x:

$$G_x = \{g\in G : g\cdot x = x\} \tag{4}$$

This is a subgroup of G, though typically not a normal one. The action of G on X is free if and only if all stabilizers are trivial.

3. Tracy-Singh product and other mathematical extensions

In this third section, we show a definition and some properties of the Tracy-Singh product. We also include a simple extension of the local stable-unstable manifold theorem and the center manifold theorem, using the tools presented in Section. These extensions are tools that will also be used in Section, where we will present the results on preservation of synchronization in nonlinear dynamical systems.

Definition 3.1. *Let λ be an eigenvalue of the $n\times n$ matrix A of multiplicity $m\leq n$. Then for $k = 1,\ldots,m$, any nonzero solution w of* [9]

$$(A-\lambda I)^k w = 0 \tag{5}$$

is called a generalized eigenvector of A.

In this case, let $w_j = u_j + v_j$ be a generalized eigenvector of the matrix A corresponding to an eigenvalue $\lambda_j = a_j + ib_j$ (note that if $b_j = 0$ then $v_j = 0$). Then, let $B = \{u_1, v_1, \ldots, u_k, v_k, \ldots, u_m, v_m\}$

be a basis of $\mathbb{R}^n$ (with $n = 2m-k$ as established by Theorems 1.7.1 and 1.7.2,). Now, we introduce the definition of Tracy-Singh product and some properties.

Definition 3.2. *If taken the matrices $A = (a_{ij})$ and $C = (c_{ij})$ of order $m \times n$ and $B = (b_{kl})$ of order $p \times q$. Let $A = (A_{ij})$ be partitioned with A_{ij} of order $m_i \times n_j$ as the (i,j) th block submatrix and $B = (B_{kl})$ of order $p_k \times q_l$ as the (k,l) th block submatrix $(\sum m_i = m, \sum n_j = n, \sum p_k = p, \sum q_l = q)$. Then, the definitions of the matrix products or sums of A and B are given as follows* [10].

Tracy-Singh product

$$A \circ B = (A_{ij} \circ B)_{ij} = \left((A_{ij} \otimes B_{kl})_{kl} \right)_{ij} \tag{6}$$

where $A_{ij} \otimes B_{kl}$ is of order $m_i p_k \times n_j q_l$, $A_{ij} \circ B$ is a Kronecker product of order $m_i p \times n_j q$, and $A \circ B$ is of order $mp \times nq$.

Tracy-Singh sum

$$A \boxplus B = A \circ I_p + I_m \circ B \tag{7}$$

where $A = (A_{ij})$ and $B = (B_{kl})$ are square matrices of respective order $m \times m$ and $p \times p$ with A_{ij} of order $m_i \times m_j$ and B_{kl} of order $p_k \times p_l$; I_p and I_m are compatibly partitioned identity matrices.

Theorem 3.1. *Let A, B, C, D, E, and F be compatibly partitioned matrices, then* [10]

1. $(A \circ B)(C \circ D) = (AC) \circ (BD)$.
2. $A \circ B \neq B \circ A$.
3. $(C \circ B = B \circ C)$ *where $C = (c_{ij})$ and c_{ij} is a scalar.*
4. $(A \circ B)' = A' \circ B'$.
5. $(A + D) \circ (B + E) = A \circ B + A \circ E + D \circ B + D \circ E$.
6. $(A \circ B) \circ F = A \circ (B \circ F)$

The next proposition presents some extensions to the local stable-unstable manifold theorem and to the center manifold theorem.

Proposition 3.1. Let E be an open subset of $\mathbb{R}^n$ containing the origin, let $f \in C^1(E)$ and ϕ_t be the flow of the nonlinear system $\dot{x} = f(x) = Ax + g(x)$. Suppose that $f(0) = 0$ and that $A = Df(0)$ have k eigenvalues with negative real part and $n-k$ eigenvalues with positive real part, that is, the origin is an hyperbolic fixed point. Then for each matrix $M \in \Lambda_U$, there exists a k

-dimensional differentiable manifold S_M tangent to the stable subspace E_M^S of the linear system $\dot{x} = MAx$ at 0 such that for all $t \geq 0$, $\phi_{M,t}(S_M) \subset S_M$ and for all $x_0 \in S_M$ [8],

$$\lim_{t \to \infty} \phi_{M,t}(x_0) = 0, \tag{8}$$

where $\phi_{M,t}$ is the flow of the nonlinear system $\dot{x} = MAx + g(x)$ and there exists an $n-k$ dimensional differentiable manifold W_M tangent to the unstable subspace E_M^W of $\dot{x} = MAx$ at 0 such that for all $t\leq 0$, $\phi_{M,t}(W_M)\subset W_M$ and for all $x_0 \in W_M$,

$$\lim_{t\to -\infty} \phi_{M,t}(x_0) = 0. \tag{9}$$

An interesting property is that Proposition 4.1 is valid for each $\overline{g}\in C^1(E)$ such that $\dot{x} = \overline{f}(x) = Ax + \overline{g}(x)$ and

$$\frac{\| \overline{g}(x) \|_2}{\| x \|_2} \to 0 \text{ as} \| x \|_2 \to 0. \tag{10}$$

In consequence, the set of matrices Λ_U generates the action of the group Λ_U on the set of the hyperbolic nonlinear systems, formally on the set of the hyperbolic vector fields $f\in C^1(E)$, $\dot{x} = \overline{f}(x) = Ax + \overline{g}(x)$ with $\overline{g}\in C^1(E)$ and

$$A\in\Omega_U \equiv \{P\in\mathbb{R}^{n\times n} : P = U^T T_P U \text{ with } T_P \text{ any upper triangular matrix}\} \tag{11}$$

Satisfying the last condition, where U is a fixed unitary matrix, the action is generated by the action of the group Λ_U on the set Ω_U. By that this first action preserves the dimension and a nonlinear systems of the stable and unstable manifolds, that is, an hyperbolic nonlinear system $\left(\dot{x} = Ax + \overline{g}(x)\right)$ is mapped in a hyperbolic nonlinear systems $\left(\dot{x} = MAx + \overline{g}(x)\right)$ and $dimS = dimS_M$ and $dimW = dimW_M$.

The proof of this Proposition 3.1 can be revised in Ref. [8].

Given a particular nonlinear system, the stable and unstable manifolds S and W are unique; then for each matrix $M\in\Lambda_U$, there exists an unique pair of manifolds (S_M, W_M) in such a way that it is possible to define a pair of functions in the following form

$$\begin{gathered} \Theta : \Lambda_U \times Man_S \to Man_S \\ \Theta(M, S) = S_M \\ \Phi : \Lambda_U \times Man_W \to Man_W \\ \Phi(M, W) = W_M \end{gathered} \tag{12}$$

Where Man_S is the set of stable manifolds and Man_W is the set of unstable manifold for autonomous nonlinear systems.

Therefore, we can say that if $A = Df(0)$ is an stable matrix A has all the n eigenvalues with negative real part, then the origin of the nonlinear system $\dot{x} = M \circ Ax + \overline{g}(x)$ is asymptotically stable; but if $A = Df(0)$ is an unstable matrix A has $n-k$ (with $n > k$) eigenvalues with positive real part, then the origin of the nonlinear system $\dot{x} = M \circ Ax + \overline{g}(x)$ is unstable.

As an extension of the local stable-unstable manifold theorem in terms of Tracy-Singh product of matrices in Λ_N and the matrix A of the vector field $f(x)$, we present the following proposition.

Proposition 3.2.

1. Let E be an open subset of $\mathbb{R}^n$ containing the origin, let $f \in C^1(E)$ and let ϕ_t be the flow of the nonlinear system $\dot{x} = f(x) = Ax + g(x)$. We suppose that $f(0) = 0$ and that $A = Df(0)$ have a k eigenvalues with negative real part and $n-k$ eigenvalues with positive real part; thus, the origin is a hyperbolic fixed point. Now, take a fixed continuously differentiable function

$$F : C^1(E) \to C^1(\overline{E}) \tag{13}$$

such that $F(g) = \hat{g}$ where $\hat{g} : \overline{E} \subset \mathbb{R}^{mn} \to \mathbb{R}^{mn}$ is a fixed continuously differentiable function with domain all $C^1(E)$; moreover, $\hat{g} \in C^1(\overline{E})$ with $\overline{E}$ an open subset of $\mathbb{R}^n$ containing the origin such that

$$\frac{\| \hat{g}(x) \|_2}{\| x \|_2} \to 0 \text{ as } \| x \|_2 \to 0. \tag{14}$$

Then, for each matrix $M \in \Lambda_U$ of $m \times m$, there exists a $mk-$ dimensional differentiable manifold $S_{M \circ A}$ tangent to the stable subspace $E^S_{M \circ A}$ of the linear system $\dot{x} = (M \circ A)x$ at 0 such that for all $t \geq 0$, $\phi_{M \circ A, t}(S_{M \circ A}) \subset S_{M \circ A}$ and for all $x_0 \in S_{M \circ A}$,

$$\lim_{t \to \infty} \phi_{M \circ A, t}(x_0) = 0, \tag{15}$$

where $\phi_{M \circ A, t}$ be the flow of the nonlinear system $\dot{x} = (M \circ A)x + \hat{g}(x)$ and there exists an $m(n-k)$ dimensional differentiable manifold $W_{\mathrm{M \circ A}}$ tangent to the unstable subspace $E^W_{M \circ A}$ of $\dot{x} = (M \circ A)x$ at 0 such that for all $t \leq 0$, $\phi_{M \circ A, t}(W_{M \circ A}) \subset W_{M \circ A}$ and for all $x_0 \in W_{M \circ A}$,

$$\lim_{t \to -\infty} \phi_{M \circ A, t}(x_0) = 0. \tag{16}$$

2. Also, there exists a function of the group Λ_N and the set of all the autonomous hyperbolic nonlinear systems of dimension n (hyperbolic vector fields of dimension n) denoted by Γ_n, to the set Γ_{mn} of all the autonomous hyperbolic nonlinear systems of dimension mn (hyperbolic vector fields of dimension mn); this function (which is similar to an action of the group Λ_N on the set Γ_n) is defined as follows

$$\begin{gathered} \vartheta : \Lambda_N \times \Gamma_n \to \Gamma_{mn} \\ \vartheta\Big(M, Ax + g(x)\Big) = (M \circ A)x + \hat{g}(x) \end{gathered} \tag{17}$$

and the new nonlinear system is

$$\begin{aligned} \dot{x} &= \vartheta\left(M, Ax + g(x)\right) \\ \dot{x} &= (M \circ A)x + \hat{g}(x)) \end{aligned} \tag{18}$$

which satisfies the following two axioms:

1. $(gh) \cdot z = g \bullet (h \cdot z)$ for all $g, h \in \Lambda_N$ and $z \in \Gamma_n$;
2. For every $z \in \Gamma_n$, there exists an unique $\hat{z} \in \Gamma_{mn}$ such that $e \cdot z = \hat{z}$ and $h \bullet \hat{z} = \mathrm{h} \cdot \mathrm{z}$ (e denotes the identity element of Λ_N, that is, is the identity matrix I_m of $m \times m$).

Where z is associated with $Ax + g(x)$ (denoted by $z \triangleq Ax + g(x)$); $h \cdot z$ means $(M_h \circ A)x + \hat{g}(x)$ (denoted by $h \cdot z \triangleq (M_h \circ A)x + \hat{g}(x)$); gh is associated with the usual product of matrices M_g, M_h, that is, $gh \triangleq M_g M_h$ and $e \cdot z$ means $(I_m \circ A)x + \hat{g}(x)$, that is, $\left(e \cdot z \triangleq (I_m \circ A)x + \hat{g}(x)\right)$ and $g \bullet (h \cdot z)$ means $(M_g \circ I_n)(M_h \circ A)x + \hat{g}(x)$ (denoted by $g \bullet (h \cdot z) \triangleq (M_g \circ I_n)(M_h \circ A)x + \hat{g}(x)$).

Proof.

1. Consider a matrix A with eigenvalues λ_i for $i = 1, 2, \ldots, n$ and the matrix M with eigenvalues μ_j for $j = 1, 2, \ldots, m$. Then, the eigenvalues of the matrix $M \circ A$ are the mn numbers $\lambda_i \mu_j$ and taking account that $\mu_j > 0$ for each $j = 1, 2, \ldots, m$. Therefore, the matrix $M \circ A$ has mk eigenvalues with negative real part and $m(n-k)$ eigenvalues with positive real part. For this, the result is a consequence of the stable-unstable manifold theorem.

2. The function $\vartheta : \Lambda_N \times \Gamma_n \to \Gamma_{mn}$ is well defined, since $F : C^1(E) \to C^1(\overline{E})$ is a fixed function; then given $g(x)$, the vector field $\hat{g}(x)$ is unique; for a fixed matrix $M_h \in \Lambda_N$, then $M_h \circ : R^{n \times n} \to \mathbb{R}^{mn \times mn}$ is a fixed function and their matrix $M_h \circ A$ is unique.

Axiom (i): Since Λ_N is a multiplicative group if $M_g, M_h \in \Lambda_N$, then $M_g M_h \in \Lambda_N$.

Then, by Theorem 3.1, we have that for all $g, h \in \Lambda_N$ and $z \in \Gamma_n$

$$(gh) \cdot z \triangleq (M_g M_h \circ A)x + \hat{g}(x) = (M_g \circ I_n)(M_h \circ A)x + \hat{g}(x) \triangleq g \bullet (h \cdot z) \tag{19}$$

Axiom (ii): For every $z \in \Gamma_n$, there exists an unique $\hat{z} \in \Gamma_{mn}$ such that $e \cdot z \triangleq (I_m \circ A)x + \hat{g}(x) = \hat{z}$, then by the Theorem 2.1

$$h \bullet \hat{z} \triangleq (M_h \circ I_n)(I_m \circ A)x + \hat{g}(x) = (M_h \circ A)x + \hat{g}(x) \triangleq h \cdot z \tag{20}$$

From what it has been said above, we can note that if $A = Df(0)$ is as stable matrix A, it has all the n eigenvalues with negative real part, then the origin of the nonlinear system $\dot{x} = (M \circ A)x + \hat{g}(x)$ is asymptotically stable; if $A = Df(0)$ is an unstable matrix A, it has n–$k(n > k)$ eigenvalues with positive real part, then the origin of the nonlinear system $\dot{x} = (M \circ A)x + \hat{g}(x)$ is unstable.

Now the following Proposition 3.2 is an extension of the center manifold theorem, similar to Proposition 3.1.

Proposition 3.3. Let be $f \in C^r(E)$ where E is an open subset of $\mathbb{R}^n$

containing the origin and $r \geq 1$. Suppose that $f(0) = 0$ and that $Df(0)$ have k eigenvalues with negative real part, j eigenvalues with positive real part and $l = n - k - j$ eigenvalues with zero real part. Then,

1. For each matrix $M \in \Lambda_U$, there exists a $m-$ dimensional differentiable center manifold $W_M^C(0)$ of class C^r tangent to the center subspace E_M^C of the linear system $\dot{x} = MAx + g(x)$ at 0 which is invariant under the flow $\phi_{M,t}$ of the nonlinear system $\dot{x} = MAx + g(x)$.

2. If taken a fixed continuously differentiable function

$$\hat{F} : C^r(E) \to C^r(\overline{E}) \tag{21}$$

such that $F(g) = \hat{g}$ where $\hat{g} : \overline{E} \subset \mathbb{R}^{mn} \to \mathbb{R}^{mn}$ is a fixed continuously differentiable function with domain all $C^r(E)$; moreover, $\hat{g} \in C^r(\overline{E})$ with $\overline{E}$ an open subset of $\mathbb{R}^n$ containing the origin such that

$$\frac{\| \hat{g}(x) \|_2}{\| x \|_2} \to 0 \text{ as } \| x \|_2 \to 0. \tag{22}$$

Then for each matrix $M \in \Lambda_N$ of $m \times m$, there exists a $ml-$ dimensional differentiable center manifold $W_{M \circ A}^C(0)$ tangent to the center subspace $E_{M \circ A}^S$ of the linear system $\dot{x} = (M \circ A)x$ at 0 which is invariant under the flow $\phi_{M \circ A,t}$ of the nonlinear system $\dot{x} = (M \circ A)x + \hat{g}(x)$.

Proof.

The proof is similar to proof of Proposition 3.1 and we make use of the center manifold theorem.

Also, there exists a similar function $\hat{\vartheta}$ to ϑ, which satisfies the axiom (i) and axiom (ii) of Proposition 3.2. However, in this case, there does not exist similar functions to Θ and Φ. due to that in general, a center manifold is not unique.

Notice that in this case, if the matrix A has $l = n - k - j \neq 0$ eigenvlues with zero real part, then the origin of the nonlinear system $\dot{x} = MAx + \hat{g}(x)$ and $\dot{x} = (M \circ A)x + \hat{g}(x)$ are not asymptotically stable.

Propositions 3.1 and 3.2 generalize Proposition 3 in Ref. [6] and give new tools for preservation of basic properties of dynamical systems and some of these properties are the stability and instability.

4. Synchronization in nonlinear dynamical system

In this section, we present that it is possible to preserve synchronization even though the dimension of the systems changes by the action of a class of transformation on the linear part to a chaotic nonlinear system. If we consider the following two n-dimensional chaotic systems,

$$\begin{aligned} \dot{x} &= Ax + g(x) \\ \dot{y} &= Ay + f(y) + u(t) \end{aligned} \tag{23}$$

Where $A \in \mathbb{R}^{n \times n}$ is a constant matrix. On the other hand,$u \in R^n$ is the control input and $f, g : R^n \rightarrow R^n$ are continuous nonlinear functions. Synchronization considered in this section is through the master and the slave system is synchronized by designing an appropriate nonlinear state-feedback control $u(t)$ attached to slave system such that $\lim_{t \to \infty} x(t) - y(t) \rightarrow 0$, where $\| \cdot \|$ is the Euclidean norm of a vector [8]. If we consider the error state vector $e = y - x \in R^n, f(y) - f(x) = L(x, y)$ and an error dynamics equation is $\dot{e} = Ae + L(x, y) + u(t)$. Taking the active control approach [5], to eliminate the nonlinear part of the error dynamics and choosing $u(t) = Bv(t) - L(x, y)$, where B is a constant gain vector which is selected such that (A, B) be controllable, we obtain:

$$\dot{e} = Ae + Bv(t) \tag{24}$$

We can see that the original synchronization problem is equivalent to stabilize the zero-input solution of the slave system through a suitable choice of the state-feedback control [8]. If the pair (A, B) is controllable, then one such suitable choice for state feedback is a linear-quadratic regulator [5], which minimizes the quadratic cost function in the next expression,

$$J\big(u(t)\big) = \int_0^\infty (e(t)^\intercal Q e(t) + v(t) R v(t)) dt \tag{25}$$

Where Q and R are positive semi-definite and positive definite weighting matrices, respectively. The state-feedback law is given by $v = -Ke$ with $K = R^{-1} B^\intercal S$ and S the solution to the *Riccati* equation

$$A^\intercal S + SA - SBR^{-1} B^\intercal + Q = 0 \tag{26}$$

This state-feedback law makes the error equation to be $\dot{e} = (A - BK)e$, with $(A - BK)$ a Hurwitz matrix.[1] The linear-quadratic regulator is a technique to obtain feedback gains [5]. It is an interesting property of (LQR) which is robustness. On the other hand, if we consider $T \in R^{m \times m}$ be a matrix with strictly positive eigenvalues, supposing that the following two nm-dimensional systems are chaotic:

$$\begin{aligned} \dot{x} &= (T \circ A)x + \hat{g}(x) \\ \dot{y} &= (T \circ A)y + \hat{f}(y) + \hat{u}(t) \end{aligned} \tag{27}$$

for some $\hat{f}, \hat{g} : R^{nm} \rightarrow R^{nm}$ continuous nonlinear functions and $\hat{u} \in R^{nm}$ is the control input. Then, for the Proposition 4.1 and the former procedure, we have that $\hat{u}(t) = \hat{\theta}(t) - \hat{L}(x, y)$ stabilizes the zero solution of the error dynamics system, where $\hat{\theta}(t) = -(BK \circ T)e$, that is, the resultant system

[1] A Hurwitz matrix is a matrix for which all its eigenvalues are such that their real part is strictly less than zero.

$$\dot{e} = (T\circ A)e + \hat{\theta}\,(t)\dot{e} = (T\circ A - T\circ BK)e \tag{28}$$

is asymptotically stable. Then, by using Lemma 2.1 and $K = -R^{-1}B^{\intercal}S$, we obtain that:

$$\begin{aligned} \dot{e} &= \Big(T\circ(A + BK)\Big)e \\ \dot{e} &= \Big(T\circ(A - BR^{-1}B^{\intercal}S)\Big)e \end{aligned} \tag{29}$$

Now, the original control $u(t) = BKe - L(x,y)$ is preserved in its linear part by the Tracy-Singh product $T\circ(\cdot)$ and the new control is given by $\hat{u}(t) = -(T\circ BK)e - \hat{L}(x,y)$. Therefore, we can interpreted the last procedure as one in which the controller $u(t)$ that achieves the synchronization in the two systems is preserved by the transformation $T\circ(\cdot)$ so that $\hat{u}(t)$ achieves the synchronization in the two resultant systems after the transformation. For that, a similar procedure is possible if we consider the transformation $(\cdot)\circ T$.

In general, under the transformation $(A,g) \rightarrow (MA,\overline{g})$ or $(A,g) \rightarrow (M\circ A,\overline{g})$ and under the hypothesis of existence of a constant state feedback $U = -Kx$, which achieves synchronization of the original chaotic systems and also that the transformed system is chaotic, then synchronization can be preserved [8]. The major contribution does not refer a better synchronization methodology; it deals that synchronization is preserved when a chaotic system changes from a lower dimension to a higher dimension.

5. Synchronization of the classical Lü system

In this section, we present the synchronization of a chaotic system. First, we propose a master and slave system. Then, from these systems, we will apply a linear transformation that allows us to preserve the synchronization. We will use the well-known Lü and Chen [11] model to show the possibility to preserve synchronization, described by

$$\begin{aligned} \dot{x}_1 &= a(x_2 - x_1) \\ \dot{x}_2 &= cx_2 - x_1x_3 \\ \dot{x}_3 &= x_1x_2 - bx_3 \end{aligned} \tag{30}$$

which has a chaotic attractor when the parameters are $a = 35, b = 3$ and $c = 14.5$. In order to observe synchronization behavior, we have a modified Lü attractor arranged as a master-slave configuration. The master and the slave systems are almost identical and the only difference is that the slave system includes an extra term which is used for the purpose of synchronization with the master system. The master system is defined by the following equations,

$$\begin{aligned} \dot{x}_1 &= 35(x_2 - x_1) \\ \dot{x}_2 &= 28x_2 - x_1x_3 \\ \dot{x}_3 &= x_1x_2 - 3x_3 \end{aligned} \tag{31}$$

and the slave system is a copy of the master system with a control function $u(t)$ to be determined in order to synchronize the two systems.

$$\begin{aligned}\dot{y}_1 &= 35(y_2 - y_1) + u_1(t) \\ \dot{y}_2 &= 28y_2 - y_1y_3 + u_2(t) \\ \dot{y}_3 &= y_1y_2 - 3y_3 + u_3(t)\end{aligned} \tag{32}$$

Now, we consider the errors $e_1 = x_1 - y_1, e_2 = x_2 - y_2$ and $e_3 = x_3 - y_3$,; then, the error dynamics can be written as:

$$\begin{aligned}\dot{e}_1 &= 35(e_2 - e_1) + u_1(t) \\ \dot{e}_2 &= 28e_2 - y_1y_3 + x_1x_3 + u_2(t) \\ \dot{e}_3 &= y_1y_2 - x_1x_2 - 3e_3 + u_3(t)\end{aligned} \tag{33}$$

If we introduce the matrices

$$A = \begin{pmatrix} -35 & 35 & 0 \\ 0 & 14.5 & 0 \\ 0 & 0 & -3 \end{pmatrix}, L(x,y) = \begin{pmatrix} 0 \\ -y_1y_3 + x_1x_3 \\ y_1y_2 - x_1x_2 \end{pmatrix}, u = \begin{pmatrix} u_1(t) \\ u_2(t) \\ u_3(t) \end{pmatrix}. \tag{34}$$

and selecting the matrix B such that (A, B) is controllable: $B = I$, the LQR controller is obtained by using weighting matrices $Q = I$ and $R = B^{\intercal}B = I$. Then, state-feedback matrix is given by

$$K = \begin{pmatrix} 0.0143 & 0.0101 & 0 \\ 0.0101 & 29.0587 & 0 \\ 0 & 0 & 0.1623 \end{pmatrix} \tag{35}$$

From the formerly said, we now present simulations made for the synchronized system of Lü and for the system also synchronized, but after the transformation of its linear part. All simulations here presented were made in *Matlab* software. In **Figure**, we show the trajectories of the master system of Lü. Each line represents one trajectory of the system along the time, taking an initial condition of $(1, 1, 1)$.

For the case of **Figure**, we show the trajectories of the slave system of Lü. As it was in the first case, each line represents one trajectory of the system along the time, taking a initial condition as $(3, 3, 3)$. **Figures** are phase space mappings of each system while maintaining the same initial condition.

On the other hand, in **Figure**, we can see the error magnitude between master and slave systems. Phase space of synchronization of the master and slave systems in **Figure** is presented. Now, we shall present a system showing modifications performed on the Lü attractor. The modified Lü master and slave systems linear and nonlinear parts may be defined as follows:

$$\begin{aligned}\dot{x} &= (T{\circ}A)x + [0 \quad -x_1x_3 \quad x_1x_2\, 0 \quad -x_4x_6 \quad x_4x_5\ ]^{\intercal} \\ \dot{y} &= (T{\circ}A)y + [0 \quad -y_1y_3 \quad y_1y_2\, 0 \quad -y_4y_6 \quad y_4y_5\ ]^{\intercal} + u\end{aligned} \tag{36}$$

Considering the error vector $e = y - x$, then the error dynamics can be written as:

$$\dot{e} = (T{\circ}A)e + L(x,y) + u \tag{37}$$

with $u = -L(x,y) + v$ and $v = -(T{\circ}BK)e$ and

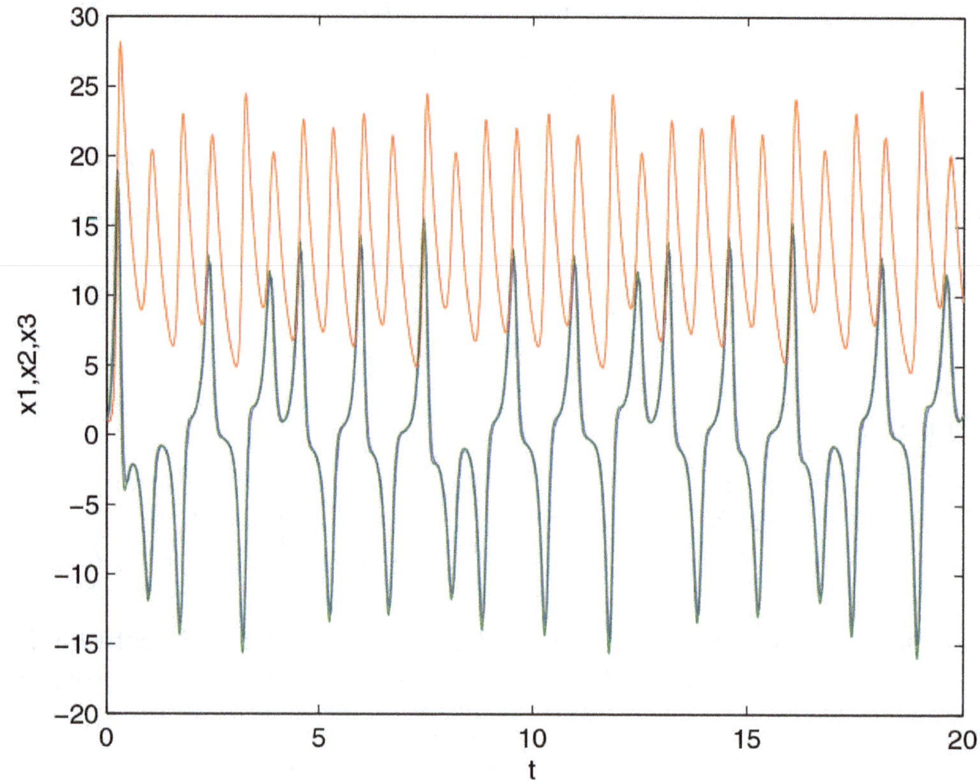

Figure. Master system of Lü.

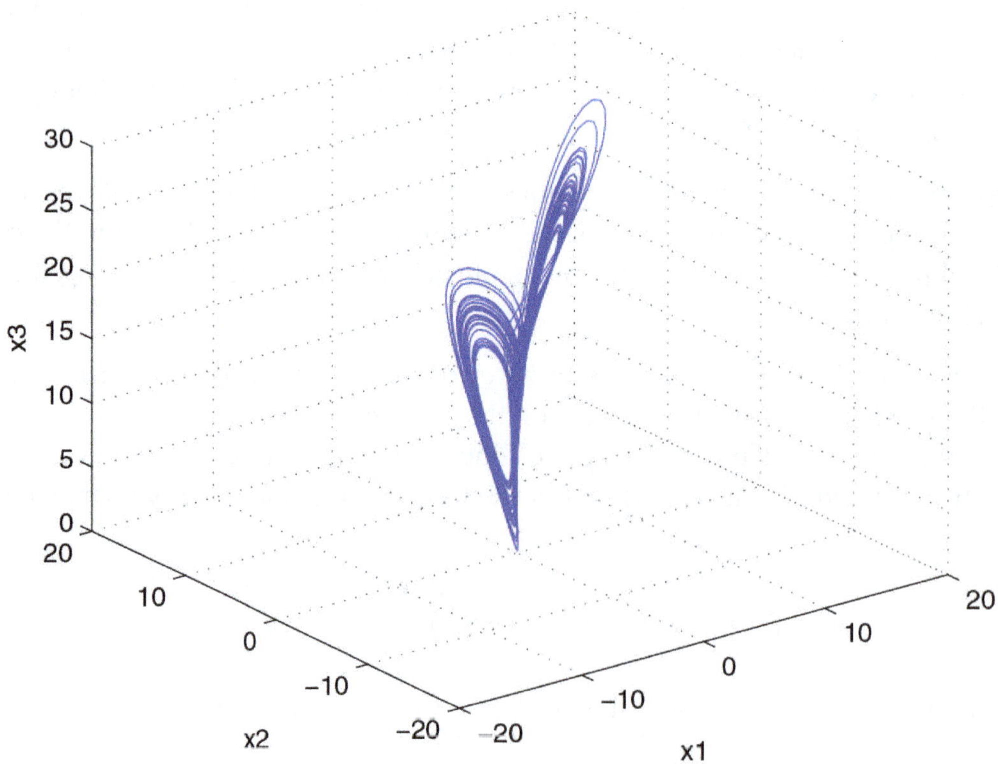

Figure. Master system of Lü.

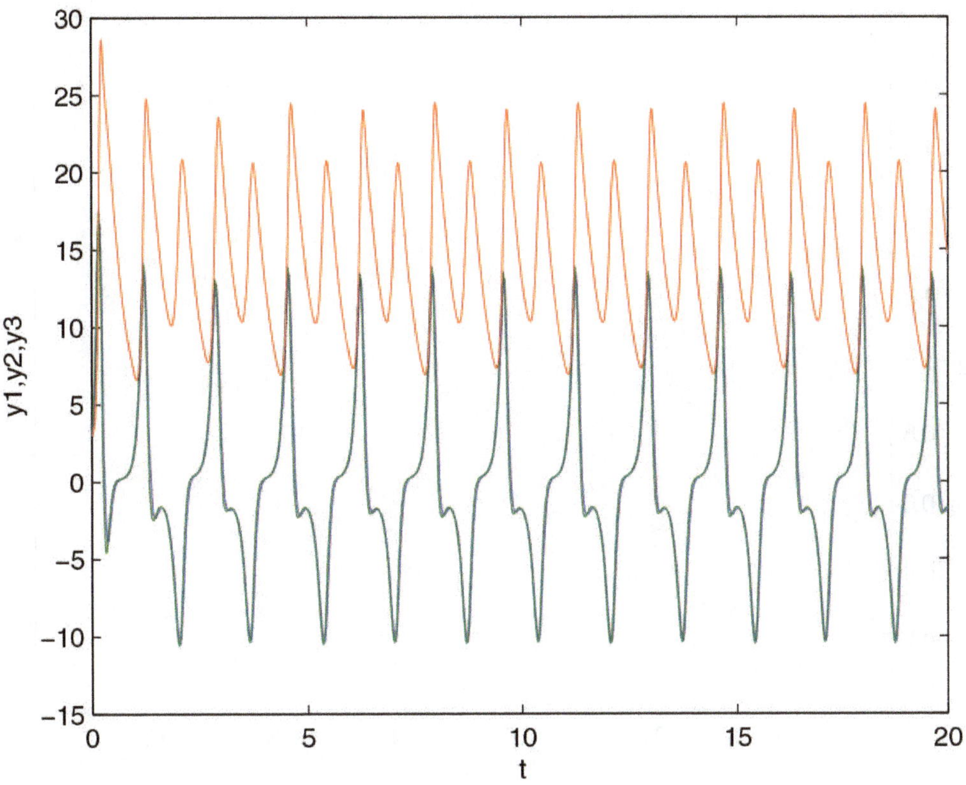

Figure. Slave system of Lü.

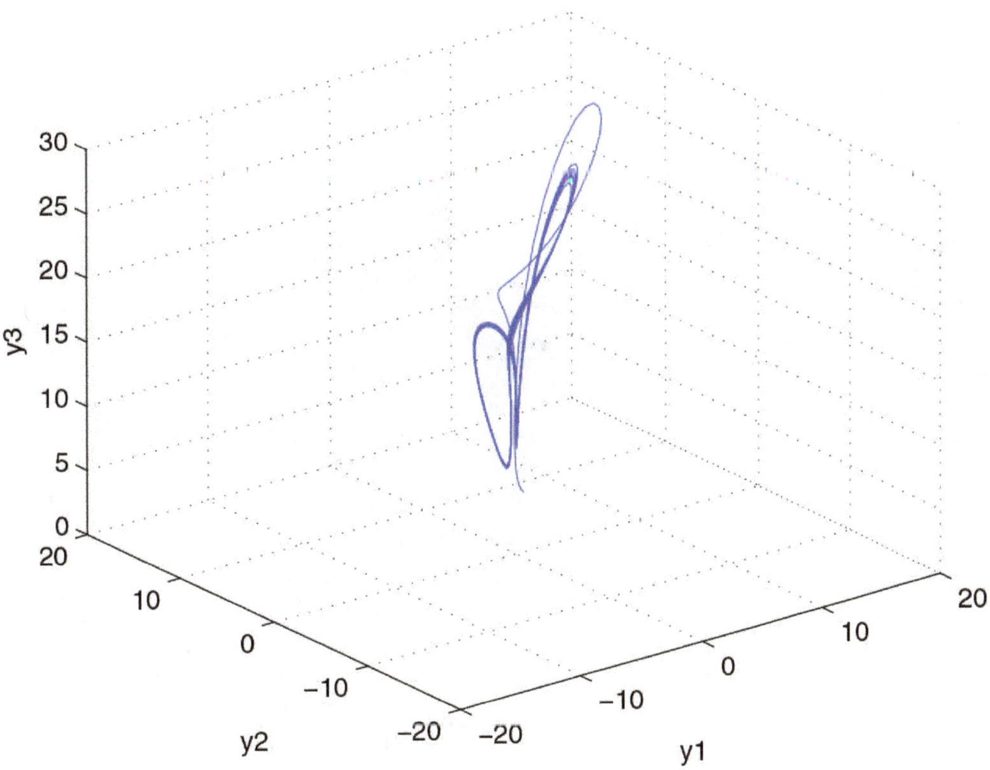

Figure. Slave system of Lü.

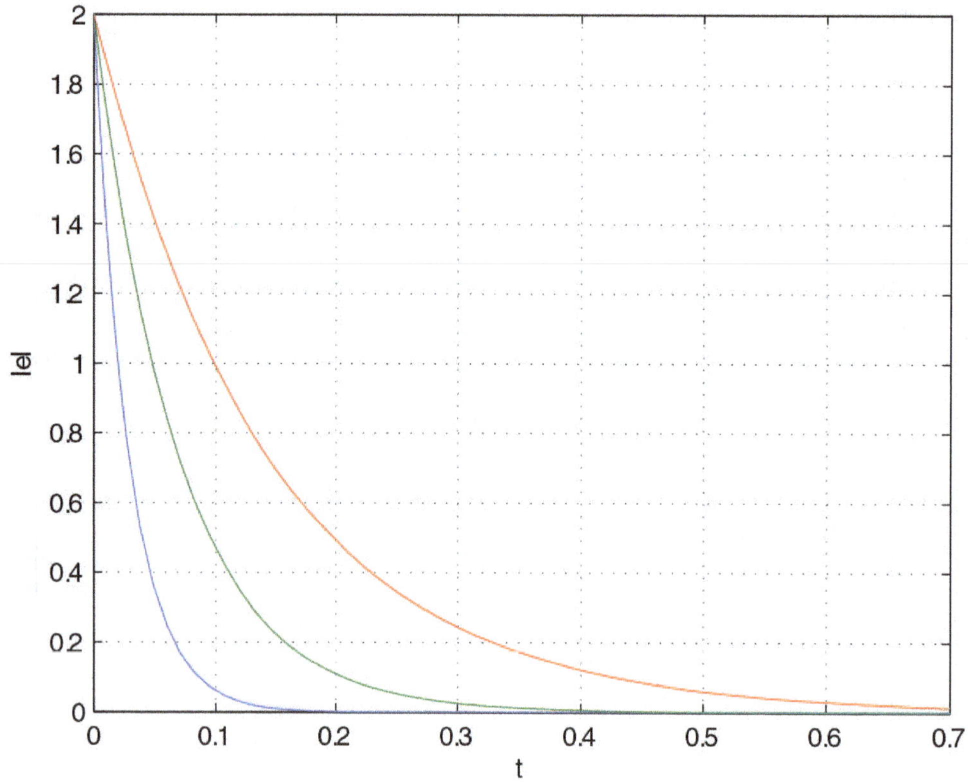

Figure. Magnitude of the error between the master and the slave systems.

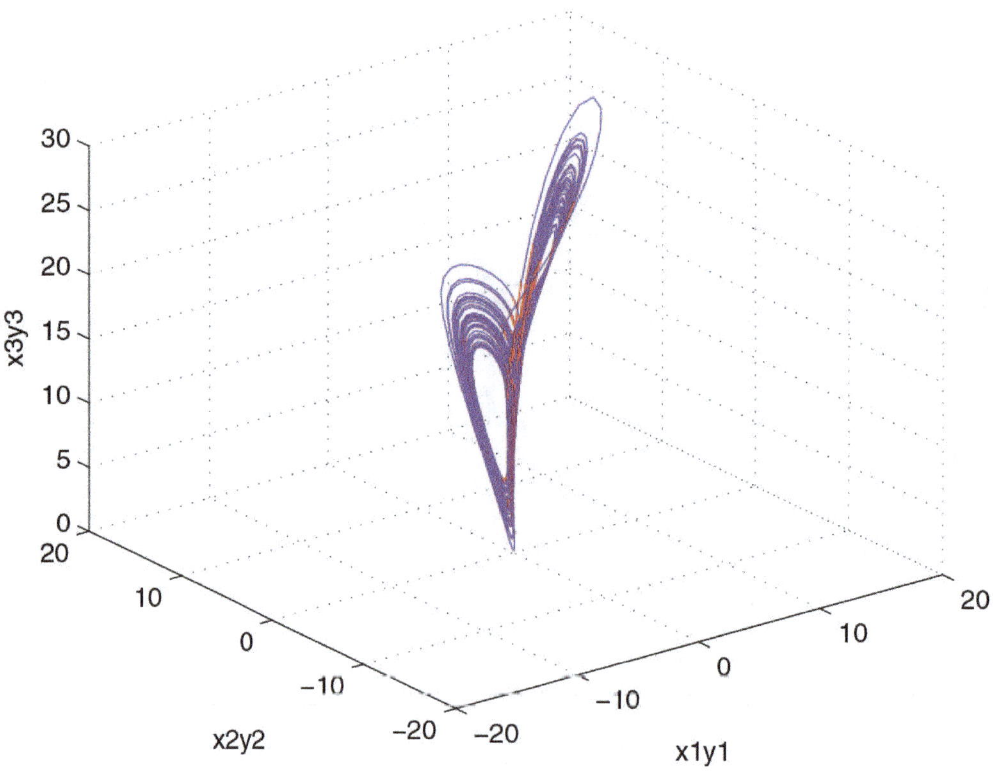

Figure. Synchronization of master and slave system of Lü.

$$A = \begin{pmatrix} -35 & 35 & 0 \\ 0 & 14.5 & 0 \\ 0 & 0 & -3 \end{pmatrix}, T = \begin{pmatrix} 1 & 1 \\ 0 & 1 \end{pmatrix}, B = [111111]^{\mathsf{T}}, \tag{38}$$
$$L(x,y) = [0 \quad -y_1y_3 + x_1x_3 \quad y_1y_2 - x_1x_2 \; 0 \quad -y_4y_6 + x_4x_6 \quad y_4y_5 - x_4x_5]^{\mathsf{T}}$$

Now, the LQR controller is obtained by using weighting matrices, $B = IQ = I$ and $R = B^{\mathsf{T}}B = I$. So the vector $L(x, y)$ takes these values because T is an upper triangular matrix and the value one on the diagonal is repeated.

$$T \circ A = \begin{pmatrix} -35 & 35 & -35 & 35 & 0 & 0 \\ 0 & 14.5 & 0 & 14.5 & 0 & 0 \\ 0 & 0 & -35 & 35 & 0 & 0 \\ 0 & 0 & 0 & 14.5 & 0 & 0 \\ 0 & 0 & 0 & 0 & -3 & -3 \\ 0 & 0 & 0 & 0 & 0 & -3 \end{pmatrix} \tag{39}$$

$$K = \begin{pmatrix} 0.0143 & 0.0101 & 0 & -0.0071 & 0.0050 & 0 \\ 0.0101 & 23.3051 & 0 & -0.0151 & 11.5941 & 0 \\ 0 & 0 & 0.1614 & 0 & 0 & -0.0757 \\ -0.0071 & -0.0151 & 0 & 0.0214 & 0.0050 & 0 \\ 0.0050 & 11.5941 & 0 & 0.0050 & 34.8411 & 0 \\ 0 & 0 & -0.0757 & 0 & 0 & 0.2324 \end{pmatrix} \tag{40}$$

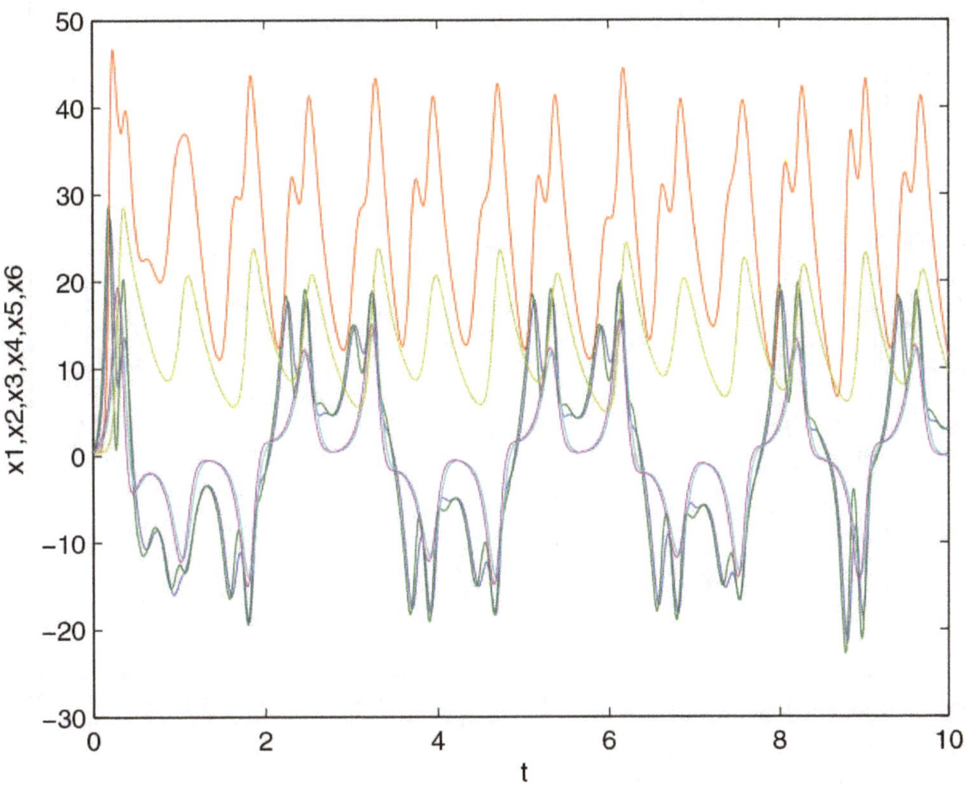

Figure. Transformation of the master system of Lü.

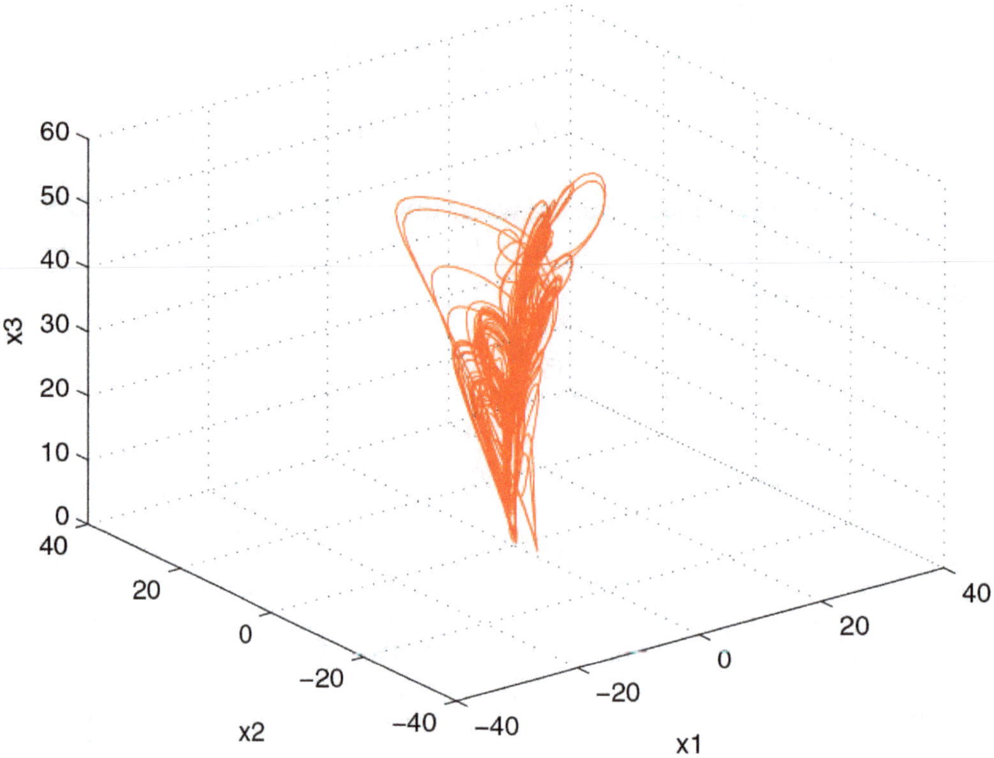

Figure. Phase space of the transformation of the master system of Lü.

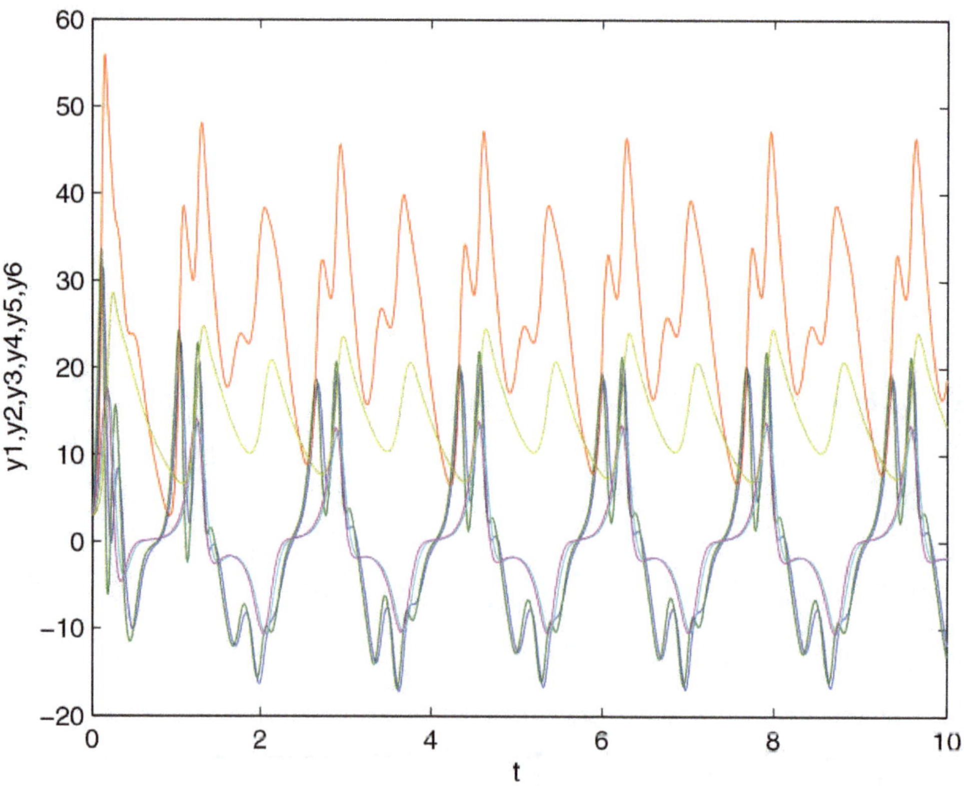

Figure. Transformation of the slave system of Lü.

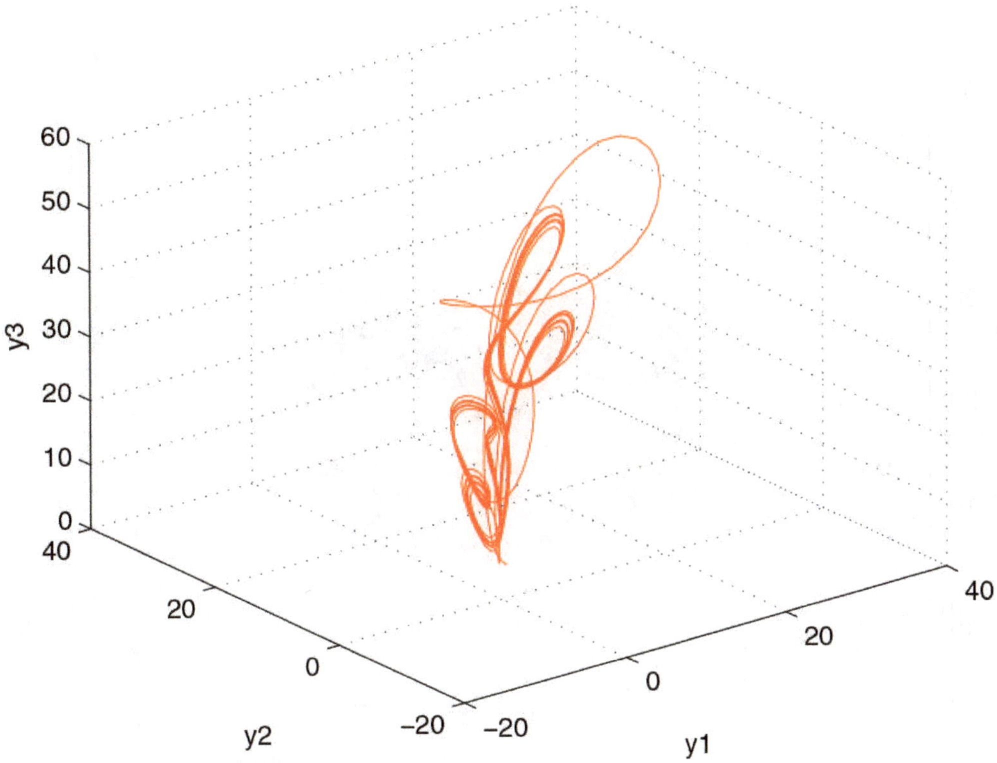

Figure. Phase space of the transformation of the slave system of Lü.

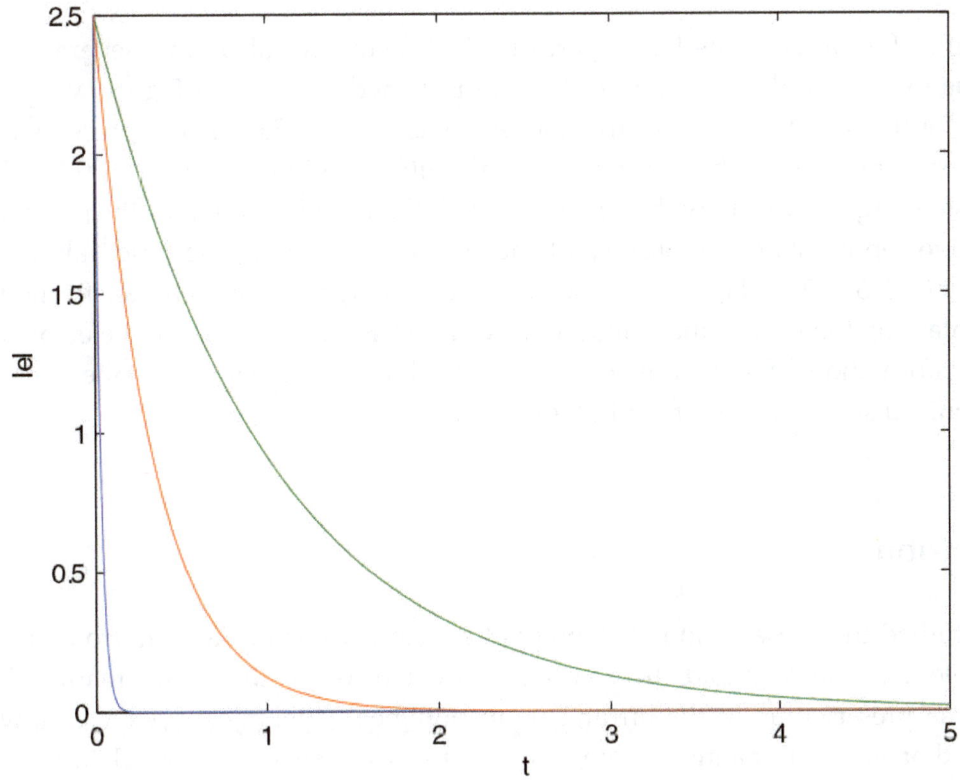

Figure. Magnitude of the error between the transformation of master and slave systems.

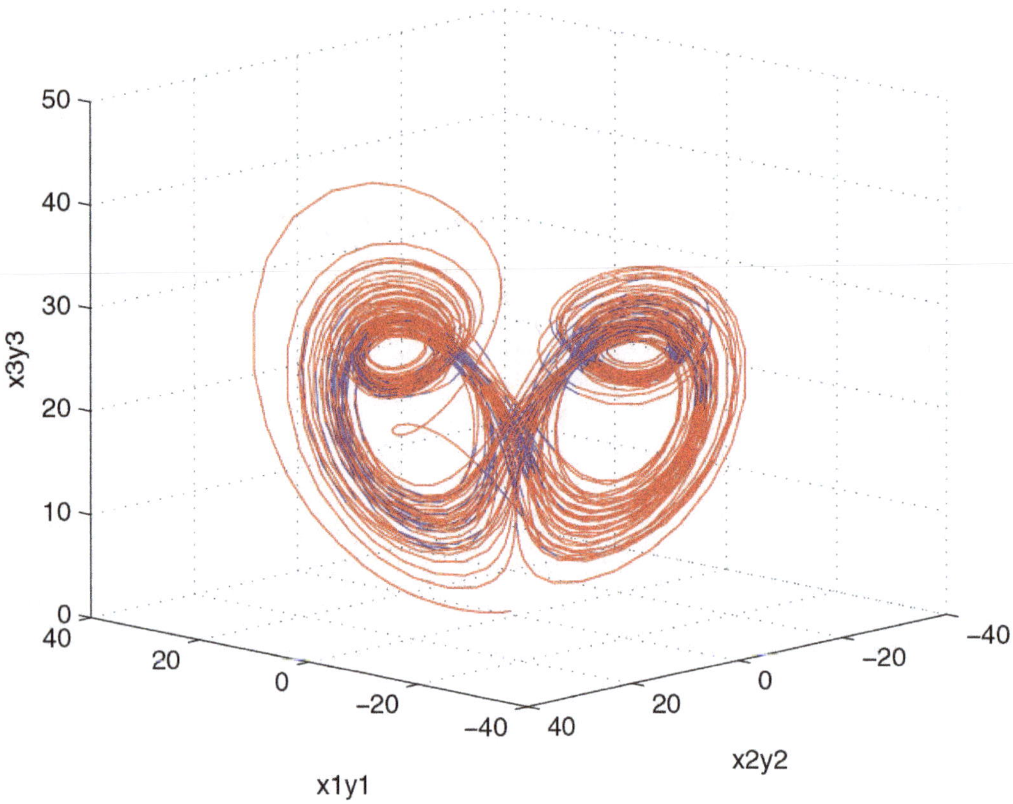

Figure. Synchronization of the transformation of the master and slave systems of Lü.

After the transformation in its linear part of Lü attractor, we also have several simulations allowing us to analyze the dynamics of the transformed system. In **Figure**, we present the trajectories of the transformation of the master system of Lü. Each line represents one trajectory of the system along with the time taking an initial condition of $(0.5, 0.5, 0.5, 0.5, 0.5, 0.5)$. For the case of **Figure**, we show the trajectories of the transformation of the slave system of Lü. Each line represents one trajectory of the system also, along the time, taking an initial condition of $(3, 3, 3, 3, 3, 3)$. **Figures** are the phase space mappings of each transformed system while maintaining the same initial condition. By last, in **Figure**, we can see the error magnitude of the transformation of synchronized system A phase space mapping of the transformation of synchronized system is presented in **Figure**.

6. Conclusion

We have studied the preservation of stability of a chaotic dynamic system, from an extension of the stable-unstable manifold theorem and an extension of the center manifold theorem based on the preservation of the linear part in nonlinear dynamical systems. However, we can check that given a chaotic system, its transformed version is also chaotic. A scheme consisting of a master-slave system for which a controller gain is obtained using a linear-quadratic regulator has been presented and synchronization is achieved and preserved even after the master-slave controller is transformed, obtaining as a consequence that the chaotic system changes to an higher dimension. It is important to note the transformation of the linear part of the chaotic system from Tracy-Singh product in which it was used to modify a Lü system, showing the effectiveness of the proposed method. The results can be extended to other techniques for feedback design, for example, adaptive control, sliding mode regulator and etcetera.

Author details

Guillermo Fernadez-Anaya[1*], Luis Alberto Quezada-Téllez[1], Jorge Antonio López-Rentería[1], Oscar A. Rosas-Jaimes[2], Rodrigo Muñoz-Vega[3], Guillermo Manuel Mallen-Fullerton[1] and José Job Flores-Godoy[4]

*Address all correspondence to: guillermo.fernandez@ibero.mx

1 Departamento de Física y Matemáticas, Universidad Iberoamericana, Ciudad de México, México

2 Facultad de Ingeniería, Universidad Autónoma del Estado de México, Toluca, Estado de, México

3 Universidad Autónoma de la Ciudad de México, Ciudad de México, México

4 Departamento de Matemática, Facultad de Ingeniería y Tecnologías, Universidad Católica del Uruguay, Uruguay

References

[1] H. Khalil. Nonlinear Systems. 3rd ed. Prentice Hall; New Jersey; 2001.

[2] D. J. Watts and S. H. Strogatz. Collective dynamics of small-world. Nature. 1998;**393**: 440–442.

[3] D. Becker-Bessudo, G. Fernandez-Anaya and J. J. Flores-Godoy. Preserving synchronization under matrix product modifications. Physica A: Statistical Mechanics and Its Applications. 2008;**387**(26): 6631–6645.

[4] T. E. Djaferis. Stability preserving maps and robust design. International Journal of Control. 2002;**75**(9): 680–690.

[5] G. Fernández-Anaya, J. C. Martínez and V. Kucera. Characterizing families of positive real matrices by matrix substitutions on scalar rational functions. Systems & Control Letters. 2006;**55**(11): 871–878.

[6] G. Fernandez-Anaya, J. J. Flores-Godoy, R. Femat and J. J. Alvarez-Ramire. Preservation of stability and synchronization in nonlinear system. Physics Letters A. 2007;**371**(3): 205–212.

[7] D. Becker-Bessudo, G. Fernández-Anaya and J. J. Flores-Godoy. Preserving synchronization using nonlinear modifications in the Jacobian matrix. Communications in Nonlinear Science and Numerical Simulation. 2011;**16**(2): 940–957.

[8] G. Fernandez-Anaya, J. J. Flores-Godoy and J. J. Alvarez-Ramirez. Synchronization preservation of dynamical networks. In: J. S. Moreno, editor. Progress in Statistical Mechanics Research. Nova Science Publishers; New York; 2008: 323–347.

[9] L. Perko. Differential Equations and Dynamical Systems. 3rd ed. Springer-Verlag; New York Inc. 2001.

[10] S. Liu. Matrix results on the Khatri-Rao and Tracy Singh products. Linear Algebra and Its Applications. 1999;**289**: 267–277.

[11] J. Lü and G. Chen. A new chaotic attractor coined. International Journal of Bifurcations and Chaos. 2002;**12**(3): 659–661.

3

Emergence of Classical Distributions from Quantum Distributions: The Continuous Energy Spectra Case

Gabino Torres-Vega

Abstract

We explore the properties of quantum states and operators that are conjugate to the Hamiltonian eigenstates and operator when the Hamiltonian spectrum is continuous, i.e., we find time-like operators $\widehat{T}$ such that $[\widehat{T}, \widehat{H}] = i\hbar$. This is a property expected for a time operator. We explicitly unfold the momentum sign degeneracy of energy states. We consider the free-particle case, and we find, among other things, that the time states are also the solution of the quantized version of the classical motion of the particle.

Keywords: time operator, time eigenstates, conjugate states, free-particle time eigenstates

1. Introduction

The problem of the time operator in quantum mechanics has been studied by numerous researchers for many years and remains a subject of current research. There are many instances in which a time variable is useful. An example of such a situation is calculating the tunneling time of a particle passing through a barrier. This time was recently measured, and it was shown to vanish [1, 2].

There are several approaches in this area that were developed by Kijowski [3], Hegerfeldt et al. [4], Weyl [5], Galapon [6], Arai and Yasumichi [7, 8], Strauss et al. [9, 10], and Hall [11], among others. The work by these authors may appear to be in four differing approaches; however, we shall show that they are simply different approaches to the same theme, approximated ones.

Some of these approaches are similar to the work of Weyl on periodic functions [5]. Weyl defined the Hermitian form

$$-i\sum_{n\neq m}\frac{(-1)^{n-m}}{n-m}x_m x_n, \tag{1}$$

where $\{x_m\}$ are the components of a vector in the basis $e^{i2\pi m/n}/\sqrt{n}$, $m = 0,1,\ldots, n-1$. Galapon, Arai et al., Straus et al., and Hall used a similar expression but with a factor of one instead of the $(-1)^{n-m}$ factor. Their results are valid in a limited region of the Hilbert space for the expression of Galapon and Arai. Strauss wanted to obtain a Lyapunov function; instead, he obtained a function that only gives the sign of time, as was shown by Hall. A different factor might result in a time operator that would be valid over the entire Hilbert space. In this chapter, we find a proper factor to obtain sensible time-like kets and operators that are valid over the entire Hilbert space, for the purely continuous energy spectrum case.

We introduce time-like kets and operators following a different route. We search for the states that are conjugate to the energy eigenstates, which is a natural approach to this subject. We find time kets and operators that are valid over the entire Hilbert space. We also find that we can make contact with the operators defined by other authors. These operators lack the oscillatory function found in this work.

Time is typically viewed as a parameter and not as a dynamical variable in classical and quantum mechanics. However, the characteristics of the time variable depend on the representation being considered. In classical mechanics, we have shown that we can talk of translations along the energy direction; in that case, the energy variable becomes a parameter, and time becomes a dynamical variable, a function of the phase-space variables [12].

For comparison, let us consider the coordinate representation of quantum mechanics. If a variable, s, with units of length is the parameter used in the shifting along the coordinate direction through the displacement operator, $e^{is\hat{P}/\hbar}$, in the momentum representation, s becomes the coordinate operator and the momentum $\hat{P}$ becomes a parameter. A similar behavior is expected when considering energy-time representations. However, the problem is to define a time representation in quantum mechanics, and we use the conjugacy concept in this chapter to find such a representation.

The basis for this work is that time is another coordinate that has to be determined. The conjugate pair coordinate-momentum is a pair of conjugate coordinates that are used to define representations of wave functions and operators. Similarly, energy and time can be used as an alternative coordinate set, but the time coordinate has to be defined. As coordinate and momentum eigenstates, the time eigenstates will also be nonnormalizable, and their peculiarities originate from the type of coordinate that energy is a semibounded quantity.

In Section, we use the rewriting of the identity operator in terms of energy eigenstates to define the states that are conjugate to the energy eigenstates and subsequently determine some of their properties and several time-like operators. We define time states for negative and positive momentum values.

Section is devoted to time-like operators and their properties. Time operators are written in three different forms. We verify that the time kets are eigenkets of the time operators. We find "evolution equations" for time kets and note that the time operators are the generators for translations along the energy direction. We also discuss how a wave packet is shifted along the energy direction.

In numerical calculations, we have to address finite regions of variables and not infinite intervals. Therefore, we focus our attention on approximate expressions for time operators in Section. We find approximate expressions of time operators that can be used in numerical calculations and are of help in the understanding of the expressions found by other authors.

The free-particle problem is analyzed in Section. We find expressions for the time kets for the free particle. The coordinate matrix elements of the time operators are also found, and we learn that the time states are also a solution to the quantum analog of the classical motion. The support of the time states embodies the classical trajectories, and as $\hbar \to 0$, we recover the classical motion.

We conclude the chapter with some concluding remarks.

2. Time eigenstates

In this section, we define the states that are conjugate to the energy eigenstates and the corresponding conjugate operator to a given quantum Hamiltonian $\widehat{H}$. We also derive some of their properties. The definition of conjugacy between the operators $\widehat{T}$ and $\widehat{H}$ that we will use here is the usual one, i.e., that these operators should comply with the constant commutator relationship $[\widehat{T}, \widehat{H}] = i\hbar$. We will consider the case of a purely continuous energy spectrum with a Hamiltonian operator $\widehat{H}$ of the form $\widehat{H} = \widehat{P}^2/2m + \widehat{V}(\widehat{Q})$, where $\widehat{P}$ is the momentum operator, $\widehat{Q}$ is the coordinate operator, and $\widehat{V}(\widehat{Q})$ is the potential energy operator. We will also consider that the sign of the momentum operator commutes with the Hamiltonian. The continuous eigenvalues of the Hamiltonian are denoted by $E \in [0, \infty)$ and correspond to the eigenkets $\{|E\rangle\}$.

We will base our definition of time states on rewriting the identity operator in terms of energy eigenstates and using the integral representation of the Dirac delta function. We assume that the Hamiltonian is self-adjoint. Thus, we will work on the span of the Hamiltonian eigenstates, denoted by

$$D = \left\{ |\psi\rangle | \, |\psi\rangle| = \int_0^\infty dE\psi(E)|E\rangle, \quad \psi(E) = \langle E|\psi\rangle \right\} \tag{2}$$

We assume that the closure relationship for the energy eigenstates holds, $\hat{I} = \int_0^{E_m} dE|E\rangle\langle E|$. The i times the derivative is self-adjoint in a finite interval and hence will work in the subspace $E \in [0, E_m]$, $E_m < \infty$, which implies that $p \in [-p_m, p_m]$, $p_m < \infty$.

We start with the rewriting of the identity operator in terms of the energy eigenkets,

$$\hat{I} = \int_0^{E_m} dE|E\rangle\langle E| = \int_0^{E_m} dE'dE\delta(E-E')|E'\rangle\langle E| = \int_0^{E_m} dE'dE\frac{1}{2\pi\hbar}\int_{-\infty}^{\infty} dt\, e^{it(E-E')/\hbar}|E'\rangle\langle E|$$
$$= \int_{-\infty}^{\infty} dt\int_0^{E_m} dE'dE\frac{e^{-itE'/\hbar}}{\sqrt{2\pi\hbar}}|E'\rangle\langle E|\frac{e^{itE/\hbar}}{\sqrt{2\pi\hbar}}, \tag{3a}$$

where we have made use of the properties of the Dirac delta function. We can separate the negative and positive momentum parts of the above expression by means of the closure relationship for the momentum states, obtaining

$$\hat{I} = \int_{-p_m}^{p_m} dp\int_0^{E_m} dE|E\rangle\langle E|p\rangle\langle p|E\rangle\langle E| = \int_{-p_m}^{p_m} dp\int_0^{E_m} dE'dE\delta(E-E')|E'\rangle\langle E'|p\rangle\langle p|E\rangle\langle E|$$
$$= \int_{-p_m}^{p_m} dp\int_0^{E_m} dE'dE\frac{1}{2\pi\hbar}\int_{-\infty}^{\infty} dte^{it(E-E')/\hbar}|E'\rangle\langle E'|p\rangle\langle p|E\rangle\langle E| \tag{3b}$$
$$= \int_{-\infty}^{\infty} dt\int_{-p_m}^{p_m} dp\int_0^{E_m} dE'dE\frac{e^{-itE'/\hbar}}{\sqrt{2\pi\hbar}}|E'\rangle\langle E'|p\rangle\langle p|E\rangle\langle E|\frac{e^{itE/\hbar}}{\sqrt{2\pi\hbar}}.$$

Thus, we define time-like kets as

$$|t\rangle := \int_0^{E_m} dE\frac{e^{-itE/\hbar}}{\sqrt{2\pi\hbar}}|E\rangle, \quad |t(p)\rangle := \int_0^{E_m} dE\frac{e^{-itE/\hbar}}{\sqrt{2\pi\hbar}}|E\rangle\langle E|p\rangle. \tag{4}$$

With these kets, the identity operator is written as

$$\hat{I} = \int_{-\infty}^{\infty} dt|t\rangle\langle t| = \int_{-\infty}^{\infty} dt\int_{-p_m}^{p_m} dp|t(p)\rangle\langle t(p)| = \hat{I}_- + \hat{I}_+, \tag{5a}$$

where

$$\hat{I}_- : = \int_{-\infty}^{\infty} dt\int_{-p_m}^{0} dp|t(p)\rangle\langle t(p)|, \quad \hat{I}_+ := \int_{-\infty}^{\infty} dt\int_0^{p_m} dp|t(p)\rangle\langle t(p)|. \tag{5b}$$

Then, the identity operator is written in terms of the time evolution of some bras and kets, which are composed of all the energy eigenstates.

Now, we define time-like operators $\widehat{T}$ and $\widehat{T}_\pm$ by introducing a factor t in the integrand of Eq. (5):

$$\widehat{T} = \int_{-\infty}^{\infty} dt\; t|t\rangle\langle t|, \tag{6a}$$

and

$$\widehat{T}_- := \int_{-\infty}^{\infty} dt\, t \int_{-p_m}^{0} dp|t(p)\rangle\langle t(p)|, \quad \widehat{T}_+ := \int_{-\infty}^{\infty} dt\, t \int_{0}^{p_m} dp|t(p)\rangle\langle t(p)|. \tag{6b}$$

The function $e^{itE/\hbar}$ exists only for $E \in [0, E_m]$, so that, for the sake of simplicity of notation, we, sometimes, will include explicitly the function $\Theta(E) - \Theta(E - E_m)$, where Θ is the step function, when necessary, otherwise we will omit this factor.

The commutator between these operators and the Hamiltonian operator is

$$\begin{aligned}
[\widehat{T}, \widehat{H}] &= \int_{-\infty}^{\infty} dt\, t \int_0^{E_m} dE' dE \frac{e^{-itE'/\hbar}}{\sqrt{2\pi\hbar}} \frac{e^{itE/\hbar}}{\sqrt{2\pi\hbar}} \left[|E'\rangle\langle E|, \widehat{H}\right] \\
&= \int_0^{E_m} dE' dE \frac{1}{2\pi\hbar} \int_{-\infty}^{\infty} dt\, t\, e^{it(E-E')/\hbar} (E - E')|E'\rangle\langle E| \\
&= \int_0^{E_m} dE' dE \frac{1}{2\pi\hbar} \int_{-\infty}^{\infty} dt \left[\left(- i\hbar \frac{\partial}{\partial E} [\Theta(E) - \Theta(E - E_m)] \right.\right. \\
&\quad \left. + i\hbar[\delta(E) - \delta(E - E_m)]\right) e^{it(E-E')/\hbar}](E - E')|E'\rangle\langle E| \\
&= \int_0^{E_m} dE' dE \frac{1}{2\pi\hbar} \int_{-\infty}^{\infty} dt\, e^{it(E-E')/\hbar} [[\Theta(E) - \Theta(E - E_m)] i\hbar \frac{\partial}{\partial E} \\
&\quad + i\hbar[\delta(E) - \delta(E - E_m)]](E - E')|E'\rangle\langle E| \\
&\quad - i\hbar \int_0^{E_m} dE' \frac{1}{2\pi\hbar} \int_{-\infty}^{\infty} dt\, e^{it(E-E')/\hbar} (E - E')|E'\rangle\langle E'|_{E=0}^{E_m} \\
&= \int_0^{E_m} dE' dE \delta(E - E') \left[i\hbar \frac{\partial}{\partial E} + i\hbar[\delta(E) - \delta(E - E_m)] \right] (E - E')|E'\rangle\langle E| \\
&\quad - i\hbar \int_0^{E_m} dE' \delta(E - E')(E - E')|E'\rangle\langle E||_{E=0}^{E_m} \\
&= \int_0^{E_m} dE' dE\, \delta(E - E') i\hbar \frac{\partial}{\partial E} (E - E')|E'\rangle\langle E| \\
&\quad + i\hbar \int_0^{E_m} dE' dE\, \delta(E - E')[\delta(E) - \delta(E - E_m)](E - E')|E'\rangle\langle E| \\
&= i\hbar \int_0^{E_m} dE' dE\, \delta(E - E') E'\rangle\langle E| \\
&\quad + \int_0^{E_m} dE' dE \delta(E - E')(E - E')|E'\rangle i\hbar \frac{\partial}{\partial E} \langle E| = i\hbar \int_0^{E_m} dE|E\rangle\langle E| \\
&= i\hbar \hat{I},
\end{aligned} \tag{7a}$$

where we have made use of the integration by parts. This is one of the properties that a time operator should comply with—the constant commutator with the Hamiltonian. We also have that

$$
\begin{aligned}
\left[\widehat{T}_-,\widehat{H}\right] &= \int_{-\infty}^{\infty} dt\, t \int_{-p_m}^{0} dp \int_{0}^{E_m} dE' dE \frac{e^{-itE'/\hbar}}{\sqrt{2\pi\hbar}} \frac{e^{itE/\hbar}}{\sqrt{2\pi\hbar}} \langle E'|p\rangle\langle p|E\rangle \left[|E'\rangle\langle E|, \widehat{H}\right] \\
&= \int_{-p_m}^{0} dp \int_{0}^{E_m} dE' dE \frac{1}{2\pi\hbar} \int_{-\infty}^{\infty} dt\, t e^{it(E-E')/\hbar} (E-E')|E'\rangle\langle E'|p\rangle\langle p|E\rangle\langle E| \\
&= \int_{-p_m}^{0} dp \int_{0}^{E_m} dE' dE \frac{1}{2\pi\hbar} \int_{-\infty}^{\infty} dt[(-i\hbar \frac{\partial}{\partial E}[\Theta(E) - \Theta(E-E_m)] \\
&\quad + i\hbar[\delta(E) - \delta(E-E_m)] e^{it(E-E')/\hbar}\Big) (E-E')|E'\rangle\langle E'|p\rangle\langle p|E\rangle\langle E| \\
&= \int_{-p_m}^{0} dp \int_{0}^{E_m} dE' dE \frac{1}{2\pi\hbar} \int_{-\infty}^{\infty} dt\, e^{it(E-E')/\hbar} \Big[[\Theta(E) - \Theta(E-E_m)\Big] i\hbar \frac{\partial}{\partial E} \\
&\quad + i\hbar[\delta(E) - \delta(E-E_m)](E-E')|E'\rangle\langle E'|p\rangle\langle p|E\rangle\langle E| \\
&\quad - i\hbar \int_{-p_m}^{0} dp \int_{0}^{E_m} dE' \frac{1}{2\pi\hbar} \int_{-\infty}^{\infty} dt\, e^{it(E-E')/\hbar} (E-E')|E'\rangle\langle E'|p\rangle\langle p|E\rangle\langle E| |_{E=0}^{E_m} \\
&= \int_{-p_m}^{0} dp \int_{0}^{E_m} dE' dE \delta(E-E') \left[i\hbar \frac{\partial}{\partial E} + i\hbar[\delta(E) - \delta(E-E_m)] \right] \\
&\quad (E-E')|E'\rangle\langle E'|p\rangle\langle p|E\rangle\langle E| \\
&\quad - i\hbar \int_{-p_m}^{0} dp \int_{0}^{E_m} dE' \delta(E-E')(E-E')|E'\rangle\langle E'|p\rangle\langle p|E\rangle\langle E|_{E=0}^{E_m} \\
&= \int_{-p_m}^{0} dp \int_{0}^{E_m} dE' dE \delta(E-E') i\hbar \frac{\partial}{\partial E} (E-E')|E'\rangle\langle E'|p\rangle\langle p|E\rangle\langle E| \\
&\quad + i\hbar \int_{-p_m}^{0} dp \int_{0}^{E_m} dE' dE \delta(E-E')[\delta(E) - \delta(E-E_m)] \\
&\quad (E-E')|E'\rangle\langle E'|p\rangle\langle p|E\rangle\langle E| \\
&= i\hbar \int_{-p_m}^{0} dp \int_{0}^{E_m} dE' dE \delta(E-E')|E'\rangle\langle E'|p\rangle\langle p|E\rangle\langle E| \\
&\quad + \int_{-p_m}^{0} dp \int_{0}^{E_m} dE' dE \delta(E-E')(E-E')|E'\rangle\langle E'|p\rangle i\hbar \frac{\partial}{\partial E} \langle p|E\rangle\langle E| \\
&= i\hbar \int_{-p_m}^{0} dp \int_{0}^{E_m} dE |E\rangle\langle E|p\rangle\langle p|E\rangle\langle E| = i\hbar \hat{I}_-,
\end{aligned} \tag{7b}
$$

$$
[\widehat{T}_+,\widehat{H}] = i\hbar \int_{0}^{p_m} dp \int_{0}^{E_m} dE |E\rangle\langle E|p\rangle\langle p|E\rangle\langle E| = i\hbar \hat{I}_+. \tag{7c}
$$

The operator

$$
-i\hbar \frac{\partial}{\partial E} + i\hbar[\delta(E) - \delta(E-E_m)] \tag{8}
$$

is a time-like operator in the energy representation, which is symmetric in the interval $[0, E_m]$ regardless of the boundary conditions at $E = 0$, E_m, when the functions exist only in the interval $[0, E_m]$.

Thus, we can say that the kets

$$|t(p)\rangle := \int_0^{E_m} dE \frac{e^{-itE/\hbar}}{\sqrt{2\pi\hbar}} |E\rangle\langle E|p\rangle = \frac{e^{-it\hat{H}/\hbar}}{\sqrt{2\pi\hbar}} |p\rangle, \tag{9a}$$

$$|t\rangle := \int_0^{E_m} dE \frac{e^{-itE/\hbar}}{\sqrt{2\pi\hbar}} |E\rangle, \tag{9b}$$

can be considered as time-like kets. We will study some of their properties in what follows.

The inner product between time states is

$$\langle t'(p')|t(p)\rangle = \frac{1}{2\pi\hbar} \langle p'|e^{-i(t-t')\hat{H}/\hbar}|p\rangle = \frac{1}{2\pi\hbar} \langle p'|p(t-t')\rangle, \tag{10a}$$

$$\begin{aligned} \langle t'|t\rangle &= \int_0^{E_m} dE \langle t'|E\rangle \langle E|t\rangle = \int_0^{E_m} dE \frac{e^{it'E/\hbar}}{\sqrt{2\pi\hbar}} \frac{e^{-itE/\hbar}}{\sqrt{2\pi\hbar}} = \frac{1}{2\pi\hbar} \int_0^{E_m} dE\, e^{i(t'-t)E/\hbar} \\ &= \frac{1}{\pi(t'-t)} e^{i(t'-t)E_m/2h} \sin\left(\frac{E_m}{2h}(t'-t)\right), \end{aligned} \tag{10b}$$

with limit

$$\lim_{E_m \to \infty} \langle t'|t\rangle = \frac{1}{2}\delta(t'-t) + \frac{i}{2\pi(t'-t)}. \tag{10c}$$

Thus, the time states are not orthogonal due to the bounded nature of the Hamiltonian operator.

2.1. Properties of the transformation function between energy and time states

The transformation function between energy and time representations is given by

$$\langle E|t\rangle = \frac{e^{-itE/\hbar}}{\sqrt{2\pi\hbar}}, \quad E \in [0, E_m], \quad t \in (-\infty, \infty). \tag{11}$$

A property of this transformation function is that it is a sort of eigenfunction of the time-like operator, $i\hbar(d/dE)[\Theta(E) - \Theta(E - E_m)] - i\hbar[\delta(E) - \delta(E - E_m)]$, when the functions exists in the interval $E \in [0, E_m]$, in the energy representation,

$$\left[i\hbar[\delta(E) - \delta(E - E_m)] - i\hbar\frac{\partial}{\partial E}[\Theta(E) - \Theta(E - E_m)]\right] \langle t|E\rangle = t[\Theta(E) - \Theta(E - E_m)] \langle t|E\rangle, \tag{12}$$

and it is also an eigenfunction of the energy operator, $i\hbar\, d/dt$,

$$i\hbar\frac{\partial}{\partial t}\langle E|t\rangle = i\hbar\frac{\partial}{\partial t}\frac{e^{-itE/\hbar}}{\sqrt{2\pi\hbar}} = E\langle E|t\rangle. \tag{13}$$

This is similar to the corresponding properties of the transformation function between coordinate and momentum representations. The squared modulus of the transformation function is constant for all values of t and E, as is desired for coordinate variables.

Time kets can be used as a coordinate system for quantum systems and are similar to coordinate or momentum eigenkets. The norm of a wave packet in the time representation is ,

$$\begin{aligned}\langle\psi|\psi\rangle &= \int_{-\infty}^{\infty} dt\langle\psi|t\rangle\ \langle t|\psi\rangle = \int_{-\infty}^{\infty} dt\int_0^{E_m} dE^{'}dE\frac{e^{-itE^{'}/\hbar}}{\sqrt{2\pi\hbar}}\langle\psi|E'\rangle\frac{e^{itE/\hbar}}{\sqrt{2\pi\hbar}}\langle E|\psi\rangle \\ &= \int_0^{E_m} dE'dE\ \langle\psi|E^{'}\rangle\ \langle E|\psi\rangle\frac{1}{2\pi\hbar}\int_{-\infty}^{\infty} dt\ e^{it(E-E^{'})/\hbar} \\ &= \int_0^{E_m} dE^{'}dE\langle\psi|E'\rangle\ \langle E|\psi\rangle\delta(E-E^{'}) \\ &= \int_0^{E_m} dE|\langle E|\psi\rangle|^2.\end{aligned} \tag{14}$$

Thus, we will obtain well-defined quantities if the wave packet $|\psi\rangle$ is normalized in the energy representation, i.e., if $\int_0^{\infty} dE|\langle E|\psi\rangle|^2 = 1$. We also note that the transformation from energy to time representations is norm preserving, i.e., it is unitary.

2.2. The time eigenstates are conjugate to the energy eigenstates

Now, the Fourier transform of the time states is

$$\begin{aligned}\int_{-\infty}^{\infty} dt\frac{e^{itE/\hbar}}{\sqrt{2\pi\hbar}}|t\rangle &= \int_{-\infty}^{\infty} dt\frac{e^{itE/\hbar}}{\sqrt{2\pi\hbar}}\int_0^{E_m} dE^{'}\frac{e^{-itE'/\hbar}}{\sqrt{2\pi\hbar}}|E^{'}\rangle = \int_0^{E_m} dE'|E'\rangle\frac{1}{2\pi\hbar}\int_{-\infty}^{\infty} dt\ e^{it(E-E')/\hbar} \\ &= \int_0^{E_m} dE'|E'\rangle\delta(E-E') = |E\rangle,\end{aligned} \tag{15}$$

Thus, the kets $|t\rangle$ and $|E\rangle$ are conjugate indeed, i.e., the definition (9) is consistent; $|t\rangle$ and $|E\rangle$ are the Fourier transforms of each other, and then an eigenstate contains all the conjugate eigenstates with the same weight.

3. Time operators

We now focus on the time operators obtained from the time kets of the previous section and on their properties. Time operators for negative, positive, and any value of the momentum are defined as

$$\hat{T}_- = \int_{-\infty}^{\infty} dt\int_{-p_m}^{0} dp|t(p)\rangle t\langle t(p)|,\ \hat{T}_+ = \int_{-\infty}^{\infty} dt\int_0^{p_m} dp|t(p)\rangle t\langle t(p)|, \tag{16a}$$

and

$$\hat{T} = \int_{-\infty}^{\infty} dt|t\rangle t\langle t|. \tag{16b}$$

The last construction was also introduced, from another perspective, by Hegerfeldt et al. [4]. Our construction is different from that of Hegerfeldt et al. because it involves all the energy eigenstates and not only those that are time reflection invariant. Our time operator exhibits the time reversal property already.

Time operators can be written in three equivalent forms in the energy representation. One form is

$$\begin{aligned}
\widehat{T}_{-} &= \int_{-\infty}^{\infty} dt \int_{-p_m}^{0} dp|t(p)\rangle t\langle t(p)| \\
&= \int_{-\infty}^{\infty} dt \int_{-p_m}^{0} dp \int_{0}^{E_m} dE'dE \frac{e^{-it\ E'/\hbar}}{\sqrt{2\pi\hbar}} |E'\rangle\langle E'|p\rangle\ t\ \langle p|E\rangle\ \langle E| \frac{e^{it\ E/\hbar}}{\sqrt{2\pi\hbar}} \\
&= \frac{1}{2\pi\hbar} \int_{-\infty}^{\infty} dt \int_{-p_m}^{0} dp \int_{0}^{E_m} dE'dE \Big(i\hbar \frac{\partial}{\partial E^{'}} [\Theta(E') - \Theta(E' - E_m)] e^{-it\ E'/\hbar} \\
&\quad -i\hbar[\delta(E') - \delta(E' - E_m)] e^{-it\ E'/\hbar} \Big) |E'\rangle\langle E'|p\rangle\ \langle p|E\rangle\langle E| e^{it\ E/\hbar} \\
&= \frac{1}{2\pi\hbar} \int_{-\infty}^{\infty} dT \int_{-p_m}^{0} dp \int_{0}^{E_m} dE'dE \Big(-[\Theta(E') - \Theta(E' - E_m)] e^{-it\ E'/h} i\hbar \frac{\partial}{\partial E^{'}} \\
&\quad -i\hbar[\delta(E') - \delta(E' - E_m)] e^{-it\ E'/\hbar} \Big) |E'\rangle\langle E'|p\rangle\langle p|E\rangle\ \langle E| e^{it\ E/\hbar} \\
&\quad + \frac{1}{2\pi\hbar} \int_{-\infty}^{\infty} dt \int_{-p_m}^{0} dp \int_{0}^{E_m} dE\ i\hbar\ [\Theta(E^{'}) \\
&\quad -\Theta(E' - E_m)] e^{-itE'/\hbar} |E'\rangle\langle E'|p\rangle\ \langle p|E\rangle\ \langle E| e^{itE/\hbar} |_{E^{'}=0}^{E_m} \\
&= \int_{-p_m}^{0} dp \int_{0}^{E_m} dE'dE\delta(E - E') \Big(-[\Theta(E') - \Theta(E' - E_m)] i\hbar \frac{\partial}{\partial E^{'}} \\
&\quad -i\hbar[\delta(E^{'}) - \delta(E^{'} - E_m)] \Big) |E^{'}\rangle\langle E^{'}|p\rangle\langle p|E\rangle\ \langle E| \\
&\quad + i\hbar \int_{-p_m}^{0} dp \int_{0}^{E_m} dE\ \delta\ (E - E')[\Theta(E') - \Theta(E' - E_m)] |E'\rangle\langle E'|p\rangle\ \langle p|E\rangle\langle E||_{E^{'}=0}^{E_m} \\
&= \int_{-p_m}^{0} dp \int_{0}^{E_m} dE[(-[\Theta(E) - \Theta(E - E_m)] i\hbar \frac{\partial}{\partial E} \\
&\quad -i\hbar\ [\delta(E) - \delta(E - E_m)]) |E\rangle\langle E|p\rangle] \langle p|E\rangle\langle E| \\
&\quad + i\hbar \int_{-p_m}^{0} dp|E'\rangle\langle E'|p\rangle\ \langle p|E'\rangle\langle E'||_{E^{'}=0}^{E_m} \\
&= \int_{-p_m}^{0} dp \int_{0}^{E_m} dE \Big(-i\hbar \frac{\partial}{\partial E} |E\rangle\langle E|p\rangle \Big) \langle p|E\rangle\langle E| - i\hbar \int_{-\infty}^{0} dp|E\rangle|\langle p|E\rangle|^2 \langle E||_{E=0}^{E_m} \\
&\quad + i\hbar \int_{-p_m}^{0} dp|E'\rangle|\langle p|E'\rangle|^2 \langle E'||_{E^{'}=0}^{E_m} \\
&= \int_{-p_m}^{0} dp \int_{0}^{E_m} dE \Big(-i\hbar \frac{\partial}{\partial E} |E\rangle\langle E|p\rangle \Big) \langle p|E\rangle\langle E|,
\end{aligned} \tag{17a}$$

where we have performed an integration by parts. We also have that

$$\hat{T}_+ = \int_{-\infty}^{\infty} dt \int_0^{p_m} dp|t(p)\rangle t\langle t(p)| = \int_0^{p_m} dp \int_0^{E_m} dE\left(-i\hbar\frac{\partial}{\partial E}|E\rangle\langle E|p\rangle\right)\langle p|E\rangle\langle E|, \tag{17b}$$

and

$$\hat{T} = \int_{-\infty}^{\infty} dt|t\rangle\, t\, \langle t| = \int_0^{E_m} dE\left(-i\hbar\frac{\partial}{\partial E}|E\rangle\right)\langle E|. \tag{17c}$$

These are the forms in which the time operators act on energy eigenkets, but they take a different form when they act on states or on both, eigenstates and wave packets.

A second energy representation of time operators is

$$\begin{aligned}
\widehat{T}_- &= \frac{1}{2\pi\hbar}\int_{-\infty}^{\infty} dt \int_{-p_m}^{0} dp \int_0^{E_m} dE'dE\, e^{-it\, E'/\hbar}|E'\rangle\langle E'|p\rangle\ \langle p|E\rangle\langle E|\Big(- i\hbar\frac{\partial}{\partial E}[\Theta(E) - \Theta(E - E_m)] \\
&\quad + i\hbar[\delta(E) - \delta(E - E_m)]\Big)e^{it\, E/\hbar} \\
&= \frac{1}{2\pi\hbar}\int_{-\infty}^{\infty} dt \int_{-p_m}^{0} dp \int_0^{E_m} dE'dE\, e^{-it\, E'/h}|E'\rangle\langle E'|p\rangle\Big[\Big(\Theta(E) \\
&\quad - \Theta(E - E_m)\Big] i\hbar\frac{\partial}{\partial E} + i\hbar[\delta(E) - \delta(E - E_m)]\big)\langle p|E\rangle\langle E|\big]e^{it\, E/\hbar} \\
&\quad - \frac{i}{2\pi}\int_{-\infty}^{\infty} dt \int_{-p_m}^{0} dp \int_0^{E_m} dE'\, e^{-it\, E'/\hbar}|E'\rangle\langle E'|p\rangle\ \langle p|E\rangle\langle E|e^{it\, E/\hbar}\Big|_{E=0}^{E_m} \\
&= \int_{-p_m}^{0} dp \int_0^{E_m} dE'dE \frac{1}{2\pi\hbar}\int_{-\infty}^{\infty} dt\, e^{it(E-E')/\hbar}|E'\rangle\ \langle E'|p\rangle\, i\hbar\frac{\partial}{\partial E}\langle p|E\rangle\langle E| \\
&\quad + i\hbar\int_{-p_m}^{0} dp \int_0^{E_m} dE'dE \frac{1}{2\pi\hbar}\int_{-\infty}^{\infty} dt\, e^{it(E-E')/\hbar}|E'\rangle\ \langle E'|p\rangle[\delta(E) \\
&\quad - \delta(E - E_m)]\langle p|E\rangle\langle E| \\
&\quad - i\hbar\int_{-p_m}^{0} dp \int_0^{E_m} dE' \frac{1}{2\pi\hbar}\int_{-\infty}^{\infty} dt\, e^{it(E-E')/\hbar}|E'\rangle\langle E'|p\rangle\ \ \langle p|E\rangle\langle E|\Big|_{E=0}^{E_m} \\
&\quad - i\hbar\int_{-p_m}^{0} dp \int_0^{E_m} dE' \frac{1}{2\pi\hbar}\int_{-\infty}^{\infty} dt\, e^{it(E-E')/\hbar}|E'\rangle\langle E'|p\rangle\ \ \langle p|E\rangle\langle E|\Big|_{E=0}^{E_m} \\
&= \int_{-p_m}^{0} dp \int_0^{E_m} dE'dE\ \ \delta(E - E')|E'\rangle\langle E'|p\rangle i\hbar\frac{\partial}{\partial E}\langle p|E\rangle\langle E| \\
&= \int_{-p_m}^{0} dp \int_0^{E_m} dE|E\rangle\ \langle E|p\rangle i\hbar\frac{\partial}{\partial E}\langle p|E\rangle\langle E|,
\end{aligned} \tag{18a}$$

$$\hat{T}_+ = \int_{-\infty}^{\infty} dt \int_0^{p_m} dp|t(p)\rangle\ t\ \langle t(p)| = \int_0^{p_m} dp \int_0^{E_m} dE|E\rangle\langle E|p\rangle i\hbar\frac{\partial}{\partial E}\langle p|E\rangle\langle E|, \tag{18b}$$

and

$$\hat{T} = \int_{-\infty}^{\infty} dt|t\rangle\, t\, \langle t| = \int_{0}^{E_m} dE|E\rangle i\hbar \frac{d}{dE}\langle E|. \tag{18c}$$

These are the various forms in which the time operators can act on states in the energy representation. The difference with the time operators when acting on energy eigenkets is a minus sign.

Other symmetric expressions for the time operators can also be obtained:

$$\begin{aligned}
\hat{T}_- &= \int_{-\infty}^{\infty} dt \int_{-p_m}^{0} dp|t(p)\rangle\, t\, \langle t(p)| \\
&= \int_{-\infty}^{\infty} dt \int_{-p_m}^{0} dp \int_{0}^{E_m} dE' dE \frac{e^{-it\,E'/\hbar}}{\sqrt{2\pi\hbar}} |E'\rangle\ \langle E'|p\rangle\, t\ \langle p|E\rangle\ \langle E| \frac{e^{itE/\hbar}}{\sqrt{2\pi\hbar}} \\
&= \int_{-p_m}^{0} dp \int_{0}^{E_m} dE' dE|E'\rangle\ \langle E'|p\rangle\ \langle p|E\rangle\ \langle E| \frac{1}{2\pi\hbar}\int_{-\infty}^{\infty} dt\, t\, e^{it(E-E')/\hbar} \\
&= -i\hbar \int_{-p_m}^{0} dp \int_{0}^{E_m} dE'dE|E'\rangle\ \langle E'|p\rangle\ \langle p|E\rangle\ \langle E|\delta'(E' - E),
\end{aligned} \tag{19a}$$

$$\hat{T}_+ = -i\hbar \int_{0}^{p_m} dp \int_{0}^{E_m} dE'dE|E'\rangle\ \langle E'|p\rangle\ \langle p|E\rangle\ \langle E|\delta'(E' - E), \tag{19b}$$

and

$$\hat{T} = -i\hbar \int_{0}^{E_m} dE'dE|E'\rangle\ \langle E|\delta'(E' - E). \tag{19c}$$

The domain of our time operators is D, defined in Eq. (2). The convergence of quantities depends on the type of wave packet that these operators act on. A wave packet of type $L^2(0, E_m)$ in the energy representation is a good choice. Thus, the domain is invariant under the action of the time operators, and the commutator between the Hamiltonian and the time operators is thus valid in the entire domain D.

3.1. Time matrix elements of the Hamiltonian

The matrix elements of the Hamiltonian in the time representation are given by

$$\begin{aligned}
\langle t'|\hat{H}|t\rangle &= \int_{0}^{E_m} dE'dE \frac{e^{it'E'/\hbar}}{\sqrt{2\pi\hbar}} \langle E'|\hat{H}|E\rangle \frac{e^{-it\,E/\hbar}}{\sqrt{2\pi\hbar}} = \frac{1}{2\pi\hbar}\int_{0}^{E_m} dE'dE\ E\, e^{it'E'/\hbar} e^{-itE/\hbar} \langle E'|E\rangle \\
&= \frac{1}{2\pi\hbar}\int_{0}^{E_m} dE\ E\, e^{i(t'-t)E/\hbar} = i\hbar \frac{d}{dt}\langle t'|t\rangle = -i\hbar \frac{d}{dt'}\langle t'|t\rangle.
\end{aligned} \tag{20}$$

This is the Schrödinger equation for time kets in the time representation.

3.2. The time ket is the eigenstate of the time operator

We can find the characteristic operator of the commutators $[\cdot,\widehat{H}]$ and $[\widehat{T},\cdot]$. Because $[\widehat{T},\widehat{H}]=i\hbar$ the commutator between the operator $e^{-i\varepsilon\widehat{T}/\hbar}$, $\mathcal{E}\in[0,E_m]$, and the Hamiltonian is

$$[e^{-i\varepsilon\hat{T}/\hbar},\hat{H}]=\sum_{n=0}^{\infty}\frac{1}{n!}\left(-i\frac{\varepsilon}{\hbar}\right)^n[\hat{T}^n,\hat{H}]=\sum_{n=1}^{\infty}\frac{1}{n!}\left(-i\frac{\varepsilon}{\hbar}\right)^n i\hbar\, n\,\hat{T}^{n-1}=\varepsilon e^{-i\varepsilon\hat{T}/\hbar}. \tag{21}$$

Similarly, the commutator between the time operator and the time propagator is

$$[\hat{T},e^{-it\hat{H}/\hbar}]=\sum_{n=0}^{\infty}\frac{1}{n!}\left(-i\frac{t}{\hbar}\right)^n[\hat{T},\hat{H}^n]=\sum_{n=1}^{\infty}\frac{1}{n!}\left(-i\frac{t}{\hbar}\right)^n i\hbar\, n\,\hat{H}^{n-1}=te^{-it\hat{H}/\hbar}, \tag{22}$$

where $t\in\mathbb{R}$.

The time ket $\hat{|}t\rangle$ is the time propagation of a zero time ket $\hat{|}0\rangle$,

$$\hat{|}t\rangle=\int_0^{\infty}dE\frac{e^{-itE/\hbar}}{\sqrt{2\pi\hbar}}|E\rangle=e^{-it\hat{H}/\hbar}\hat{|}0\rangle,\ \hat{|}0\rangle=\frac{1}{\sqrt{2\pi\hbar}}\int_0^{\infty}dE|E\rangle. \tag{23}$$

Thus, according to Eq. (22), we can say that the time ket is an eigenstate of the time operator

$$\widehat{T}|t\rangle=\widehat{T}e^{-it\hat{H}/\hbar}|0\rangle=e^{-it\hat{H}/\hbar}\widehat{T}|0\rangle+t\,e^{-it\hat{H}/\hbar}|0\rangle=t|t\rangle, \tag{24}$$

where we have set $\widehat{T}|0\rangle=0$ because $|0\rangle$ is the zero-time state.

An "evolution equation" for the energy eigenstate is,

$$\begin{aligned}\widehat{T}|E\rangle&=\int_{-\infty}^{\infty}dt\frac{e^{itE/\hbar}}{\sqrt{2\pi\hbar}}\widehat{T}|t\rangle=\int_{-\infty}^{\infty}dt\frac{e^{itE/\hbar}}{\sqrt{2\pi\hbar}}t|t\rangle\\&=\int_{-\infty}^{\infty}dt\left(-i\hbar\frac{\partial}{\partial E}[\Theta(E)-\Theta(E-E_m)]\frac{e^{itE/\hbar}}{\sqrt{2\pi\hbar}}+i\hbar[\delta(E)-\delta(E-E_m)]\frac{e^{itE/\hbar}}{\sqrt{2\pi\hbar}}\right)|t\rangle\\&=\left(-i\hbar\frac{d}{dE}[\Theta(E)-\Theta(E-E_m)]+i\hbar[\delta(E)-\delta(E-E_m)]\right)|E\rangle\\&=[\Theta(E)-\Theta(E-E_m)]\left(-i\hbar\frac{d}{dE}\right)|E\rangle.\end{aligned} \tag{25}$$

Thus, the time operator is the generator of translations along the energy direction. All quantities are well defined as long as E and t belong to the allowed set of values for them. For other values of E and $E+\mathcal{E}$, we will get a linear combination of the energy eigenstates [14].

3.3. Shifting of operators

The shifting of the Hamiltonian along the energy direction is

$$\widehat{H}(\mathcal{E}) := e^{-i\varepsilon\hat{T}/\hbar}\widehat{H}e^{i\varepsilon\hat{T}/\hbar} = (\widehat{H}\, e^{-i\varepsilon\hat{T}/\hbar} + \mathcal{E}\, e^{-i\varepsilon\hat{T}/\hbar})e^{i\varepsilon\hat{T}/\hbar} = \widehat{H} + \mathcal{E}, \tag{26}$$

where $0 \leq E + \varepsilon$. For the translation of the time operator (22)), we have

$$\hat{T}(t) := e^{it\widehat{H}/\hbar}\hat{T}e^{-it\widehat{H}/\hbar} = e^{it\widehat{H}/\hbar}(e^{-it\widehat{H}/\hbar}\hat{T} + te^{-it\widehat{H}/\hbar}) = \hat{T} + t. \tag{27}$$

These operations are well defined as long as $E + \varepsilon \geq 0$ [6, 14]. The derivative with respect to t of the time-shifted operator (27) is

$$\frac{d}{dt}\widehat{T}(t) = \hat{I}, \tag{28}$$

that is, in the energy-time representations, t is the value that the time operator $\widehat{T}$ can take and not simply a parameter. Similarly, in the case of a translation of the Hamiltonian operator by the time operator, i.e., Eq. (26), we find that

$$\frac{d}{d\varepsilon}\widehat{H}(\varepsilon) = \hat{I}. \tag{29}$$

Therefore, in the energy-time representations, ε is not simply a parameter, but it is related to the values that the Hamiltonian $\widehat{H}$ can take.

Thus, the use of energy and time eigenkets and operators instead of coordinate and momentum eigenkets and operators is similar to going from a parametric representation of curves, with time being the parameter of evolution, to a nonparametric representation in which time is now one of the coordinates.

4. Approximate expressions

In this section, we make contact with other expressions that have been used by other authors. Other works have not made use of the Sa(x;1) factor that appears in our results. The results in this section will allow us to obtain a better understanding of previous results.

4.1. Approximating the integral in an infinite interval

As an approximation, we replace the integral in an infinite interval $(2\pi)^{-1}\int_{-\infty}^{\infty} dt$ with the integral in the finite interval $t \in [-T/2, T/2]$, $\lim_{T\to\infty}(1/T)\int_{-T/2}^{T/2} dt$. Then,

$$
\begin{aligned}
\widehat{T}_{-} &= \int_{-\infty}^{\infty} dt \int_{-p_m}^{0} dp|t(p)\rangle t\langle t(p)| \\
&\cong \frac{1}{T}\int_{-T/2}^{T/2} dt \int_{-p_m}^{0} dp \int_{0}^{E_m} dE^{'} dE\, e^{-itE^{'}/\hbar}|E^{'}\rangle\ \langle E^{'}|p\rangle t\langle p|E\rangle\ \langle E|e^{itE/\hbar} \\
&= \int_{-p_m}^{0} dp \int_{0}^{E_m} dE^{'} dE|E^{'}\rangle\ \langle E^{'}|p\rangle\ \langle p|E\rangle\ \langle E|\frac{1}{T}\int_{-T/2}^{T/2} dt\, t\, e^{it(E-E^{'})/\hbar} \qquad (30a) \\
&= \int_{-p_m}^{0} dp \int_{0}^{E_m} dE^{'} dE|E^{'}\rangle\ \langle E^{'}|p\rangle\ \langle p|E\rangle\ \langle E|\frac{i\hbar}{E-E^{'}}\mathrm{Sa}\left(\frac{T}{2\hbar}(E-E^{'});1\right) \\
&= \int_{0}^{E_m} dE^{'} dE \frac{i\hbar}{E-E^{'}}\mathrm{Sa}\left(\frac{T}{2\hbar}(E-E^{'});1\right)|E^{'}\rangle\ \langle E^{'}|\hat{I}_{-}|E\rangle\ \langle E|,
\end{aligned}
$$

$$
\widehat{T}_{+} \cong \int_{0}^{E_m} dE^{'} dE \frac{i\hbar}{E-E^{'}}\mathrm{Sa}\left(\frac{T}{2\hbar}(E-E^{'});1\right)|E^{'}\rangle\ \langle E^{'}|\hat{I}_{+}|E\rangle\ \langle E|, \qquad (30b)
$$

and

$$
\widehat{T} \cong \int_{0}^{E_m} dE^{'} dE|E^{'}\rangle\ \langle E|\frac{i\hbar}{E-E^{'}}\mathrm{Sa}\left(\frac{T}{2\hbar}(E-E^{'});1\right), \qquad (30c)
$$

where the Sa function of type one is defined as

$$
\mathrm{Sa}(x;1) := \frac{\sin(x)}{x} - \cos(x). \qquad (31)
$$

A plot of this function can be found in **Figure**. This function is zero at $x = 0$ and oscillates between $\approx \pm 1$. The limit $T \to \infty$ of the integral of $\mathrm{Sa}(Tx/2;1)/Tx$ times a function $f(x)$ gives an approximation to the derivative of the latter at $x = 0$.

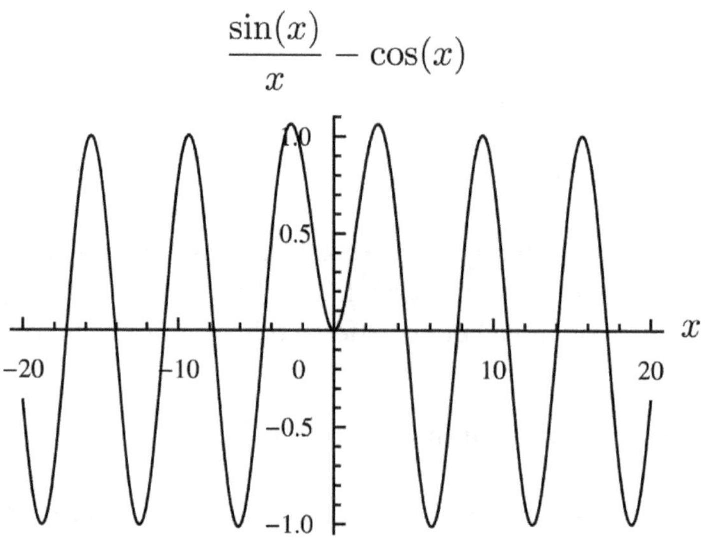

Figure. A plot of the function $\mathrm{Sa}(x;1) := \sin(x)/x - \cos(x)$.

Expressions that resemble Eq. (30c), but without the Sa factor, were used by other authors as a function that gives the sign of time in the continuous energy spectrum case [9–11].

5. The free particle

As an example of the time kets provided by our method, let us apply the derived results to the free-particle system. We find expressions for time eigenkets, including the case when a distinction of the sign of the momentum is needed. In this model, the momentum operator $\hat{P}$ commutes with the Hamiltonian operator $\hat{H}$, indicating a symmetry, allowing for some simplifications.

A set of energy eigenfunctions, in the coordinate representation, for the free-particle model is

$$\langle q|E_{\pm}\rangle = \frac{e^{\pm i\sqrt{2mE}q/\hbar}}{\sqrt{2\pi\hbar}}, \quad E \in [0,\infty)\,. \tag{32}$$

The subscripts in these functions indicate the sign of the momentum of the particle.

Thus, the zero-time eigenstate for the free particle is given as

$$\begin{aligned}\langle q|0_{\pm}\rangle &:= \int_0^{\infty} dE \frac{1}{\sqrt{2\pi\hbar}}\langle q|E_{\pm}\rangle = \frac{1}{\sqrt{2\pi\hbar}}\int_0^{\infty} dE \frac{e^{\pm i\sqrt{2mE}\;q/\hbar}}{\sqrt{2\pi\hbar}} = \frac{1}{\sqrt{2\pi\hbar}}\int_0^{\infty}\frac{p\,dp}{m}\frac{e^{\pm ip\;q/h}}{\sqrt{2\pi\hbar}} \\ &= \frac{1}{m}\left(\mp i\,\hbar\frac{d}{dq}\right)\frac{1}{2\pi\hbar}\int_0^{\infty} dp\;\; e^{\pm i\,p\;q/h} = \mp i\frac{\hbar}{m}\frac{d}{dq}\left(\frac{\delta(q)}{2} \pm \frac{i}{2\pi q}\right),\end{aligned} \tag{33}$$

where we have made the change in variable $E = p^2/2m$. The unit of the last ket is time^{-1}. Various other authors have used kets obtained by direct quantization of the classical expression for the time variable and have obtained a time ket with units of time$^{1/2}$. However, our kets exhibit the properties discussed in this chapter.

Figure shows a three-dimensional plot of the approximation of the squared modulus of the time states $\langle q|t_{-}\rangle$ and $\langle q|t_{+}\rangle$, obtained by not integrating from $E = 0$ to ∞ but up to a finite, large, value of E. They start highly localized at the origin and subsequently they move away from it and spread with time. The support of these functions resembles the classical motion curve $mq = pt$.

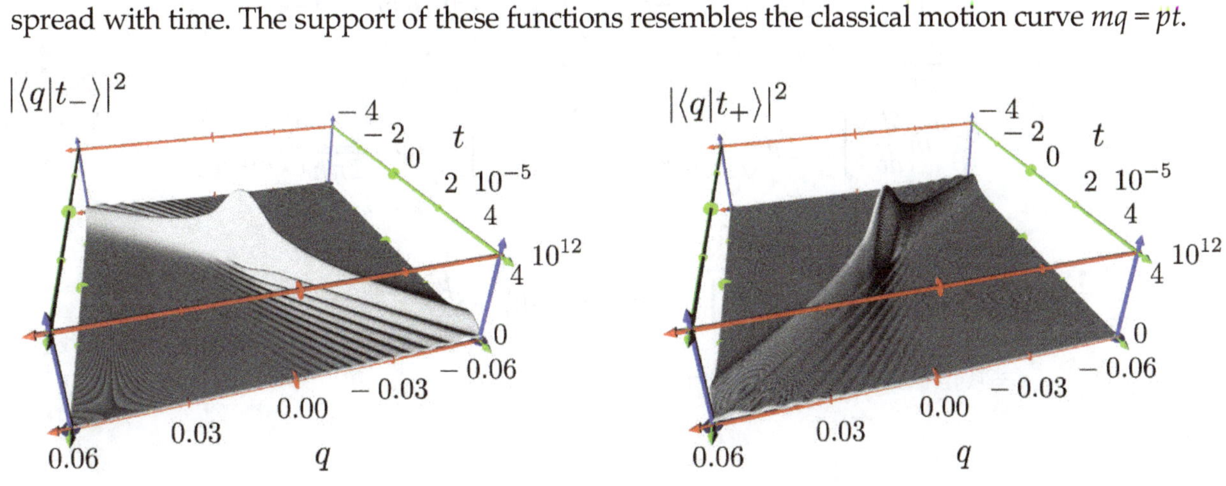

Figure. Three-dimensional plots of the squared modulus of the approximate time kets $|\langle q|t_{-}\rangle|^2$ and $|\langle q|t_{+}\rangle|^2$ for the free-particle model. The density is initially a highly localized density at $q = 0$ but subsequently it spreads and moves away from the origin. Dimensionless units.

For the sake of completeness, we write down the matrix elements of the time operators in the coordinate representation. They are

$$\langle q'|\widehat{T}_\pm|q\rangle = \int_0^\infty dE\langle q'|E_\pm\rangle\left(i\hbar\frac{\partial}{\partial E}\right)\langle E_\pm|q\rangle = \int_0^\infty dE\frac{e^{\pm i\sqrt{2mE}q'/\hbar}}{\sqrt{2\pi\hbar}}\left(i\hbar\frac{\partial}{\partial E}\right)\frac{e^{\mp i\sqrt{2mE}q/\hbar}}{\sqrt{2\pi\hbar}}$$
$$= \pm\int_0^\infty dE\frac{e^{\pm i\sqrt{2mE}q'/\hbar}}{\sqrt{2\pi\hbar}}\frac{m}{\sqrt{2mE}}q\frac{e^{\mp i\sqrt{2mE}q/\hbar}}{\sqrt{2\pi\hbar}} = \pm\frac{1}{2\pi\hbar}q\int_0^\infty dp\, e^{\pm ip(q'-q)/\hbar} \tag{34}$$
$$= \pm q\left[\frac{\delta(q-q')}{2}+\frac{i}{2\pi(q-q')}\right].$$

5.1. Solution to the quantized version of the classical motion of a free particle

The following calculation shows that the time states can also be the solution to the quantized classical expression for the motion of a free particle initially located at $q = 0$, i.e., the quantization of $mq = pt$. Let us rewrite the product $mq\langle q|t_-\rangle$ as follows:

$$mq\langle q|t_-\rangle = mq\int_0^{E_m} dE\frac{e^{-itE/\hbar}}{\sqrt{2\pi\hbar}}\langle q|E_-\rangle = mq\int_0^{E_m} dE\frac{e^{-itE/\hbar}}{\sqrt{2\pi\hbar}}\frac{e^{-i\sqrt{2mE}q/\hbar}}{\sqrt{2\pi\hbar}}$$
$$= mq\int_{-p_m}^0 dp\frac{p}{m}\frac{e^{-itp^2/2m\hbar}}{\sqrt{2\pi\hbar}}\frac{e^{ipq/\hbar}}{\sqrt{2\pi\hbar}}$$
$$= m\int_{-p_m}^0 dp\frac{p}{m}\frac{e^{-itp^2/2m\hbar}}{\sqrt{2\pi\hbar}}\left(-i\hbar\frac{\partial}{\partial p}\right)\frac{e^{i\,pq/\hbar}}{\sqrt{2\pi\hbar}}$$
$$= m\int_{-p_m}^0 dp\frac{e^{ipq/\hbar}}{\sqrt{2\pi\hbar}}\left(i\hbar\frac{\partial}{\partial p}\right)\frac{p}{m}\frac{e^{-itp^2/2m\hbar}}{\sqrt{2\pi\hbar}} - i\hbar m\frac{p}{m}\frac{e^{-itp^2/2m\hbar}}{\sqrt{2\pi\hbar}}\frac{e^{ipq/\hbar}}{\sqrt{2\pi\hbar}}\Big|_{p=-p_m}^0$$
$$= i\hbar\, m\int_{-p_m}^0 dp\frac{e^{ipq/\hbar}}{\sqrt{2\pi\hbar}}\left(\frac{1}{m}-i\frac{p}{m}\frac{2tp}{2m\hbar}\right)\frac{e^{-itp^2/2m\hbar}}{\sqrt{2\pi\hbar}}$$
$$+ i\hbar\frac{p_m}{2\pi\hbar}e^{-itp_m{}^2/2m\hbar}e^{ip_mq/\hbar} \tag{35a}$$
$$= i\hbar\, m\int_{-p_m}^0 dp\frac{e^{ipq/\hbar}}{\sqrt{2\pi\hbar}}\frac{1}{m}\frac{e^{-itp^2/2m\hbar}}{\sqrt{2\pi\hbar}} + t\int_{-p_m}^0 dp\frac{e^{ipq/\hbar}}{\sqrt{2\pi\hbar}}\frac{p^2}{m}\frac{e^{-itp^2/2m\hbar}}{\sqrt{2\pi\hbar}}$$
$$+ i\frac{p_m}{2\pi}e^{-itp_m^2/2m\hbar}e^{i\,p_mq/\hbar}$$
$$= t\left(-i\hbar\frac{d}{dq}\right)\int_{-p_m}^0 dp\frac{e^{ipq/\hbar}}{\sqrt{2\pi\hbar}}\frac{p}{m}\frac{e^{-itp^2/2m\hbar}}{\sqrt{2\pi\hbar}} + i\hbar\int_{-p_m}^0 dp\frac{e^{ipq/\hbar}}{\sqrt{2\pi\hbar}}\frac{e^{-itp^2/2m\hbar}}{\sqrt{2\pi\hbar}}$$
$$+ i\frac{p_m}{2\pi}e^{-itp_m^2/2m\hbar}e^{ip_mq/\hbar}$$
$$= t\left(-i\hbar\frac{d}{dq}\right)\langle q|t_-\rangle + i\hbar\int_{-p_m}^0 dp\langle q|p\rangle\,\langle p|E\rangle + i\frac{p_m}{2\pi}e^{-itp_m^2/2m\hbar}e^{i\,p_mq/\hbar}$$
$$= t\langle q|\hat{P}|t_-\rangle + i\hbar(\langle q|\hat{I}_-|E\rangle + \langle q|\hat{P}|p_m\rangle\,\langle p_m|E\rangle),$$

$$mq\langle q|t_+\rangle = t\langle q|\hat{P}|t_+\rangle + i\hbar(\langle q|\hat{I}_+|E\rangle - \langle q|\hat{P}|p_m\rangle\,\langle p_m|E\rangle). \tag{35b}$$

We can think of the last two terms in the above equations as quantum corrections to the classical trajectory of a free particle. These correction terms seem to vanish when $\hbar \to 0$.

On the other hand, the straightforward solution to the quantized version of the classical expression for the motion of a free particle gives a quite different function. The solution to the differential equation

$$mq\ f(q;\ t) = t\left(-i\hbar\frac{d}{dq}\right)f(q;\ t) \tag{36}$$

is

$$f(q;\ t) = N\ e^{i\ mq^2/2\hbar t}, \tag{37}$$

where N is a normalization constant. The squared modulus of this function is constant for all q and for all t. The squared modulus of the corresponding momentum function,

$$f(p;\ t) = N\sqrt{i\frac{t}{m}}\ e^{-i\ tp^2/2m\hbar}, \tag{38}$$

is not a localized function either; it actually is proportional to the transformation function between energy and time representations, in momentum representation. Thus, the route of forming conjugate states to the energy eigenstates seems to be a better path for obtaining appropriate time eigenstates.

6. Conclusions

We have introduced time-like states and time-like operators that are conjugate to the energy eigenstates and Hamiltonian operator, respectively. We have also given an interpretation of the obtained states and operators, and we have found that expressions obtained via other approaches to finding time eigenstates can be related to our expressions. However, the oscillatory Sa factor that we use solves many difficulties found in previous treatments. We have found the form of the time states for the free particle and a time operator that is valid for any L^2-type wave functions.

The approximation to time operators that we have introduced in this chapter uses expressions that can be adapted to the case of discrete energy spectra. We will explore this possibility in a later paper. From the literature on time operators, it might be believed that the treatment for a continuous energy spectrum is different from that for discrete energy spectrum systems. But, the results of this study suggest that both types of systems can be addressed in a similar manner.

Finally, we have found that the spectral measure $M(d\tau)$ of $\hat{T}$ is a nonorthogonal resolution of the identity defined by

$$\langle E^{'}|\widehat{M_I}(d\tau)|E\rangle = \frac{e^{it(E^{'}-E)/\hbar}}{2\pi\hbar}d\tau. \tag{39}$$

This measure exhibits the covariance property, as was previously stated by Holevo [13].

Author details

Gabino Torres-Vega

Address all correspondence to: gabino@fis.cinvestav.mx

Physics Department, Cinvestav, México

References

[1] Galapon EA: Only above barrier energy components contribute to barrier traversal time. Phys Rev Lett. 2012;**108**:170402. DOI: 10.1103/PhysRevLett.108.170402

[2] Eckle P, Pfeiffer AN, Cirelli C, Staudte A, Dörner R, Muller HG, Büttiker M, Keller U: Attosecond ionization and tunnelling delay time measurements in Helium. Science. 2008;**322**:1524–1529. DOI: 10.1126/science.1163439

[3] Kijowski J: On the time operator in quantum mechanics and the Heisenberg uncertainty relation for energy and time. Rep Math Phys. 1974;**6**:361–386.

[4] Hegerfeldt GC, Muga JG, Muñoz J: Manufacturing time operators: Covariance, selection criteria, and examples. Phys Rev A. 2010;**82**:012113. DOI: 10.1103/PhysRevA.82.012113

[5] Weyl H: The Theory of Groups and Quantum Mechanics. 2nd ed. USA: Dover Publications, Inc.; 1950. 447 p.

[6] Galapon EA: Self-adjoint time operator is the rule for discrete semi-bounded Hamiltonians. Proc R Soc Lond A. 2002;**458**:2671–2690. DOI: 10.1098/rspa.2002.0992

[7] Arai A, Yasumichi M: Time operators of a Hamiltonian with purely discrete spectrum. Rev Math Phys. 2008;**20**:951–978.

[8] Arai A: Necessary and sufficient conditions for a Hamiltonian with discrete eigenvalues to have time operators. Lett Math Phys. 2009;**87**:67. DOI: 10.1007/sl 1005-008-0286-z

[9] Strauss Y, Silman J, Machnes S, Horwitz LP: An arrow of time operator for standard Quantum Mechanics. 2008. quant-ph 0802.2448.

[10] Strauss Y: Forward and backward time observables for quantum evolution and quantum stochastic processes – I: The time observables. 2007. math-ph 0706.0268v1.

[11] Hall MJW: Comment on "An arrow of time operator for standard quantum mechanics" (a sign of the time!). 2008. quant-ph 0802.2682.

[12] Torres-Vega G: Conjugate dynamical systems: Classical analogue of the quantum energy translation. J Phys A: Math Theor. 2012;**45**:215302. DOI: 10.1088/1751-8113/45/21/215302

[13] Holevo AS: Estimation of shift parameters of a quantum state. Rep Math Phys. 1978;**13**:379–399.

[14] Martínez-Pérez A, Torres-Vega G: Translations in quantum mechanics revisited. The point spectrum case. Can J Phys. 2016;**94**:1365.

4

Oscillation Criteria for Second-Order Neutral Damped Differential Equations with Delay Argument

Said R. Grace and Irena Jadlovská

Abstract

The chapter is devoted to study the oscillation of all solutions to second-order nonlinear neutral damped differential equations with delay argument. New oscillation criteria are obtained by employing a refinement of the generalized Riccati transformations and integral averaging techniques.

2010 Mathematics Subject Classification: 34C10, 34K11.

Keywords: neutral differential equation, damping, delay, second-order, generalized Riccati technique, oscillation

1. Introduction

In the chapter, we are mainly concerned with the oscillatory behavior of solutions to second-order nonlinear neutral damped differential equations with delay argument of the form

$$\left(r(t)\left(z'(t)\right)^{\alpha}\right)' + p(t)\left(z'(t)\right)^{\alpha} + q(t)f\left(x(\sigma(t))\right) = 0, \quad t \geq t_0, \tag{1}$$

where $\alpha \geq 1$ is a quotient of positive odd integers and

$$z(t) = x(t) + a(t)x(\tau(t)). \tag{2}$$

Throughout, we suppose that the following hypotheses hold:

i. $r,\ p,\ q \in C(\mathcal{J}, \mathbb{R}^+)$, where $\mathcal{J} = [t_0, \infty)$ and $\mathbb{R}^+ = (0,\infty)$;

ii. $a \in C(\mathcal{J}, \mathbb{R})$, $0 \leq a(t) \leq 1$;

iii. $\tau \in C(\mathcal{I}, \mathbb{R})$, $\tau(t) \leq t$, $\tau(t) \to \infty$ as $t \to \infty$;

iv. $\sigma \in C^1(\mathcal{I}, \mathbb{R})$, $\sigma(t) \leq t$, $\sigma'(t) \geq 0$, $\sigma(t) \to \infty$ as $t \to \infty$;

v. $f \in C(\mathbb{R}, \mathbb{R})$, such that $xf(x) > 0$ and $f(x)/x^{\beta} \geq k > 0$ for $x \neq 0$, where k is a constant and β is the ratio of odd positive integers.

By a solution of Eq. (1), we mean a nontrivial real-valued function $x(t)$, which has the property $z(t) \in C^1([T_x, \infty))$, $r(t)\left(z'(t)\right)^{\alpha} \in C^1([T_x, \infty))$, $T_x \geq t_0$, and satisfies Eq. (1) on $[T_x, \infty)$. In the sequel, we will restrict our attention to those solutions $x(t)$ of Eq. (1) that satisfy the condition

$$\sup\{|x(t)| : T \leq t < \infty\} > 0 \quad \text{for} \quad T \geq T_x. \tag{3}$$

We make the standing hypothesis that Eq. (1) admits such a solution. As is customary, a solution of Eq. (1) is said to be oscillatory if it is neither eventually positive nor eventually negative on $[T_x, \infty)$ and otherwise, it is termed nonoscillatory. The equation itself is called oscillatory if all its solutions are oscillatory.

Remark 1. All the functional inequalities considered in the sequel are assumed to hold eventually, that is, they are satisfied for all t large enough.

Oscillation theory was created in 1836 with a paper of Jacques Charles François Sturm published in *Journal des Mathematiqués Pures et Appliqueés*. His long and detailed memoir [1] was one of the first contributions in Liouville's newly founded journal and initiated a whole new research into the qualitative analysis of differential equations. Heretofore, the theory of differential equations was primarily about finding solutions of a given equation and so was very limited. Contrarily, the main idea of Sturm was to obtain geometric properties of solutions (such as sign changes, zeros, boundaries, and oscillation) directly from the differential equation, without benefit of solutions themselves.

Henceforth, the oscillation theory for ordinary differential equations has undergone a significant development. Nowadays, it is considered as coherent, self-contained domain in the qualitative theory of differential equations that is turning mainly toward the study of solution properties of functional differential equations (FDEs).

The problem of obtaining sufficient conditions for asymptotic and oscillatory properties of different classes of FDEs has experienced long-term interest of many researchers. This is caused by the fact that differential equations, especially those with deviating argument, are deemed to be adequate in modeling of the countless processes in all areas of science. For a summary of the most significant efforts and recent findings in the oscillation theory of FDEs and vast bibliography therein, we refer the reader to the excellent monographs [2–6].

In a neutral delay differential equation the highest-order derivative of the unknown function appears both with and without delay. The study of qualitative properties of solutions of such equations has, besides its theoretical interest, significant practical importance. This is due to the fact that neutral differential equations arise in various phenomena including problems concerning electric networks containing lossless transmission lines (as in high-speed computers where such lines are used to interconnect switching circuits), in the study of vibrating masses attached to an elastic bar or in the solution of variational problems with time delays. We refer the reader to the monograph [7] for further applications in science and technology.

So far, most of the results obtained in the literature has centered around the special *undamped* form of Eq. (1), i.e., when $p(t) = 0$. For instance, in one of the pioneering works on the subject, Grammatikopoulos et al. [8] studied the second-order neutral differential equation with constant delay of the form

$$(x(t) + a(t)x(t-\tau)^{''} + q(t)x(t-\tau) = 0 \tag{4}$$

and proved that Eq. (4) is oscillatory if

$$\int_{t_0}^{\infty} q(s)\left(1-a(s-\tau)\right) \mathrm{d}s = \infty. \tag{5}$$

Later on, Grace and Lalli [9] extended the results from [8] to the more general equation

$$\left(r(t)(x(t) + a(t)x(t-\tau)^{'}\right)^{'} + q(t)f\left(x(t-\tau)\right) = 0, \tag{6}$$

with

$$\frac{f(x)}{x} \geq k, \quad k > 0 \quad \text{and} \quad \int_{t_0}^{\infty} \frac{\mathrm{d}s}{r(s)} = \infty \tag{7}$$

and showed that Eq. (6) is oscillatory if there exists a continuously differentiable function $\rho(t)$ such that

$$\int_{t_0}^{\infty} \left(\rho(s)q(s)(1-a(s-\tau)) - \frac{\left(\rho^{'}(s)\right)^2 r(s-\tau)}{4k\rho(s)} \right) \mathrm{d}s = \infty. \tag{8}$$

In Ref. [10], Dong has involved to study the oscillation problem for a half-linear case of Eq. (1) and by defining a sequence of continuous functions has obtained various kinds of better results. Afterward, his approach has been further developed by several authors. However, it appears that very little is known regarding the oscillation of Eq. (1) with $p(t) \neq 0$ and $\alpha \neq \beta$. Motivated by the results of Ref. [10], this chapter presents some new oscillation criteria, which are applicable on Eq. (1).

On the other hand, Eq. (1) can be considered as a natural generalization of the second-order delay differential equation of the form

$$\left(r(t)\left(x^{'}(t)\right)^{\alpha}\right) + p(t)\left(x^{'}(t)\right)^{\alpha} + q(t)f\left(x(\sigma(t))\right) = 0. \tag{9}$$

Very recently, the authors of [19] studied the oscillation problem of Eq. (9) with $p(t) = 0$ and $\alpha = \beta$. Their ideas, which are based on careful investigation of classical techniques covering

Riccati transformations and integral averages, will be extended to the more general equation (1).

2. Main results

For the simplicity and without further mention, we use the following notations:

$$A(t) = \exp\left(-\int_{t_0}^{t} \frac{p(s)}{r(s)}\mathrm{d}s\right), \quad Q(t) = kq(t)\left(1-a(\sigma(t))\right)^{\beta}, \tag{10}$$

$$R(t) = \int_{t}^{\infty} \left(\frac{A(s)}{r(s)}\right)^{\frac{1}{\alpha}}\mathrm{d}s, \quad \tilde{Q}(t) = q(t)\left(1-a(\sigma(t))\frac{R(\tau(\sigma(t)))}{R(\sigma(t))}\right)^{\beta}, \tag{11}$$

$$P(t) = \frac{\phi'(t)}{\phi(t)} - \frac{p(t)}{r(t)}, \quad \tilde{q}(t) = Q(t) + \frac{p(t)A(t)}{r(t)}\int_{t}^{\infty}\frac{Q(s)}{A(s)}\mathrm{d}s, \tag{12}$$

where $\phi(t) \in C^1(\mathcal{J}, \mathbb{R})$ is a given function and will be specified later.

The organization of this chapter is as follows. Before stating our main results, we present two lemmas that ensure that any solution $x(t)$ of Eq. (1) satisfies the condition

$$z(t) > 0, \quad z'(t) > 0, \quad \left(r(t)\left(z'(t)\right)^{\alpha}\right)' < 0, \tag{13}$$

for t sufficiently large. Next, we get our main oscillation results for Eq. (1) by employing the generalized Riccati transformations and integral averaging techniques. We base our arguments on the assumption that the function $P(t)$ is positive or negative.

Lemma 1. Assume that

$$\int_{t_0}^{\infty} \left(\frac{A(s)}{r(s)}\right)^{\frac{1}{\alpha}}\mathrm{d}s = \infty \tag{14}$$

holds and Eq. (1) has a positive solution $x(t)$ on $\mathcal{J}$. Then there exists a $T \in \mathcal{J}$, sufficiently large, such that

$$z(t) > 0, \quad z'(t) > 0, \quad \left(r(t)\left(z'(t)\right)^{\alpha}\right)' < 0, \tag{15}$$

on $[T, \infty)$.

Proof. Since, $x(t)$ is a positive solution of Eq. (1) on $\mathcal{J}$, then, by the assumptions (iii) and (iv), there exists a $t_1 \in \mathcal{J}$ such that $x(\tau(t)) > 0$ and $x(\sigma(t)) > 0$ on $[t_1, \infty)$. Define the function $z(t)$ as in Eq. (2). Then it is easy to see that $z(t) \geq x(t) > 0$, for $t \geq t_1$, and at the same time, from Eq. (1), we get

$$\left(r(t)\left(z'(t)\right)^{\alpha}\right)' + p(t)\left(z'(t)\right)^{\alpha} = -q(t)f\left(x(\sigma(t))\right) < 0. \tag{16}$$

We assert that $\frac{r(t)}{A(t)}\left(z'(t)\right)^{\alpha}$ is decreasing. Clearly, by writing the left-hand side of Eq. (16) in the form

$$\left(r(t)\left(z'(t)\right)^{\alpha}\right)' + \frac{p(t)}{r(t)}r(t)\left(z'(t)\right)^{\alpha} < 0, \tag{17}$$

we get

$$\left(\frac{r(t)}{A(t)}\left(z'(t)\right)^{\alpha}\right)' = -\frac{q(t)}{A(t)}f(x(\sigma(t))) < 0 \tag{18}$$

and so the assertion is proved.

Now, we claim that $z'(t) > 0$ on $[t_1, \infty)$. If not, then there exists $t_2 \in [t_1, \infty)$ such that $z'(t_2) < 0$. Using the fact that $\frac{r(t)}{A(t)}\left(z'(t)\right)^{\alpha}$ is decreasing, we obtain, for $t \geq t_2$,

$$\frac{r(t)}{A(t)}\left(z'(t)\right)^{\alpha} < c := \frac{r(t_2)}{A(t_2)}\left(z'(t_2)\right)^{\alpha} < 0. \tag{19}$$

Integrating the above inequality from t_2 to t, we find that

$$z(t) < z(t_2) + c^{\frac{1}{\alpha}}\int_{t_2}^{t}\left(\frac{A(s)}{r(s)}\right)^{\frac{1}{\alpha}}\mathrm{d}s \tag{20}$$

for $t \geq t_2$. By condition (14), $z(t)$ approaches to $-\infty$ as $t \to \infty$, which contradicts the fact that $z(t)$ is eventually positive. Therefore, $z'(t) > 0$ and from Eq. (1), we have that $\left(r(t)\left(z'(t)\right)^{\alpha}\right)' < 0$. The proof is complete.

Lemma 2. Assume that

$$\int_{t_0}^{\infty}\left(\frac{A(u)}{r(u)}\int_{t_0}^{u}\frac{\tilde{Q}(s)R^{\beta}(\sigma(s))}{A(s)}\mathrm{d}s\right)^{\frac{1}{\alpha}}\mathrm{d}u = \infty, \tag{21}$$

holds and Eq. (1) has a positive solution $x(t)$ on $\mathcal{I}$. Then there exists $T \in \mathcal{I}$, sufficiently large, such that

$$z(t) > 0, \quad z'(t) > 0, \quad \left(r(t)\left(z'(t)\right)^{\alpha}\right)' < 0, \tag{22}$$

on $[T, \infty)$.

Proof. Similarly to the proof of Lemma 1, we assume that there exists $t_2 \in \mathscr{I}$ such that $z'(t) < 0$ on $[t_2, \infty)$. Taking Eq. (18) into account, we have

$$z'(s) \le \left(\frac{r(t)}{A(t)}\frac{A(s)}{r(s)}\right)^{\frac{1}{\alpha}} z'(t), \tag{23}$$

for $s \ge t \ge t_2$. Integrating the above inequality from t to t', $t' \ge t \ge t_2$, we get

$$z(t') \le z(t) + \left(\frac{r(t)}{A(t)}\right)^{\frac{1}{\alpha}} z'(t) \int_t^{t'} \left(\frac{r(s)}{A(s)}\right)^{-\frac{1}{\alpha}} ds. \tag{24}$$

Letting $t' \to \infty$, we have

$$z(t) \ge -R(t)\left(\frac{r(t)}{A(t)}\right)^{\frac{1}{\alpha}} z'(t), \tag{25}$$

which yields

$$\left(\frac{z(t)}{R(t)}\right)' \ge 0 \tag{26}$$

and hence we see that $\frac{z(t)}{R(t)}$ is nondecreasing. By Eq. (2) and (iii), we have

$$\begin{aligned} x(t) &= z(t) - a(t)x(\tau(t)) \\ &\ge z(t) - a(t)z(\tau(t)) \\ &\ge \left(1 - a(t)\frac{R(\tau(t))}{R(t)}\right) z(t), \end{aligned} \tag{27}$$

which together with Eq. (1) and the assumption (v) yields

$$\begin{aligned} \left(r(t)\left(z'(t)\right)^{\alpha}\right)' + p(t)\left(z'(t)\right)^{\alpha} &\le -kq(t)\left(1 - a(\sigma(t))\frac{R(\tau(\sigma(t)))}{R(\sigma(t))}\right)^{\beta} z^{\beta}(\sigma(t)) \\ &= -k\tilde{Q}(t) z^{\beta}(\sigma(t)). \end{aligned} \tag{28}$$

On the other hand, from Eq. (23), we have

$$\frac{r(t)\left(z'(t)\right)^{\alpha}}{A(t)} \le \frac{r(t_2)\left(z'(t_2)\right)^{\alpha}}{A(t_2)}, \tag{29}$$

that is,

$$\frac{r(t)}{A(t)}\left(-z'(t)\right)^{\alpha} \ge \frac{r(t_2)}{A(t_2)}\left(-z'(t_2)\right)^{\alpha} := \gamma^{\alpha} \tag{30}$$

for some positive constant γ. Setting Eq. (30) into Eq. (25), we obtain

$$z(t) \geq \gamma R(t) \tag{31}$$

and so, Eq. (28) becomes

$$\left(r(t)\left(z'(t)\right)^{\alpha}\right)' + p(t)\left(z'(t)\right)^{\alpha} \leq -\tilde{\gamma}\tilde{Q}(t)R^{\beta}(\sigma(t)), \tag{32}$$

where $\tilde{\gamma} := k\gamma^{\beta}$. Now, if we define the function

$$U(t) = r(t)\left(-z'(t)\right)^{\alpha} > 0, \tag{33}$$

then

$$U'(t) + \frac{p(t)}{r(t)}U(t) \geq \tilde{\gamma}\tilde{Q}(t)R^{\beta}(\sigma(t)), \tag{34}$$

or, equally

$$\left(\frac{U(t)}{A(t)}\right)' \geq \tilde{\gamma}\frac{\tilde{Q}(t)R^{\beta}(\sigma(t))}{A(t)}. \tag{35}$$

Integrating the above inequality from t_2 to t, we get

$$U(t) \geq \tilde{\gamma}A(t)\int_{t_2}^{t}\frac{\tilde{Q}(s)R^{\beta}(\sigma(s))}{A(s)}ds \tag{36}$$

or

$$r(t)\left(-z'(t)\right)^{\alpha} \geq \tilde{\gamma}A(t)\int_{t_2}^{t}\frac{\tilde{Q}(s)R^{\beta}(\sigma(s))}{A(s)}ds. \tag{37}$$

It follows from this last inequality that

$$0 < z(t) \leq z(t_2) - \tilde{\gamma}\int_{t_2}^{t}\left(\frac{A(u)}{r(u)}\int_{t_2}^{u}\frac{\tilde{Q}(s)R^{\beta}(\sigma(s))}{A(s)}ds\right)^{\frac{1}{\alpha}}du \tag{38}$$

for $t \geq t_2$. As $t \to \infty$, then by condition Eq. (21), $z(t)$ approaches to $-\infty$, which contradicts the fact that $z(t)$ is eventually positive. Therefore, $z'(t) > 0$ and from Eq. (1), we have $\left(r(t)\left(z'(t)\right)^{\alpha}\right)' < 0$. The proof is complete.

Lemma 3. Assume that

$$\int_{t_0}^{\infty}\left(\frac{A(u)}{r(u)}\int_{t_0}^{u}\frac{\tilde{Q}(s)}{A(s)}ds\right)^{\frac{1}{\alpha}}du = \infty, \tag{39}$$

holds and Eq. (1) has a positive solution $x(t)$ on $\mathcal{I}$. Then there exists $T \in \mathcal{I}$, sufficiently large, such that either

$$z(t) > 0, \quad z'(t) > 0, \quad \left(r(t)\left(z'(t)\right)^{\alpha}\right)' < 0, \tag{40}$$

on $[T, \infty)$ or $\lim_{t\to\infty} x(t) = 0$.

Proof. As in the proof of Lemma 1, we assume that there exists $t_2 \in \mathcal{I}$ such that $z'(t) < 0$ on $[t_2, \infty)$. So, $z(t)$ is decreasing and

$$\lim_{t\to\infty} z(t) =: b \geq 0 \tag{41}$$

exists. Therefore, there exists $t_3 \in [t_2, \infty)$ such that

$$z(\sigma(t)) > z(t) \geq b > 0. \tag{42}$$

As in the proof of Lemma 2, we obtain Eq. (27), i.e.,

$$\begin{aligned} x(\sigma(t)) &\geq \left(1 - a(\sigma(t))\frac{R(\tau(\sigma(t)))}{R(\sigma(t))}\right) z(\sigma(t)) \\ &\geq b\left(1 - a(\sigma(t))\frac{R(\tau(\sigma(t)))}{R(\sigma(t))}\right), \quad \text{for} \quad t \geq t_3. \end{aligned} \tag{43}$$

Thus,

$$\begin{aligned} \left(r(t)\left(z'(t)\right)^{\alpha}\right)' + p(t)\left(z'(t)\right)^{\alpha} &\leq -\tilde{b}\, q(t)\left(1 - a(\sigma(t))\frac{R(\tau(\sigma(t)))}{R(\sigma(t))}\right)^{\beta} \\ &= -\tilde{b}\tilde{Q}(t), \end{aligned} \tag{44}$$

where $\tilde{b} := kb^{\beta}$.

Define the function $U(t)$ as in Eq. (103). Then Eq. (44) becomes

$$\left(\frac{U(t)}{A(t)}\right)' \geq \tilde{b}\frac{\tilde{Q}(t)}{A(t)}. \tag{45}$$

Integrating the above inequality twice from t_3 to t, one gets

$$0 < z(t) \leq z(t_3) - \tilde{b}\int_{t_3}^{t}\left(\frac{A(u)}{r(u)}\int_{t_3}^{u}\frac{\tilde{Q}(s)}{A(s)}ds\right)^{\frac{1}{\alpha}} du, \tag{46}$$

for $t \geq t_3$. As $t \to \infty$, then by condition (39), $z(t)$ approaches to $-\infty$, which contradicts the fact that $z(t)$ is eventually positive. Thus, $b = 0$ and from $0 \leq x(t) \leq z(t)$, we see that $\lim_{t\to\infty} x(t) = 0$. The proof is complete.

Using results of Lemmas 1 and 2, we can obtain the following oscillation criteria for Eq. (1).

Theorem 1. Let conditions (i)–(v) and one of the conditions (14) or (21) hold. Furthermore, assume that there exists a positive continuously differentiable function $\phi(t)$ such that, for all sufficiently large, T, $T_1 \geq T$,

$$P(t)\geq 0 \tag{47}$$

on $[T, \infty)$ and

$$\limsup_{t\to\infty}\left\{\phi(t)A(t)\int_t^{\infty}\frac{Q(s)}{A(s)}ds+\int_{T_1}^{t}\left[\phi(s)Q(s)-\frac{\alpha^{\alpha}}{(\alpha+1)^{\alpha+1}}\frac{\phi(s)r(\sigma(s))\left(P(s)\right)^{\alpha+1}}{\left(\beta\sigma'(s)\psi(s)\right)^{\alpha}}\right]ds\right\}=\infty, \tag{48}$$

where

$$\psi(t)=\begin{cases} c_1, & c_1 \text{ is some positive constant if } \beta>\alpha \\ 1, & \text{if } \beta=\alpha \\ c_2\left(\int_T^t r^{-\frac{1}{\alpha}}(s)ds\right)^{\frac{\beta-\alpha}{\alpha}}, & c_2 \text{ is some positive constant if } \beta<\alpha. \end{cases} \tag{49}$$

Then, Eq. (1) is oscillatory.

Proof. Suppose to the contrary that $x(t)$ is a nonoscillatory solution of Eq. (1). Then, without loss of generality, we may assume that there exists $T\in\mathcal{I}$ large enough, so that $x(t)$ satisfies the conclusions of Lemma 1 or 2 on $[T, \infty)$ with

$$x(t)>0,\quad x(\tau(t))>0,\quad x(\sigma(t))>0 \tag{50}$$

on $[T, \infty)$. In particular, we have

$$z(t)>0,\quad z'(t)>0,\quad \left(r(t)\left(z'(t)\right)^{\alpha}\right)'<0,\quad \text{for}\quad t\geq T. \tag{51}$$

By Eq. (2) and the assumption (iii), we get

$$\begin{aligned} x(t) &= z(t)-a(t)x(\tau(t)) \\ &\geq z(t)-a(t)z(\tau(t)) \\ &\geq (1-a(t))z(t), \end{aligned} \tag{52}$$

which together with Eq. (1) implies

$$\begin{aligned} \left(r(t)\left(z'(t)\right)^{\alpha}\right)'+\frac{p(t)}{r(t)}\left(z'(t)\right)^{\alpha} &\leq -kq(t)\left(1-a(\sigma(t))\right)^{\beta}z^{\beta}(\sigma(t)) \\ &= -Q(t)z^{\beta}(\sigma(t)). \end{aligned} \tag{53}$$

We consider the generalized Riccati substitution

$$w(t)=\phi(t)\frac{r(t)\left(z'(t)\right)^{\alpha}}{z^{\beta}(\sigma(t))}>0,\quad \text{for}\quad t\geq T. \tag{54}$$

As in the proof of Lemma 1, we get Eq. (18), which in view of the assumption (v) yields

$$\left(\frac{r(t)}{A(t)}\left(z'(t)\right)^{\alpha}\right)' \leq -\frac{Q(t)}{A(t)}z^{\beta}(\sigma(t)). \tag{55}$$

Integrating Eq. (55) from t to ∞ and using the fact that $z(t)$ is increasing, we have

$$\begin{aligned}\frac{r(t)}{A(t)}\left(z'(t)\right)^{\alpha} &\geq \int_t^{\infty}\frac{Q(s)}{A(s)}z^{\beta}(\sigma(s))\mathrm{d}s\\ &\geq z^{\beta}(\sigma(t))\int_t^{\infty}\frac{Q(s)}{A(s)}\mathrm{d}s.\end{aligned} \tag{56}$$

So it follows from Eq. (56) and the definition (54) of $w(t)$ that

$$w(t) = \phi(t)\frac{r(t)\left(z'(t)\right)^{\alpha}}{z^{\beta}(\sigma(t))} \geq \phi(t)A(t)\int_t^{\infty}\frac{Q(s)}{A(s)}\mathrm{d}s. \tag{57}$$

By Eq. (53) we can easily prove that

$$\begin{aligned}w'(t) &= \left(r(t)\left(z'(t)\right)^{\alpha}\right)'\frac{\phi(t)}{z^{\beta}(\sigma(t))} + \left(\frac{\phi(t)}{z^{\beta}(\sigma(t))}\right)' r(t)\left(z'(t)\right)^{\alpha}\\ &\leq -\frac{\phi(t)}{z^{\beta}(\sigma(t))}\left(p(t)\left(z'(t)\right)^{\beta} + Q(t)z^{\beta}(\sigma(t))\right)\\ &\quad + r(t)\left(z'(t)\right)^{\alpha}\left(\frac{\phi'(t)}{z^{\beta}(\sigma(t))} - \frac{\phi(t)\left(z^{\beta}(\sigma(t))\right)'}{z^{\beta+1}(\sigma(t))}\right)\\ &\leq -\phi(t)Q(t) + w(t)\left(\frac{\phi'(t)}{\phi(t)} - \frac{p(t)}{r(t)}\right)\\ &\quad -\beta\phi(t)\frac{r(t)(z'(t))^{\beta}z'(\sigma(t))\sigma'(t)}{z^{\beta+1}(\sigma(t))}.\end{aligned} \tag{58}$$

On the other hand, since $r(t)\left(z'(t)\right)^{\alpha}$ is decreasing, we have

$$\frac{z'(\sigma(t))}{z'(t)} \geq \left(\frac{r(t)}{r(\sigma(t))}\right)^{\frac{1}{\alpha}} \tag{59}$$

and thus Eq. (58) becomes

$$\begin{aligned}w'(t) \leq\ & -\phi(t)Q(t) + P(t)w(t)\\ & -\frac{\beta\phi(t)\sigma'(t)}{r^{\frac{1}{\alpha}}(\sigma(t))}\left(\frac{w(t)}{\phi(t)}\right)^{\frac{\alpha+1}{\alpha}} z^{\frac{\beta-\alpha}{\alpha}}(\sigma(t)).\end{aligned} \tag{60}$$

Now, we consider the following three cases:

Case I: $\beta > \alpha$.

In this case, since $z'(t) > 0$ for $t \geq T$, then there exists $T_1 \geq T$ such that $z(\sigma(t)) \geq c$ for $t \geq T_1$. This implies that

$$z^{\frac{\beta-\alpha}{\alpha}}(\sigma(t)) \geq c^{\frac{\beta-\alpha}{\alpha}} := c_1 \tag{61}$$

Case II: $\beta = \alpha$.

In this case, we see that $z^{\frac{\beta-\alpha}{\alpha}}(\sigma(t)) = 1$.

Case III: $\beta < \alpha$.

Since $r(t)\left(z'(t)\right)^{\alpha}$ is decreasing, there exists a constant d such that

$$r(t)\left(z'(t)\right)^{\alpha} \leq d \tag{62}$$

for $t \geq T$. Integrating the above inequality from T to t, we have

$$z(t) \leq z(T) + \int_T^t \left(\frac{d}{r(s)}\right)^{\frac{1}{\alpha}} \mathrm{ds}. \tag{63}$$

Hence, there exists $T_1 \geq T$ and a constant d_1 depending on d such that

$$z(t) \leq d_1 \int_T^t r^{-\frac{1}{\alpha}}(s) \mathrm{ds}, \quad \text{for} \quad t \geq T_1 \tag{64}$$

and thus

$$z^{\frac{\beta-\alpha}{\alpha}}(\sigma(t)) \geq d_1^{\frac{\beta-\alpha}{\alpha}} \left(\int_T^t r^{-\frac{1}{\alpha}}(s) \mathrm{ds}\right)^{\frac{\beta-\alpha}{\alpha}} = d_2 \left(\int_T^t r^{-\frac{1}{\alpha}}(s) \mathrm{ds}\right)^{\frac{\beta-\alpha}{\alpha}} \tag{65}$$

for some positive constant d_2.

Using these three cases and the definition of $\psi(t)$, we get

$$w'(t) \leq -\phi(t)Q(t) + P(t)w(t) - \frac{\beta\sigma'(t)\psi(t)}{\left(\phi(t)r(\sigma(t))\right)^{\frac{1}{\alpha}}} w^{\frac{1+\alpha}{\alpha}}(t) \tag{66}$$

for $t \geq T_1 \geq T$. Setting

$$A := P(t), \tag{67}$$

$$B := \frac{\beta\sigma'(t)\psi(t)}{\left(\phi(t)r(\sigma(t))\right)^{\frac{1}{\alpha}}}, \tag{68}$$

and using the inequality

$$Au - Bu^{\frac{1+\alpha}{\alpha}} \leq \frac{\alpha^{\alpha}}{(\alpha+1)^{\alpha+1}} \frac{A^{\alpha+1}}{B^{\alpha}}, \tag{69}$$

we obtain

$$w^{'}(t) \leq -\phi(t)Q(t) + \frac{\alpha^{\alpha}}{(\alpha+1)^{\alpha+1}} \frac{\phi(t)r(\sigma(t))\left(P(t)\right)^{\alpha+1}}{\left(\beta\sigma^{'}(t)\psi(t)\right)^{\alpha}}. \tag{70}$$

Integrating the above inequality from T_1 to t, we have

$$w(t) \leq w(T_1) - \int_{T_1}^{t} \left(\phi(s)Q(s) - \frac{\alpha^{\alpha}}{(\alpha+1)^{\alpha+1}} \frac{\phi(s)r(\sigma(s))\left(P(s)\right)^{\alpha+1}}{\left(\beta\sigma^{'}(s)\psi(s)\right)^{\alpha}} \right) ds. \tag{71}$$

Taking Eq. (57) into account, we get

$$\begin{aligned} w(T_1) \quad & \geq \phi(t)A(t) \int_{t}^{\infty} \frac{Q(s)}{A(s)} ds \\ & + \int_{T_1}^{t} \left(\phi(s)Q(s) - \frac{\alpha^{\alpha}}{(\alpha+1)^{\alpha+1}} \frac{\phi(s)r(\sigma(s))\left(P(s)\right)^{\alpha+1}}{\left(\beta\sigma^{'}(s)\psi(s)\right)^{\alpha}} \right) ds. \end{aligned} \tag{72}$$

Taking the lim sup on both sides of the above inequality as $t \to \infty$, we obtain a contradiction to the condition (48). This completes the proof.

Remark 2. Note that the presence of the term $\phi(t)A(t)\int_{t}^{\infty} \frac{Q(t)}{A(s)} ds$ in Eq. (57) improves a number of related results in, e.g., [9, 13–18, 20].

Setting $\phi(t) = t$ in Eq. (57), then the following corollary becomes immediate.

Corollary 1. Let conditions (i)–(v) and one of the conditions (14) or (21) hold. Assume that, for all sufficiently large, T, $T_1 \geq T$,

$$tp(t) \leq r(t) \tag{73}$$

on $[T, \infty)$ and

$$\lim_{t\to\infty} sup \left\{ tA(t) \int_{t}^{\infty} \frac{Q(s)}{A(s)} ds \quad + \int_{T_1}^{t} \left[sQ(s) - \frac{\alpha^{\alpha}}{(\alpha+1)^{\alpha+1}} \frac{sr(\sigma(s))\left(\frac{1}{s} - \frac{p(s)}{r(s)}\right)^{\alpha+1}}{\left(\beta\sigma^{'}(s)\psi(s)\right)^{\alpha}} \right] ds \right\} = \infty, \tag{74}$$

where $\psi(t)$ is as in Theorem 1. Then Eq. (1) is oscillatory.

Corollary 2. Assume that the conditions (39) and (74) hold. Then Eq. (1) is oscillatory or $\lim_{t\to\infty} x(t) = 0$.

Next, we present some complementary oscillation results for Eq. (1) by using an integral averaging technique due to Philos. We need the class of functions F. Let

$$\mathcal{D}_0 = \{(t, s) : t > s \geq t_0\} \quad \text{and} \quad \mathcal{D} = \{(t, s) : t > s \geq t_0\} \tag{75}$$

The function $H(t, s) \in C(\mathcal{D}, \mathbb{R})$ is said to belong to a class F if

(a) $H(t, t) = 0$ for $t \geq T$, $H(t, s) > 0$ for $(t, s) \in \mathcal{D}_0$

(b) $H(t, s)$ has a continuous and nonpositive partial derivative on $\mathcal{D}_0$ with respect to the second variable such that

$$\frac{\partial}{\partial s}\Big(H(t, s)\phi(s)\Big) - H(t, s)\frac{\phi(s)p(s)}{r(s)} = -h(t, s)\Big(H(t, s)\phi(s)\Big)^{\frac{\alpha}{\alpha+1}} \tag{76}$$

for all $(t, s) \in \mathcal{D}_0$.

Theorem 2. Let conditions (i)–(v) and one of the conditions (14) or (21) hold. Furthermore, assume that there exist functions $H(t, s)$, $h(t, s) \in F$ such that, for all sufficiently large, T, for $T_1 \geq T$,

$$\limsup_{t\to\infty} \frac{1}{H(t, T_1)} \int_{T_1}^{t} \Big(H(t, s)\Big(\phi(s)Q(s) + \rho(s)\phi(s)p(s)\Big) - \frac{\alpha^\alpha}{(\alpha+1)^{\alpha+1}} \frac{h^{\alpha+1}(t, s)r(\sigma(s))}{\beta^\alpha \Big(\sigma'(s)\psi(s)\Big)^\alpha}\Big) ds = \infty \tag{77}$$

where $\phi(t)$ and $\rho(t)$ are continuously differentiable functions and $\psi(t)$ is as in Theorem 1. Then Eq. (1) is oscillatory.

Proof. Suppose to the contrary that $x(t)$ is a nonoscillatory solution of Eq. (1). Then, without loss of generality, we may assume that there exists $T \in \mathcal{I}$ large enough, so that $x(t)$ satisfies the conclusions of Lemma 1 or 2 on $[T, \infty)$ with

$$x(t) > 0, \quad x(\tau(t)) > 0, \quad x(\sigma(t)) > 0 \tag{78}$$

on $[T, \infty)$. In particular, we have

$$z(t) > 0, \quad z'(t) > 0, \quad \Big(r(t)\big(z'(t)\big)^\alpha\Big)' < 0, \quad \text{for} \quad t \geq T. \tag{79}$$

Define the function $w(t)$ as

$$w(t) = \phi(t)r(t)\left(\frac{\big(z'(t)\big)^\alpha}{z^\beta(\sigma(t))} + \rho(t)\right) \geq \phi(t)r(t)\rho(t), \tag{80}$$

where $\rho(t) \in C^1(\mathcal{I}, \mathbb{R})$. Similarly to the proof of Theorem 1, we obtain the inequality

$$\begin{aligned} w'(t) \le & -\phi(t)Q(t) + \phi(t)\Big(r(t)\rho(t)\Big)' + \left(\frac{\phi'(t)}{\phi(t)} - \frac{p(t)}{r(t)}\right)w(t) \\ & - \frac{\beta\sigma'(t)\psi(t)}{\Big(\phi(t)r(\sigma(t))\Big)^{\frac{1}{\alpha}}}\Big(w(t) - \phi(t)r(t)\rho(t)\Big)^{\frac{1+\alpha}{\alpha}}. \end{aligned} \tag{81}$$

Multiplying Eq. (81) by $H(t, s)$, integrating with respect to s from T_1 to t for $t \ge T_1 \ge T$, and using (a) and (b), we find that

$$\begin{aligned} & \int_{T_1}^{t} H(t, s)\phi(s)\Big(Q(s) - \Big(r(s)\rho(s)\Big)\Big)' \mathrm{d}s \\ \le & -\int_{T}^{t} H(t, s)w'(s)\mathrm{d}s + \int_{T_1}^{t} H(t, s)\left(\frac{\phi'(s)}{\phi(s)} - \frac{p(s)}{r(s)}\right)w(s)\mathrm{d}s \\ & -\int_{T_1}^{t} \frac{\beta H(t, s)\sigma'(s)\psi(s)}{\Big(\phi(s)r(\sigma(s))\Big)^{\frac{1}{\alpha}}}\Big(w(s) - \phi(s)r(s)\rho(s)\Big)^{\frac{1+\alpha}{\alpha}} \mathrm{d}s \\ = & \, H(t, s)w(s)\Big|^{T_1 t} + \int_{T_1}^{t}\left(\frac{\partial}{\partial s}H(t, s) + H(t, s)\left(\frac{\phi'(s)}{\phi(s)} - \frac{p(s)}{r(s)}\right)\right)w(s)\mathrm{d}s \\ & -\int_{T_1}^{t} \frac{\beta H(t, s)\sigma'(s)\psi(s)}{\Big(\phi(s)r(\sigma(s))\Big)^{\frac{1}{\alpha}}}\Big(w(s) - \phi(s)r(s)\rho(s)\Big)^{\frac{1+\alpha}{\alpha}} \mathrm{d}s \\ = & \, H(t, T_1)w(T_1) + \int_{T_1}^{t} -\frac{h(t, s)}{\phi(s)}\Big(H(t, s)\phi(s)\Big)^{\frac{\alpha}{\alpha+1}} w(s)\mathrm{d}s \\ & -\int_{T_1}^{t} \frac{\beta H(t, s)\sigma'(s)\psi(s)}{\Big(\phi(s)r(\sigma(s))\Big)^{\frac{1}{\alpha}}}\Big(w(s) - \phi(s)r(s)\rho(s)\Big)^{\frac{1+\alpha}{\alpha}} \mathrm{d}s \end{aligned} \tag{82}$$

Setting

$$A := -\frac{h(t, s)}{\phi(s)}[H(t, s)\phi(s)]^{\frac{\alpha}{\alpha+1}}, \quad B := \frac{\beta H(t, s)\sigma'(s)\psi(s)}{\Big(\phi(s)r(\sigma(s))\Big)^{\frac{1}{\alpha}}} \tag{83}$$

and

$$C := \phi(s)r(s)\rho(s) \tag{84}$$

and using the inequality

$$Au - B(u - C)^{\frac{1+\alpha}{\alpha}} \le AC + \frac{\alpha^{\alpha}}{(\alpha+1)^{\alpha+1}}\frac{A^{\alpha+1}}{B^{\alpha}}, \tag{85}$$

we obtain

$$\begin{aligned} & \int_{T_1}^{t} H(t, s)\phi(s)\Big(Q(s) - \Big(r(s)\rho(s)\Big)'\Big)\mathrm{d}s \\ \le & \, H(t, T_1)w(T_1) + \int_{T_1}^{t} -h(t, s)r(s)\rho(s)[H(t, s)\phi(s)]^{\frac{\alpha}{\alpha+1}}\mathrm{d}s \\ & + \int_{T_1}^{t} \frac{\alpha^{\alpha}}{(\alpha+1)^{\alpha+1}} \frac{h^{\alpha+1}(t, s)r(\sigma(s))}{\beta^{\alpha}\Big(\sigma'(s)\psi(s)\Big)^{\alpha}}\mathrm{d}s \end{aligned} \tag{86}$$

Thus,

$$\begin{aligned} H(t, T_1)w(T_1) \geq & \int_{T_1}^{t} H(t, s)\phi(s)\left(Q(s)-\left(r(s)\rho(s)\right)'\right)ds \\ & + \int_{T_1}^{t} h(t, s)r(s)\rho(s)[H(t, s)\phi(s)]^{\frac{\alpha}{\alpha+1}}ds \\ & - \int_{T_1}^{t} \frac{\alpha^{\alpha}}{(\alpha+1)^{\alpha+1}} \frac{h^{\alpha+1}(t, s)r(\sigma(s))}{\beta^{\alpha}\left(\sigma'(s)\psi(s)\right)^{\alpha}} ds. \end{aligned} \tag{87}$$

That is,

$$\begin{aligned} & H(t, T_1)w(T_1) \\ & \geq \int_{T_1}^{t} H(t, s)\phi(s)\left(Q(s)-\left(r(s)\rho(s)\right)'\right)ds \\ & + \int_{T_1}^{t} -r(s)\rho(s)\left(\frac{\partial}{\partial s}\left(H(t, s)\phi(s)\right)-H(t, s)\frac{\phi(s)p(s)}{r(s)}\right)ds \\ & - \int_{T_1}^{t} \frac{\alpha^{\alpha}}{(\alpha+1)^{\alpha+1}} \frac{h^{\alpha+1}(t, s)r(\sigma(s))}{\beta^{\alpha}\left(\sigma'(s)\psi(s)\right)^{\alpha}} ds \\ & = \int_{T_1}^{t} H(t, s)\left(\phi(s)Q(s) + \rho(s)\phi(s)p(s)\right)ds \\ & - H(t, s)\phi(s)r(s)\rho(s)\Big|_{T_1}^{t} - \int_{T_1}^{t} \frac{\alpha^{\alpha}}{(\alpha+1)^{\alpha+1}} \frac{h^{\alpha+1}(t, s)r(\sigma(s))}{\beta^{\alpha}\left(\sigma'(s)\psi(s)\right)^{\alpha}} ds \end{aligned} \tag{88}$$

It follows that

$$\begin{aligned} & \int_{T_1}^{t} H(t, s)\left(\phi(s)Q(s) + \rho(s)\phi(s)p(s)\right)ds \\ & - \int_{T_1}^{t} \frac{\alpha^{\alpha}}{(\alpha+1)^{\alpha+1}} \frac{h^{\alpha+1}(t, s)r(\sigma(s))}{\beta^{\alpha}\left(\sigma'(s)\psi(s)\right)^{\alpha}} ds \\ & \leq H(t, T_1)\left(w(T_1)-\phi(T_1)r(T_1)\rho(T_1)\right), \end{aligned} \tag{89}$$

which is a contradiction to Eq. (77). The proof is complete.

Remark 3. Authors in [15, 20] studied a partial case of Eq. (1) by employing the generalized Riccati substitution (80). Note that the function $\rho(t)$ used in the generalized Riccati substitution (80) finally becomes unimportant. Thus, we can put $\rho(t) = 0$ and obtain similar results to those from [15, 20].

In the next part, we provide several oscillation results for Eq. (1) under the assumption that the function $P(t)$ is nonpositive. These results generalize those from [10] for Eq. (1) in such sense that $\alpha \neq \beta$ and $p(t) \neq 0$.

Theorem 3. Let conditions (i)–(v) and one of the conditions (14) or (21) hold. Furthermore, assume that there exists a continuously differentiable function $\phi(t)$ such that, for all sufficiently large, T, $T_1 \geq T$,

$$P(t)\leq 0 \tag{90}$$

on $[T,\ \infty)$ and

$$\limsup_{t\to\infty}\left[\phi(t)A(t)\int_t^{\infty}\frac{Q(s)}{A(s)}\mathrm{d}s+\int_{T_1}^{t}\phi(s)\left(Q(s)-A(s)P(s)\int_s^{\infty}\frac{Q(u)}{A(u)}\mathrm{d}u\right)\mathrm{d}s\right]=\infty. \tag{91}$$

Then Eq. (1) is oscillatory.

Proof. Suppose to the contrary that $x(t)$ is a nonoscillatory solution of Eq. (1). Then, without loss of generality, we may assume that there exists $T\in\mathcal{I}$ large enough, so that $x(t)$ satisfies the conclusions of Lemma 1 or 2 on $[T, \infty)$ with

$$x(t)>0,\quad x(\tau(t))>0,\quad x(\sigma(t))>0 \tag{92}$$

on $[T, \infty)$. In particular, we have

$$z(t)>0,\quad z'(t)>0,\quad \left(r(t)\left(z'(t)\right)^{\alpha}\right)'<0,\quad \text{for}\quad t\geq T. \tag{93}$$

Proceeding as in the proof of Theorem 1, we obtain the inequality (66), i.e.,

$$w'(t)\leq\ -\phi(t)Q(t)+P(t)w(t)-\frac{\beta\sigma'(t)\psi(t)}{\left(\phi(t)r(\sigma(t))\right)^{\frac{1}{\alpha}}}w^{\frac{1+\alpha}{\alpha}}(t) \tag{94}$$

for $t\geq T_1\geq T$. Using Eq. (90), and setting Eq. (57) in Eq. (94), we get

$$\begin{aligned}w'(t)\ &\leq-\phi(t)Q(t)+\phi(t)A(t)P(t)\int_t^{\infty}\frac{Q(s)}{A(s)}\mathrm{d}s\\&\quad-\frac{\beta\sigma'(t)\psi(t)}{\left(\phi(t)r(\sigma(t))\right)^{\frac{1}{\alpha}}}w^{\frac{1+\alpha}{\alpha}}(t)\\&\leq-\phi(t)Q(t)+\phi(t)A(t)P(t)\int_t^{\infty}\frac{Q(s)}{A(s)}\mathrm{d}s,\end{aligned} \tag{95}$$

that is,

$$w'(t)+\phi(t)Q(t)-\phi(t)A(t)P(t)\int_t^{\infty}\frac{Q(s)}{A(s)}\mathrm{d}s\leq 0. \tag{96}$$

Integrating the above inequality from T_1 to t, we have

$$\begin{aligned}w(T_1)\ &\geq w(t)+\int_{T_1}^{t}\left(\phi(s)Q(s)-\phi(s)A(s)P(s)\int_s^{\infty}\frac{Q(u)}{A(u)}\mathrm{d}u\right)\mathrm{d}s\\&\geq\phi(t)A(t)\int_t^{\infty}\frac{Q(s)}{A(s)}\mathrm{d}s+\int_{T_1}^{t}\left(\phi(s)Q(s)-\phi(s)A(s)P(s)\int_s^{\infty}\frac{Q(u)}{A(u)}\mathrm{d}u\right)\mathrm{d}s\end{aligned} \tag{97}$$

Taking the lim sup on both sides of the above inequality as $t\to\infty$, we obtain a contradiction to condition Eq. (91). This completes the proof.

Setting $\phi(t) = 1$, we have the following consequence.

Corollary 3. Let conditions (i)–(v) and one of the conditions (14) or (21) hold. Assume that

$$\limsup_{t\to\infty}\left[A(t)\int_t^{\infty}\frac{Q(s)}{A(s)}\mathrm{d}s + \int_{T_1}^{t}\tilde{q}(s)\mathrm{d}s\right] = \infty, \tag{98}$$

for all sufficiently large T, for $T_1 \geq T$. Then Eq. (1) is oscillatory.

Define a sequence of functions $\{y_n(t)\}_{n=0}^{\infty}$ as

$$y_0(t) = \int_t^{\infty}\tilde{q}(s)\mathrm{d}s, \quad t \geq T \tag{99}$$

$$y_n(t) = \int_t^{\infty}\frac{\beta\sigma'(s)\psi(s)}{r^{\frac{1}{\alpha}}(\sigma(s))}\left(y_{n-1}(s)\right)^{\frac{1+\alpha}{\alpha}}\mathrm{d}s + y_0(t), \quad t \geq T, \quad n = 1, 2, 3, \ldots, \tag{100}$$

for $T \geq t_0$ sufficiently large.

By induction, we can see that $y_n \leq y_{n+1}$, $n = 1, 2, 3, \ldots$.

Lemma 4. Let conditions (i)–(v) and one of the conditions (14) or (21) hold. Assume that $x(t)$ is a positive solution of Eq. (1) on $\mathcal{I}$. Then there exists $T \in \mathcal{I}$, sufficiently large, such that

$$w(t) \geq y_n(t), \tag{101}$$

where $w(t)$ and $y_n(t)$ are defined as Eqs. (54) and (100), respectively. Furthermore, there exists a positive function $y(t)$ on $[T_1, \infty)$, $T_1 \geq T$, such that

$$\lim_{n\to\infty} y_n(t) = y(t) \tag{102}$$

and

$$y(t) = \int_t^{\infty}\frac{\beta\sigma'(s)\psi(s)}{r^{\frac{1}{\alpha}}(\sigma(s))}\left(y(s)\right)^{\frac{1+\alpha}{\alpha}}\mathrm{d}s + y_0(t). \tag{103}$$

Proof. Similarly to the proof of Theorem 3, we obtain Eq. (95). Setting $\phi(t) = 1$ in Eq. (95), we get

$$w'(t) + Q(t) + \frac{p(t)A(t)}{r(t)}\int_t^{\infty}\frac{Q(s)}{A(s)}\mathrm{d}s + \frac{\beta\sigma'(t)\psi(t)}{r^{\frac{1}{\alpha}}(\sigma(t))}w^{\frac{1+\alpha}{\alpha}}(t) \leq 0 \tag{104}$$

for $t \geq T_1 \geq T$. Integrating Eq. (104) from t to t', we get

$$w(t') - w(t) + \int_t^{t'}\tilde{q}(s)\mathrm{d}s + \int_t^{t'}\frac{\beta\sigma'(s)\psi(s)}{r^{\frac{1}{\alpha}}(\sigma(s))}w^{\frac{1+\alpha}{\alpha}}(s)\mathrm{d}s \leq 0 \tag{105}$$

or

$$w(t') - w(t) + \int_t^{t'}\frac{\beta\sigma'(s)\psi(s)}{r^{\frac{1}{\alpha}}(\sigma(s))}w^{\frac{1+\alpha}{\alpha}}(s)\mathrm{d}s \leq 0. \tag{106}$$

We assert that

$$\int_t^\infty \frac{\beta\sigma'(s)\psi(s)}{r^{\frac{1}{\alpha}}(\sigma(s))} w^{\frac{1+\alpha}{\alpha}}(s)ds < \infty. \tag{107}$$

If not, then

$$w(t') \le w(t) - \int_t^{t'} \frac{\beta\sigma'(s)\psi(s)}{r^{\frac{1}{\alpha}}(\sigma(s))} w^{\frac{1+\alpha}{\alpha}}(s)ds \to -\infty \tag{108}$$

as $t' \to \infty$, which contradicts to the positivity of $w(t)$ and thus the assertion is proved. By Eq. (104), we see that $w(t)$ is decreasing that means

$$\lim_{t\to\infty} w(t) = k, \quad k \ge 0. \tag{109}$$

By virtue of Eq. (107), we have $k = 0$. Thus, letting $t' \to \infty$ in Eq. (105), we get

$$\begin{aligned} w(t) &\ge \int_t^\infty \tilde{q}(s)ds + \int_t^\infty \frac{\beta\sigma'(s)\psi(s)}{r^{\frac{1}{\alpha}}(\sigma(s))} w^{\frac{1+\alpha}{\alpha}}(s)ds \\ &= y_0(t) + \int_t^\infty \frac{\beta\sigma'(s)\psi(s)}{r^{\frac{1}{\alpha}}(\sigma(s))} w^{\frac{1+\alpha}{\alpha}}(s)ds, \end{aligned} \tag{110}$$

that is,

$$w(t) \ge \int_t^\infty \tilde{q}(s)ds = y_0(t). \tag{111}$$

Moreover, by induction, we have that

$$w(t) \ge y_n(t), \quad \text{for} \quad t \ge T_1, \quad n = 1, 2, 3, \ldots. \tag{112}$$

Thus, since the sequence $\{y_n(t)\}_{n=0}^\infty$ is monotone increasing and bounded above, it converges to $y(t)$. Letting $n \to \infty$ and using Lebesgue monotone convergence theorem in Eq. (100), we get Eq. (103). The proof is complete.

Theorem 4. Let conditions (i)–(v) and one of the conditions (14) or (21) hold. If

$$\lim_{t\to\infty} inf \left(\frac{1}{y_0(t)} \int_t^\infty \frac{\beta\sigma'(s)\psi(s)}{r^{\frac{1}{\alpha}}(\sigma(s))} \left(y_0(s)\right)^{\frac{1+\alpha}{\alpha}} ds \right) > \frac{\alpha}{(\alpha+1)^{\frac{1+\alpha}{\alpha}}}, \tag{113}$$

where $\psi(t)$ is as in Theorem 1, then Eq. (1) is oscillatory.

Proof. Suppose to the contrary that $x(t)$ is a nonoscillatory solution of Eq. (1). Then, without loss of generality, we may assume that there exists $T \in \mathcal{I}$ large enough, so that $x(t)$ satisfies the conclusions of Lemma 1 or 2 on $[T, \infty)$ with

$$x(t) > 0, \quad x(\tau(t)) > 0, \quad x(\sigma(t)) > 0 \tag{114}$$

on $[T, \infty)$. In particular, we have

$$z(t) > 0, \quad z'(t) > 0, \quad \left(r(t)\left(z'(t)\right)^{\alpha}\right)' < 0, \quad \text{for} \quad t \geq T. \tag{115}$$

By Eq. (113), there exists a constant $\gamma > \frac{\alpha}{(\alpha+1)^{\frac{1+\alpha}{\alpha}}}$ such that

$$\lim_{t\to\infty} inf \frac{1}{y_0(t)} \int_t^{\infty} \frac{\beta\sigma'(s)\psi(s)}{r^{\frac{1}{\alpha}}(\sigma(s))} \left(y_0(s)\right)^{\frac{1+\alpha}{\alpha}} ds > \gamma. \tag{116}$$

Proceeding as in the proof of Lemma 4, we obtain Eq. (110) and from that, we have

$$\frac{w(t)}{y_0(t)} \geq 1 + \frac{1}{y_0(t)} \int_t^{\infty} \frac{\beta\sigma'(s)\psi(s)}{r^{\frac{1}{\alpha}}(\sigma(s))} \left(y_0(s)\right)^{\frac{1+\alpha}{\alpha}} \left(\frac{w(s)}{y_0(s)}\right)^{\frac{1+\alpha}{\alpha}} ds \tag{117}$$

Let

$$\lambda = \inf_{t \geq t_1} \frac{w(t)}{y_0(t)}. \tag{118}$$

Then it is easy to see that $\lambda \geq 1$ and

$$\lambda \geq 1 + \lambda^{\frac{1+\alpha}{\alpha}} \gamma, \tag{119}$$

which contradicts the admissible value of λ and γ, and thus completes the proof.

Theorem 5. Let conditions (i)–(v), one of the conditions (14) or (21) hold, and $y_n(t)$ be defined as in Eq. (100). If there exists some $y_n(t)$ such that, for T sufficiently large,

$$\lim_{t\to\infty} sup \; y_n(t) \left(\int_T^{\sigma(t)} r^{-\frac{1}{\alpha}}(s) ds\right)^{\alpha} > \frac{1}{\psi(t)}, \tag{120}$$

where $\psi(t)$ is as in Theorem 1, then Eq. (1) is oscillatory.

Proof. Suppose to the contrary that $x(t)$ is a nonoscillatory solution of Eq. (1). Then, without loss of generality, we may assume that there exists $T \in \mathcal{I}$ large enough, so that $x(t)$ satisfies the conclusions of Lemma 1 or 2 on $[T, \infty)$ with

$$x(t) > 0, \quad x(\tau(t)) > 0, \quad x(\sigma(t)) > 0 \tag{121}$$

on $[T, \infty)$. In particular, we have

$$z(t) > 0, \quad z'(t) > 0, \quad \left(r(t)\left(z'(t)\right)^{\alpha}\right)' < 0, \quad \text{for} \quad t \geq T. \tag{122}$$

Proceeding as in the proof of Theorem 3 and using defining $w(t)$ as in Eq. (54), for $T_1 \geq T$, we get

$$
\begin{aligned}
\frac{1}{w(t)} &= \frac{z^{\beta}(\sigma(t))}{r(t)\left(z'(t)\right)^{\alpha}} \\
&\geq \frac{\psi(t)}{r(t)}\left(\frac{z(\sigma(t))}{z'(t)}\right)^{\alpha} \\
&= \frac{\psi(t)}{r(t)}\left(\frac{z(T_1)+\int_{T_1}^{\sigma(t)} r^{-\frac{1}{\alpha}}(s) r^{\frac{1}{\alpha}}(s) z'(s)\mathrm{d}s}{z'(t)}\right)^{\alpha} \\
&\geq \psi(t)\left(\int_{T_1}^{\sigma(t)} r^{-\frac{1}{\alpha}}(s)\mathrm{d}s\right)^{\alpha}
\end{aligned}
\tag{123}
$$

Thus,

$$
w(t)\left(\int_{T}^{\sigma(t)} r^{-\frac{1}{\alpha}}(s)\mathrm{d}s\right)^{\alpha} \leq \frac{1}{\psi(t)}\left(\frac{\int_{T}^{\sigma(t)} r^{-\frac{1}{\alpha}}(s)\mathrm{d}s}{\int_{T_1}^{\sigma(t)} r^{-\frac{1}{\alpha}}(s)\mathrm{d}s}\right)^{\alpha} \tag{124}
$$

And therefore,

$$
\lim_{t\to\infty} sup\ w(t)\left(\int_{T}^{\sigma(t)} r^{-\frac{1}{\alpha}}(s)\mathrm{d}s\right)^{\alpha} \leq \frac{1}{\psi(t)}, \tag{125}
$$

which contradicts Eq. (120). The proof is complete.

Theorem 6. Let conditions (i)–(v), one of the conditions (14) or (21) hold, and $y_n(t)$ be defined as in Eq. (100). If there exists some $y_n(t)$ such that

$$
\int_{T_1}^{\infty} \tilde{q}(t)\exp\left(\int_{T_1}^{t} \frac{\beta\sigma'(s)\psi(s)}{r^{\frac{1}{\alpha}}(\sigma(s))} y_n^{\frac{1}{\alpha}}(s)\mathrm{d}s\right)\mathrm{d}t = \infty \tag{126}
$$

or

$$
\int_{T_1}^{\infty} \frac{\beta\sigma'(t)\psi(t) y_n^{\frac{1}{\alpha}}(t) y_0(t)}{r^{\frac{1}{\alpha}}(\sigma(t))}\exp\left(\int_{T_1}^{t} \frac{\beta\sigma'(s)\psi(s)}{r^{\frac{1}{\alpha}}(\sigma(s))} y_n^{\frac{1}{\alpha}}(s)\mathrm{d}s\right)\mathrm{d}t = \infty, \tag{127}
$$

for T sufficiently large and $T_1 \geq T$, where $\psi(t)$ is as in Theorem 1, then Eq. (1) is oscillatory.

Proof. Suppose to the contrary that $x(t)$ is a nonoscillatory solution of Eq. (1). Then, without loss of generality, we may assume that there exists $T \in \mathcal{I}$ large enough, so that $x(t)$ satisfies the conclusions of Lemma 1 or 2 on $[T, \infty)$ with

$$
x(t) > 0, \quad x(\tau(t)) > 0, \quad x(\sigma(t)) > 0 \tag{128}
$$

on $[T, \infty)$. In particular, we have

$$z(t) > 0, \quad z^{'}(t) > 0, \quad \left(r(t)\left(z^{'}(t)\right)^{\alpha}\right)^{'} < 0, \quad \text{for} \quad t \geq T. \tag{129}$$

From Eq. (103), we have

$$y^{'}(t) = -\frac{\beta\sigma^{'}(t)\psi(t)}{r^{\frac{1}{\alpha}}(\sigma(t))}\left(y(t)\right)^{\frac{1+\alpha}{\alpha}} - \tilde{q}(t), \tag{130}$$

for all $t \geq T_1 \geq T$. Since $y(t) \geq y_n(t)$, Eq. (130) yields

$$y^{'}(t) \leq -\frac{\beta\sigma^{'}(t)\psi(t)}{r^{\frac{1}{\alpha}}(\sigma(t))} y_n^{\frac{1}{\alpha}}(t) y(t) - \tilde{q}(t). \tag{131}$$

Multiplying the above inequality by the integration factor

$$\exp\left(\int_{T_1}^{t} \frac{\beta\sigma^{'}(s)\psi(s)}{r^{\frac{1}{\alpha}}(\sigma(s))} y_n^{\frac{1}{\alpha}}(s) \mathrm{d}s\right), \tag{132}$$

one gets

$$\begin{aligned} y(t) \leq \; & \exp\left(-\int_{T_1}^{t} \frac{\beta\sigma^{'}(s)\psi(s)}{r^{\frac{1}{\alpha}}(\sigma(s))} y_n^{\frac{1}{\alpha}}(s) \mathrm{d}s\right) \\ & \left(y(t_1) - \int_{T_1}^{t} \tilde{q}(s) \exp\left(\int_{T_1}^{s} \frac{\beta\sigma^{'}(u)\psi(u)}{r^{\frac{1}{\alpha}}(\sigma(u))} y_n^{\frac{1}{\alpha}}(u) \mathrm{d}u\right) \mathrm{d}s\right), \end{aligned} \tag{133}$$

from which we have that

$$\int_{T_1}^{t} \tilde{q}(s) \exp\left(\int_{T_1}^{s} \frac{\beta\sigma^{'}(u)\psi(u)}{r^{\frac{1}{\alpha}}(\sigma(u))} y_n^{\frac{1}{\alpha}}(u) \mathrm{d}u\right) \mathrm{d}s \leq y(T_1) < \infty. \tag{134}$$

This is a contradiction with Eq. (126).

Now denote

$$u(t) = \int_{t}^{\infty} \frac{\beta\sigma^{'}(s)\psi(s)}{r^{\frac{1}{\alpha}}(\sigma(s))}\left(y(s)\right)^{\frac{1+\alpha}{\alpha}} \mathrm{d}s \tag{135}$$

Taking the derivative of $u(t)$, one gets

$$
\begin{aligned}
u'(t) &= -\frac{\beta\sigma'(t)\psi(t)}{r^{\frac{1}{\alpha}}(\sigma(t))}\left(y(t)\right)^{\frac{1+\alpha}{\alpha}} \\
&\leq -\frac{\beta\sigma'(t)\psi(t)}{r^{\frac{1}{\alpha}}(\sigma(t))} y_n^{\frac{1}{\alpha}}(t)y(t) \\
&= \frac{\beta\sigma'(t)\psi(t)}{r^{\frac{1}{\alpha}}(\sigma(t))} y_n^{\frac{1}{\alpha}}(t)\left(u(t) + y_0(t)\right)
\end{aligned} \tag{136}
$$

Proceeding in a similar manner to that above, we conclude that

$$
\int_{T_1}^{\infty} \frac{\beta\sigma'(t)\psi(t)}{r^{\frac{1}{\alpha}}(\sigma(t))} y_n^{\frac{1}{\alpha}}(t)y_0(t)\exp\left(\int_{T_1}^{t} \frac{\beta\sigma'(s)\psi(s)}{r^{\frac{1}{\alpha}}(\sigma(s))} y_n^{\frac{1}{\alpha}}(s)\mathrm{d}s\right)\mathrm{d}t < \infty, \tag{137}
$$

which contradicts to Eq. (127). The proof is complete.

Author details

Said R. Grace[1]* and Irena Jadlovská[2]

*Address all correspondence to: saidgrace@yahoo.com

1 Department of Engineering Mathematics, Faculty of Engineering, Cairo University, Giza, Egypt

2 Department of Mathematics and Theoretical Informatics, Faculty of Electrical Engineering and Informatics, Technical University of Košice, Košice, Slovakia

References

[1] C. Sturm. Memoir on linear differential equations of second-order. *J. Math. Pures Appl.*, 1 (1936), 106–186.

[2] R. P. Agarwal, S. R. Grace, and D. O'Regan. *Oscillation Theory for Second Order Linear, Half-Linear, Superlinear and Sublinear Dynamic Equations*, Kluwer Academic, Dordrchet, 2002.

[3] R. P. Agarwal, S. R. Grace, and D. O'Regan. *Oscillation Theory for Second Order Dynamic Equations*, Taylor & Francis, London and New York, 2003.

[4] R. P. Agarwal, M. Bohner, and W. T. Li. *Nonoscillation and Oscillation: Theory for Functional Differential Equations*, Marcel Dekker Inc., New York, 2004.

[5] I. Györi and G. Ladas. *Oscillation Theory of Delay Differential Equations with Applications*, Clarendon Press, Oxford, 1991.

[6] L. H. Erbe, Q. Kong, and B. G. Zhang. *Oscillation Theory for Functional Differential Equations*, Marcel Dekker Inc., New York, 1995.

[7] J. K. Hale. *Functional Differential Equations. Analytic Theory of Differential Equations*. Springer, Berlin Heidelberg, 1971.

[8] M. K. Grammatikopoulos, G. Ladas, and A. Meimaridou. Oscillation of second order neutral delay differential equations, *Rat. Math.*, 1 (1985), 267–274.

[9] S. R. Grace and B. S. Lalli. Oscillation of nonlinear second order neutral differential equations. *Rat. Math.*. 3 (1987), 77–84.

[10] J. G. Dong. Oscillation behavior of second order nonlinear neutral differential equations with deviating arguments. *Comput. Math. Appl.*, 59 (2010), 3710–3717.

[11] B. Baculíková and J. Džurina. Oscillation theorems for second order neutral differential equations. *Comput. Math. Appl.*, 61 (2011), 94–99.

[12] B. Baculíková and J. Džurina. Oscillation theorems for second-order nonlinear neutral differential equations. *Comput. Math. Appl.* 62 (2011), 4472–4478.

[13] T. Li, Z. Han, C. Zhang, and H. Li. Oscillation criteria for second-order superlinear neutral differential equations. *Abstr. Appl. Anal.*, 2011 (2011). Hindawi Publishing Corporation. pp 1–17

[14] T. Li, Y. V. Rogovchenko, and C. Zhang. Oscillation results for second-order nonlinear neutral differential equations. *Adv. Differ. Eqs.*, 2013 (2013), 1–13.

[15] Yu. V. Rogovchenko and F. Tuncay. Oscillation criteria for second-order nonlinear differential equations with damping. *Nonlinear Anal.*, 69 (2008), 208–221.

[16] T. Li and Y. V. Rogovchenko. Oscillation theorems for second-order nonlinear neutral delay differential equations. *Abstr. Appl. Anal.*, 2014 (2014), pp 1–5.

[17] T. Li and Y. V. Rogovchenko. Oscillatory behavior of second-order nonlinear neutral differential equations. *Abstr. Appl. Anal.*, 2014 (2014), pp. 1–8.

[18] T. Li, E. Thandapani, J.R. Graef, and E. Tunç. Oscillation of second-order Emden-Fowler neutral differential equations. *Nonlinear Stud.*, 20 (2013), 1, 1–8.

[19] H. Wu, L. Erbe, and A. Peterson. Oscillation of solution to second-order half-linear delay dynamic equations on time scales. *Electr. J. Differ. Eqs*, 71 (2016), 1–15.

[20] T. Li, Y. V. Rogovchenko, and S. Tang. Oscillation of second-order nonlinear differential equations with damping. *Math. Slov.*, 64.5 (2014), 1227–1236.

5

Relationship between Interpolation and Differential Equations: A Class of Collocation Methods

Francesco Aldo Costabile, Maria Italia Gualtieri and Anna Napoli

Abstract

In this chapter, the connection between general linear interpolation and initial, boundary and multipoint value problems is explained. First, a result of a theoretical nature is given, which highlights the relationship between the interpolation problem and the Fredholm integral equation for high-order differential problems. After observing that the given problem is equivalent to a Fredholm integral equation, this relation is used in order to determine a general procedure for the numerical solution of high-order differential problems by means of appropriate collocation methods based on the integration of the Fredholm integral equation. The classical analysis of the class of the obtained methods is carried out. Some particular cases are illustrated. Numerical examples are given in order to illustrate the efficiency of the method.

Keywords: boundary value problem, initial value problem, collocation methods, interpolation, Birkhoff, Lagrange, Peano, Fredholm

1. Introduction

The relationship between interpolation and differential equations theories has already been considered. In Ref. ([1], p. 72), Davis observed that the Peano kernel in the interpolation problem

$$y(a) = \alpha, \quad y(b) = \beta, \qquad a, b, \alpha, \beta \in \mathbb{R}, \tag{1}$$

is the Green's function of the differential problem

$$\phi''(x) = f(x)$$
$$\phi(a) = \phi(b) = 0$$

where $\phi(x) = y(x) - P_1[y](x)$, being $P_1[y](x)$ the unique interpolatory polynomial for Eq. (1).

He observed that *"these remarks indicate the close relationship between Peano kernels and Green's functions, and hence between interpolation theory and the theory of linear differential equations. Unfortunately, we shall not be able to pursue this relationship"* [1].

Later, Agarwal ([2], p. 2), Agarwal and Wong ([3], pp. 21, 151, 186) considered some separate boundary value problems and the related Fredholm integral equation, using only polynomial interpolation, without taking into account the related Peano kernel. They used Fredholm integral equation in order to obtain existence and uniqueness results for the solution of the considered boundary value problems.

Linear interpolation has an important role also in the numerical solution of differential problems. For example, finite difference methods approximate the solution $y(x)$ of a boundary value problem by a sequence of overlapping polynomials which interpolate $y(x)$ in a set of grid points. This is obtained by replacing the differential equation with finite difference equations on a mesh of points that covers the range of integration. The resultant algebraic system of equations is often solved with iterative pro-cesses, such as relaxation methods.

Many authors used linear interpolation with spline functions for the numerical solution of boundary value problems.

Here, we take into account a more general nonlinear initial/boundary/multipoint value problems for high-order differential equations

$$\begin{cases} y^{(r)}(x) = f\left(x, \mathbf{y}(x)\right), & x \in I = [a,b], \quad r \geq 1 \\ L_i[y](x) = w_i, \quad i = 0,\ldots,r-1, \quad x \in I \end{cases} \tag{2}$$

where $\mathbf{y}(x) = (y(x), y'(x), \ldots, y^{(q)}(x))$, $0 \leq q < r$, $y \in \mathscr{C}^r(I)$, and L_i are r linearly independent functionals on $\mathscr{C}^r(I)$. Moreover, we suppose that the function $f : [a,b] \times \mathbb{R}^{q+1} \to \mathbb{R}$ is continuous at least in the interior of the domain of interest, and it satisfies a uniform Lipschitz condition in $\mathbf{y}$, which means that there exists a nonnegative constant Λ, such that, whenever $(x, y_0, y_1, \ldots, y_q)$ and $(x, \overline{y}_0, \overline{y}_1, \ldots, \overline{y}_q)$ are in the domain of f, the following inequality holds

$$\left| f(x, y_0, y_1, \ldots, y_q) - f(x, \overline{y}_0, \overline{y}_1, \ldots, \overline{y}_q) \right| \leq \Lambda \sum_{k=0}^{q} \left| y_k - \overline{y}_k \right|. \tag{3}$$

If $L_i[y] = \Phi\left(\mathbf{y}(a)\right), i = 0, \ldots, r-1$, then (2) is an initial value problem (IVP); if $L_i[y] = \Phi\left(\mathbf{y}(a), \mathbf{y}(b)\right), i = 0, \ldots, r-1$, then (2) is a boundary value problem (BVP); if $L_i[y] = \Phi\left(\mathbf{y}(x_j)\right), i = 0, \ldots, r-1, j = 0, \ldots, m \geq 2$, then (2) is a multipoint value problem (MVP).

In this chapter,

- we assume that the conditions for the existence and uniqueness of solution of problem (2) in a certain appropriate domain of $[a, b] \times R^{q+1}$ are satisfied and that the solution $y(x)$ is differentiable with continuity up to what is necessary;
- we get the Fredholm integral equation related to problem (2), by polynomial interpolation and the Peano kernel of the linear interpolation problem $L_i[y](x) = w_i,\ i = 0, \ldots, r-1$. In this way, we point out the close relationship between Green's function and Peano kernel;
- then, we construct a class of spectral collocation (pseudospectral) methods which are derived by a linear interpolation process.

The reason for which we prefer collocation methods is their superior accuracy for problems whose solutions are sufficiently smooth functions. Recently, Boyd ([11], p. 8) observed that *"When many decimal places of accuracy are needed, the contest between pseudospectral algorithms and finite difference and finite element methods is not an even battle but a rout: pseudospectral methods win hands-down."*

2. The Fredholm integral equation for problem (2)

We consider the general differential problem (2), and we prove that it is equivalent to a Fredholm integral equation.

Proposition 1 *[1, p. 35] The linear interpolation problem*

$$L_i[P](x) = w_i, \qquad w_i, \in \mathbb{R},\ i = 0, \ldots, r-1,\ P \in P_{r-1},\ x \in I \tag{4}$$

with L_i, $i = 0, \ldots, r-1$, linearly independent functionals on $\mathscr{C}^r(I)$, has the unique solution

$$P_{r-1}(t) = -\frac{1}{G} \begin{vmatrix} 0 & 1 & t & \cdots & \cdots & t^{r-1} \\ w_0 & & & & & \\ w_1 & & & & & \\ \vdots & & & L_i[x^j] & & \\ \vdots & & & & & \\ w_{r-1} & & & & & \end{vmatrix}, \quad G = |L_i[x^j]|_{i,j=0,\ldots,r-1}. \tag{5}$$

Proof. Since the L_i, $i = 0, \ldots, r-1$ are linearly independent, the result follows from the general linear interpolation theory.

Proposition 2 *If $y \in \mathscr{C}^r(I)$ and $L_i[y](x) = w_i$, $i = 0, \ldots, r-1$, $x \in I$, then*

$$y(x) = P_{r-1}[y](x) + \int_a^b K_r^x(x, t)\, y^{(r)}(t)\, dt, \qquad \forall x \in I \text{ fixed}, \tag{6}$$

with $L_i[y] = L_i[P_{r-1}]$, $i = 0, \ldots, r-1$, $P_{r-1}[y](x) = P_{r-1}(x)$, *and*

$$K_r^x(x,t) = \frac{1}{(r-1)!}\left[(x-t)_+^{r-1} - P_{r-1}\left[(x-t)_+^{r-1}\right](x)\right], \tag{7}$$

where index x means that $K_r^x(x,t)$ is considered as a function of x.

Proof. It follows by observing that $P_{r-1}[(x)_+^j](t) = (t)_+^j, j = 0,\ldots,r-1$ and from Peano kernel Theorem [1].

Theorem 1 *With the above notations and under the mentioned hypothesis, problem (2) is equivalent to the Fredholm integral equation*

$$y(x) = P_{r-1}[y](x) + \int_a^b K_r^x(x,t) f\Big(t, y(t)\Big)\, dt. \tag{8}$$

Proof. The result follows from the uniqueness of the Peano kernel and from Propositions 1 and 2.

Corollary 1 *It results* $L_i[K_r^x] = 0,\ i = 0,\ldots,r-1$.

From Theorem 1, general results on the existence and uniqueness of solution of problem (2) by standard techniques [2, 3] can be obtained. In the following, we will not linger over them, but we will outline the close relationship between interpolation and differential equations. Particularly, we will use linear interpolation in order to determine a class of collocation methods for the numerical solution of problem (2).

3. A class of Birkhoff-Lagrange collocation methods

Integral Eq. (8) allows to determine a very wide class of numerical methods for Eq. (2), which we call *methods of collocation for integration.*

Let $\{x_i\}_{i=1}^m$ be m distinct points in $[a,b]$ and denoted by $l_i(t),\ i = 1,\ldots,m,$ the fundamental Lagrange polynomials on the nodes x_i, that is

$$l_i(t) = \frac{\omega_m(t)}{(t-x_i)\omega'_m(x_i)}, \quad \text{where } \omega_m(t) = \prod_{k=1}^{m}(t-x_k). \tag{9}$$

Theorem 2 *If the solution $y(x)$ of Eq. (8) is in $\mathscr{C}^{r+m}(I)$, then*

$$y(x) = P_{r-1}[y](x) + \sum_{i=1}^{m} p_{r,i,m}(x) f\Big(x_i, \mathbf{y}(x_i)\Big) + T_{r,m}(y,x), \tag{10}$$

where

$$p_{r,i,m}(x) = \int_a^b K_r^x(x,t) l_i(t)\, dt, \quad i = 1,\ldots,m, \tag{11}$$

and the remainder term $T_{r,m}(y,x)$ is given by:

$$T_{r,m}(y,x) = \frac{1}{m!}\int_a^b K_r^x(x,t)\omega_m(t)y^{(r+m)}(\xi_x)\,dt, \tag{12}$$

being ξ_x a suitable point of the smallest interval containing x and all x_i, $i = 1,\dots,m$.

Proof. From Lagrange interpolation

$$y^{(r)}(x) = \sum_{i=1}^{m} l_i(x)y^{(r)}(x_i) + \overline{R}_m(y,x) \tag{13}$$

where

$$\overline{R}_m(y,x) = \frac{1}{m!}\omega_m(t)y^{(r+m)}(\xi_x) \tag{14}$$

is the remainder term. From (2), $f(x,y(x)) = \sum_{i=1}^{m} l_i(x)y^{(r)}(x_i) + \overline{R}_m(y,x)$. Then, from Theorem 1, inserting Eq. (13) into (8), we obtain Eq. (10).

Theorem 2 suggests to consider the implicitly defined polynomial

$$y_{r,m}(x) = P_{r-1}[y_{r,m}](x) + \sum_{i=1}^{m} p_{r,i,m}(x)f\left(x_i, y_{r,m}(x_i)\right). \tag{15}$$

For polynomial (15), the following theorem holds.

Theorem 3 (The main Theorem). *Polynomial (15), of degree $r+m-1$, satisfies the relations*

$$\begin{aligned} L_i[y_{r,m}](x) &= w_i, \qquad i = 0,\dots,r-1,\ x \in I,\ w_i \in \mathbb{R} \\ y_{r,m}^{(r)}(x_j) &= f\left(x_j, y_{r,m}(x_j)\right) \qquad j = 1,\dots,m, \end{aligned} \tag{16}$$

that is, $y_{r,m}(x)$ is a collocation polynomial for Eq. (2) at nodes x_j, $j = 1,\dots,m$.

Proof. From (15), Corollary 1 and the linearity of operators L_i, we get $L_i[y_{r,m}](x) = w_i$, $i = 0,\dots,$ $r-1$. By Theorems 1 and 2, we obtain $y^{(r)}(x_i) = y_{r,m}^{(r)}(x_i)$, and from Eq. (11), $p_{r,i,m}^{(r)}(x) = l_i(x)$. Hence, relations (16) follow.

Remark 1 *(Hermite-Birkhoff-type interpolation). Theorem 3 is equivalent to the general Hermite-Birkhoff interpolation problem [12]: given $w_i \in \mathbb{R}$, $i = 0,\dots,r-1$, and $\alpha_j \in \mathbb{R}$, $j = 1,\dots,m$, determine, if there exists, the polynomial $Q(x) \in \mathscr{P}_{m+r-1}$ such that*

$$\begin{aligned} L_i[Q] &= w_i, \qquad i = 0,\dots,r-1 \\ Q^{(r)}(x_j) &= \alpha_j, \quad j = 1,\dots,m,\ x_j \in I. \end{aligned} \tag{17}$$

Remark 2 *In the case of IVPs, for each method (15), we can derive the corresponding implicit Runge-Kutta method. For example, for $r = 2$, let $b = x_0 + h$ and $x_i = x_0 + c_i h$ with $c_i \in [0,1]$. With the change of coordinates $x = x_0 + th$, $t \in [0,1]$, we can write*

$$p_{r,i,m}(x) = p_{r,i,m}(x_0 + th) = h^2 \int_0^t \int_0^r l_i(s)\, ds\, dr, \qquad l_i(s) = \prod_{\substack{k=1 \\ k \neq i}}^{m} \frac{s-c_k}{c_i-c_k}. \tag{18}$$

Putting $f(x_i, y_{r,m}(x_i)) = y''_{r,m}(x_i) \equiv K_i$, $a_{i,j} = p_{r,j}(x_i) = h^2 \int_0^{c_i} (c_i - s) l_j(s)\, ds$, *we have*

$$K_i = f\left(x_0 + c_i h, y_0 + y'_0 th + \sum_{j=1}^{m} a_{i,j} K_j \right) \tag{19}$$

and

$$\begin{cases} y_1(t) \equiv y_{r,m}(x_0 + th) = y_0 + y'_0 th + h^2 \sum_{i=1}^{m} p_{r,i,m}(x_0 + th) K_i \\ y'_1(t) \equiv y'_{r,m}(x_0 + th) = y'_0 h + h^2 \sum_{i=1}^{m} p'_{r,i,m}(x_0 + th) K_i. \end{cases} \tag{20}$$

Eqs. (19) and (20) are the well-known continuous Runge-Kutta method for second-order differential equations. Particularly, for $t = 1$, *we have the implicit Runge-Kutta-Nystrom method.*

3.1. A-priori estimation of error

In what follows, we consider the norm

$$\|f\| = \max_{a \le t \le b} \sum_{k=0}^{q} |f^{(k)}(t)|, \qquad \forall f \in \mathscr{C}^{(q)}(I). \tag{21}$$

Moreover, we define

$$Q_m = \sum_{i=1}^{m} \|p_{r,i,m}\|, \quad F(x) = \int_a^b K_r^x(x,t) dt, \quad H = \max_{a \le t \le b} |\overline{R}_m(y,t)|, \tag{22}$$

where $\overline{R}_m(y,t)$ is defined as in (14).

Theorem 4 *With the previous notations, if* $\Lambda Q_m < 1$, *then*

$$\|y - y_{r,m}\| \le \frac{H\|F\|}{1 - \Lambda Q_m}. \tag{23}$$

Proof. By deriving Eqs. (10) and (15), s times, $s = 0,\ldots,q$, we get

$$y^{(s)}(x) - y^{(s)}_{r,m}(x) = \sum_{i=1}^{m} p^{(s)}_{r,i,m}(x) \left[f\left(x_i, \mathbf{y}(x_i)\right) - f\left(x_i, \mathbf{y}_{r,m}(x_i)\right) \right] + \frac{\partial^s}{\partial x^s} \int_a^b K_r^x(x,t) \overline{R}_m(y,t) dt. \tag{24}$$

It follows that

$$\begin{aligned} |y^{(s)}(x)-y^{(s)}_{r,m}(x)| &\le \sum_{i=1}^{m} |p^{(s)}_{r,i,m}(x)| \Lambda \sum_{k=0}^{q} |y^{(k)}(x_i)-y^{(k)}_{r,m}(x_i)| + H\,|F^{(s)}(x)| \\ &\le \Lambda \| y-y_{r,m} \| \sum_{i=1}^{m} |p^{(s)}_{r,i,m}(x)| + H|F^{(s)}(x)|. \end{aligned} \tag{25}$$

From this, we obtain inequality (23).

4. Algorithms and implementation

To calculate the approximate solution of problem (2) by polynomial $y_{r,m}(x)$ at $x \in I$, we need the values $y^{(s)}_{r,m}(x_k)$, $k = 1,\dots,m$, $s = 0,\dots,q$. In order to get these values, we propose the following algorithm:

- Put $y^{(s)}_k = y^{(s)}_{r,m}(x_k)$, $k = 1,\dots,m$, $s = 0,\dots,q$ and consider the following system

$$y^{(s)}_k = P^{(s)}_{r-1}[y_k](x_k) + \sum_{i=1}^{m} p^{(s)}_{r,i}(x_k) f(x_i, \mathrm{y}_i), \tag{26}$$

$k = 1,\dots,m$, $s = 0,\dots,q$, where $\mathrm{y}_i = (y_i, y'_i, \dots, y^{(q)}_i)$.

System (26) can be written in the form

$$Y - AF(Y) = C \tag{27}$$

where

$$A = \begin{pmatrix} A_0 & 0 & \cdots & 0 \\ 0 & \ddots & & \vdots \\ \vdots & & \ddots & 0 \\ 0 & \cdots & 0 & A_q \end{pmatrix}_{m(q+1)\times m(q+1)} \tag{28}$$

with

$$A_s = \begin{pmatrix} \tilde{a}^{(s)}_{1,1} & \cdots & \tilde{a}^{(s)}_{1,m} \\ \vdots & & \vdots \\ \tilde{a}^{(s)}_{m,1} & \cdots & \tilde{a}^{(s)}_{m,m} \end{pmatrix}_{m\times m} \qquad \tilde{a}^{(s)}_{i,j} = p^{(s)}_{r,j}(x_i), \quad s = 0,\dots,q, \tag{29}$$

$$Y = (\overline{Y}_0, \dots, \overline{Y}_q)^T_{m(q+1)\times 1}, \qquad \overline{Y}_s = \left(y^{(s)}_1, \dots, y^{(s)}_m\right), \tag{30}$$

$$F(Y) = (\underbrace{F_m, \dots, F_m}_{q})^T, \qquad F_m = (f_1, \dots, f_m)^T, \qquad f_i = f(x_i, \mathrm{y}_i), \tag{31}$$

$$B_s = \left(P^{(s)}_{r-1}[y_1](x_1), \dots, P^{(s)}_{r-1}[y_m](x_m)\right), \qquad C = (B_0, \dots, B_q)^T_{m(q+1)\times 1}. \tag{32}$$

From Eq. (27), we get

$$Y = AF(Y) + C\,, \tag{33}$$

or, putting $G(Y) = AF(Y) + C$,

$$Y = G(Y)\,. \tag{34}$$

For the existence and uniqueness of solution of system (34), we can prove, with standard technique, the following theorem.

Theorem 5 *If* $T = \Lambda\|A\|_\infty < 1$, *system (34) has a unique solution which can be calculated by an iterative method*

$$(Y_m)_{v+1} = G\Big((Y_m)_{(v)}\Big), \qquad v \geq 0 \tag{35}$$

with a fixed $(Y_m)_0 \in \mathbb{R}^{m(q+1)}$ *and* $G(Y_m) = AF(Y_m) + C$.

Moreover, if Y *is the exact solution,*

$$\|(Y_m)_{v+1} - Y\|_\infty \leq \frac{T^v}{1-T}\|\,(Y_m)_1 - (Y_m)_0\|_\infty\,. \tag{36}$$

Remark 3 *If f is linear, then system (27) is a linear system which can be solved by a more suitable method.*

Remark 4 *System (27) can be considered as a discrete method for the numerical solution of (2).*

Remark 5 *Method (15) can generate the polynomial sequence*

$$(y_{r,m}(x))_v = P_{r-1}[y_{r,m}](x) + \sum_{i=1}^{m} p_{r,i,m}(x) f(x_i, (y_{r,m}(x_i))_{v-1}), \quad (y_{r,m})_0 = P_{r-1}[y](x) \tag{37}$$

which is equivalent to the discretization of Picard method for differential equations.

4.1. Numerical computation of the entries of matrix A

To calculate the elements $\tilde{a}_{i,k}^{(s)}$ of the matrix A in Eq. (27), we have to compute the integrals

$$p_{r,k}^{(s)}(x) = \frac{d^s}{dx^s}\int_a^b K_r^x(x,t) l_i(t)\,dt \tag{38}$$

for $x = x_i$. Integrating by parts, it remains to solve the problem of the computation of

$$F_{i1}(x_j) = \int_a^{x_j} l_i(t)dt, \quad F_{ik}(x_j) = \int_a^{x_j} F_{i,k-1}(t)dt \qquad k = 2,\ldots,n \tag{39}$$

$$M_{i1}(x_j) = \int_{x_j}^b l_i(t)dt, \quad M_{ik}(x_j) = \int_{x_j}^b M_{i,k-1}(t)dt \qquad k = 2,\ldots,n \tag{40}$$

$i,j = 1,\ldots m$. To this aim, it suffices to compute

$$\int_c^{x_j=t_k}\int_c^{t_{k-1}}\cdots\int_c^{t_1} r_{m,i}(t)\,dt\,dt_1\cdots dt_{k-1} \tag{41}$$

where $c=a$ or $c=b$, $r_{0,0}(t)=1$,

$$r_{m,i}(t)=(t-x_1)\cdots(t-x_{i-1})(t-x_{i+1})\cdots(t-x_m) \qquad i=1,2,\ldots,m\,. \tag{42}$$

For the computation of the integral (41), we use the recursive algorithm introduced in Ref. [13]: for each $i=1,\ldots,m$, let us consider the new points $z_j^{(i)}=x_j$ if $j<i$, and $z_j^{(i)}=x_{j+1}$ if $j\geq i$. Moreover, let us define $g_{0,1,c}^{(i)}(x)=x-c$ and for $s=1,\ldots,m-1$

$$g_{s,j,c}^{(i)}(x)=\int_c^{x=t_j}\int_c^{t_{j-1}}\cdots\int_c^{t_1}\left(t-z_1^{(i)}\right)\left(t-z_2^{(i)}\right)\cdots\left(t-z_s^{(i)}\right)dt\,dt_1\cdots dt_{j-1}. \tag{43}$$

We can easily compute $g_{0,j,c}^{(i)}(x)=\frac{(x-c)^j}{j!}$. For the computation of Eq. (43), the following recurrence formula [13] holds

$$g_{s,j,c}^{(i)}(x)=\left(x-z_s^{(i)}\right)g_{s-1,j,c}^{(i)}(x)-j g_{s-1,j+1,c}^{(i)}(x). \tag{44}$$

Thus, if $W_i=\prod_{k=1,\,k\neq i}^{m}(x_i-x_k)$, then

$$F_{ik}(x_j)=\frac{g_{m-1,k,a}^{(i)}(x_j)}{W_i}, \qquad M_{ik}(x_j)=(-1)^k\frac{g_{m-1,k,b}^{(i)}(x_j)}{W_i}. \tag{45}$$

Remark 6 *An alternative approach for the exact computation of integrals (39) and (40) is to use a quadrature formula with a suitable degree of precision.*

4.2. Outline of the method

Summarizing the proposed method consists of the following steps:

1. determine the interpolation polynomial $P_{r-1}[y](x)$ satisfying the boundary conditions and compute the Peano remainder;
2. approximate $y^{(r)}(x)$ by Lagrange interpolation on a set of fixed nodal point;
3. compute the elements of matrix A (28) and solve system (26);
4. obtain polynomial (15).

5. Some particular cases

Now we consider some special cases of problem (2), and for each case, we determine $P_{r-1}[y](x)$ and $K_r^x(x,t)$. We also show how the proposed class of methods includes the methods presented in previous works [12–24].

5.1. Initial value problems

In the case of initial value problems, in Refs. [13, 17, 25], problem

$$y^{(r)}(x)=f(x,y(x)) \tag{46}$$

has been considered, while in Ref. [23], the authors introduced the more general equation

$$y^{(r)}(x)=f\left(x,y(x),y'(x),\ldots,y^{(r)}(x)\right), \qquad q\leq r-1. \tag{47}$$

In both cases

$$P_{r-1}[y](x)=\sum_{i=0}^{r-1}\frac{(x-a)^i}{i!}y^{(i)}(a) \tag{48}$$

and

$$K_r^x(x,t)=\frac{1}{(r-1)!}(x-t)_+^{r-1}. \tag{49}$$

If $\{x_i\}_{i=1}^m$ are the zeros of Chebyshev polynomials of first and second kind, the explicit expression for polynomials $p_{r,i,m}(x)$ can be obtained [13, 17, 25] for some values of r.

Particularly, for $r=1$ and $r=2$, in the case of zeros of Chebyshev polynomials of first kind, we get

$$\begin{aligned}p_{1,i,m}(x)=&\frac{1}{m}\sum_{k=2}^{m-1}\left\{\left[\frac{T_{k+1}(x)}{k+1}-\frac{T_{k-1}(x)}{k-1}+2\frac{(-1)^{k-1}}{k^2-1}\right]\cos\left(\frac{2i-1}{2m}k\pi\right)\right\}\\&+\frac{1}{m}\left[x+1+\cos\left(\frac{2i-1}{2m}\pi\right)(x^2-1)\right]\end{aligned} \tag{50}$$

where $T_{k-1}(x)$ and $T_{k+1}(x)$ are the Chebyshev polynomials of the first kind and degree $k-1$ and $k+1$, respectively, and

$$\begin{aligned}p_{2,i,m}(x)=&\frac{1}{m}\left\{\frac{(x+1)^2}{2}+\frac{x^3-3x-2}{3}\left(\frac{\cos\frac{\pi(2i-1)}{2m}+x\cos\pi(2i-1)}{m}\right)\right.\\&+\frac{1}{2}\sum_{k=3}^{m-1}\cos\frac{k\pi(2i-1)}{2m}\left[\frac{T_{k+2}(x)}{(k+1)(k+2)}-2\frac{T_k(x)}{k^2-1}\right.\\&\left.\left.+\frac{T_{k-2}(x)}{(k-1)(k-2)}-\frac{12k(-1)^k}{k(k^2-1)(k^2-4)}-\frac{4(-1)^k}{k^2-1}(x+1)\right]\right\}.\end{aligned} \tag{51}$$

In the case of zeros of Chebyshev polynomials of second kind

$$p_{1,i,m}(x) = \frac{2}{m+1} \sin \frac{\pi i}{m+1} \sum_{k=0}^{m-1} \sin \frac{(k+1)\pi i}{m+1} \frac{1}{k+1} \left[T_{k+1}(x) + (-1)^k \right] \tag{52}$$

and

$$\begin{aligned} p_{2,i,m}(x) &= \frac{1}{m+1} \sin \frac{\pi i}{m+1} \left\{ \sin \frac{\pi i}{m+1} (x+1)^2 \right. \\ &\left. + \sum_{k=2}^{m} \frac{1}{k} \sin \frac{k\pi i}{m+1} \left[\frac{T_{k+1}(x)}{k+1} - \frac{T_{k-1}(x)}{k-1} - 2\left(x + \frac{k^2}{k^2-1} \right) (-1)^k \right] \right\} \end{aligned} \tag{53}$$

In Refs. [13, 25], the authors presented the corresponding implicit Runge-Kutta methods too.

In Ref. [26], Coleman and Booth used also a polynomial interpolant of degree n for y'', but they started from an identity different to Eq. (8) and derived a collocation method for which the nodes $\{x_i\}_{i=1}^m$ are the zeros of Chebyshev polynomials of second kind.

5.2. Boundary value problems

5.2.1. Case $r = 2n$

For $n = 1$, for the exact solution $y(x)$ of the second-order BVP

$$y''(x) = f(x, y(x), y'(x)), \quad y(-1) = y_0,\ y(1) = y_1 \tag{54}$$

$x \in [-1, 1]$, it is known that

$$y(x) = \frac{y_1 + y_0}{2} + x\frac{y_1 - y_0}{2} + \int_{-1}^{1} K_2^x(x,t) f(x, y(x), y'(x)) dt \tag{55}$$

where

$$K_2^x(x,t) = \begin{cases} \dfrac{(t+1)(x-1)}{2} & t \le x \\ \dfrac{(x+1)(t-1)}{2} & x < t\,. \end{cases} \tag{56}$$

By applying method (15), we get [16]

$$y_{2,m}(x) = \frac{y_1 + y_0}{2} + x\frac{y_1 - y_0}{2} + \sum_{i=1}^{m} p_{r,i,m}(x) f\left(x_i, y(x_i), y'(x_i)\right) \tag{57}$$

with $p_{r,i,m}(x) = \int_{-1}^{1} K_2^x(x,t) l_i(t) dt$.

If $x_i = \cos \frac{\pi i}{m+1}$, $i = 1, \ldots, m$, we obtain explicitly the expression of $p_{r,i,m}(x)$ [18]

$$p_{r,i,m}(x) = \frac{1}{m+1} \sin\frac{\pi i}{m+1}\left[\sum_{k=2}^{m}\frac{G_k(x)}{k}\sin\frac{k\pi i}{m+1} + (x^2-1)\sin\frac{\pi i}{m+1}\right] \tag{58}$$

where

$$G_k(x) = \frac{T_{k+1}(x)}{k+1} - \frac{T_{k-1}(x)}{k-1} + \begin{cases} \frac{2x}{k^2-1} & \text{even } k \\ k3\frac{2}{k^2-1} & \text{odd } k. \end{cases} \tag{59}$$

The same method has been presented in Ref. [24], where also stability has been studied.

Now assume $[a,b] = [0,1]$ and $r > 2$. Several types of boundary conditions can be considered.

-Hermite boundary conditions [22]:

$$y^{(h)}(0) = \alpha_h, \quad y^{(h)}(1) = \beta_h, \qquad h = 0,\ldots,n-1 \tag{60}$$

with α_h, β_h, $h = 0,\ldots,n-1$ real constants.

In this case, P_{r-1} is the Hermite polynomial of degree $2n-1$

$$P_{2n-1}[y](x) = \sum_{i=0}^{n-1}(y^{(i)}(0)H_{i1}(x) + y^{(i)}(1)H_{i2}(x)) \tag{61}$$

with

$$\begin{aligned} H_{i1}(x) &= \frac{x^i(1-x)^n}{i!}\sum_{s=0}^{n-i-1}\binom{n+s-1}{n-1}x^s \\ H_{i2}(x) &= \frac{x^n(1-x)^i}{i!}\sum_{s=0}^{n-i-1}\binom{n+s-1}{n-1}(1-x)^s. \end{aligned} \tag{62}$$

The kernel is

$$K_{2n}^x(x,t) = \begin{cases} \sum_{i=0}^{n-1}\frac{(-t)^{2n-i-1}}{(2n-i-1)!}H_{i1}(x) & 0 \le t \le x \\ -\sum_{i=0}^{n-1}\frac{(1-t)^{2n-i-1}}{(2n-i-1)!}H_{i2}(x) & x \le t \le 1\,. \end{cases} \tag{63}$$

-Lidstone boundary conditions [19]:

$$y^{(2h)}(0) = \alpha_h, \quad y^{(2h)}(1) = \beta_h, \qquad h = 0,\ldots,n-1 \tag{64}$$

where α_h, β_h, $h = 0,\ldots,n$ are real constants.

In this case, P_{r-1} is the Lidstone interpolating polynomial [3] of degree $2n-1$

$$P_{2n-1}[y](x) = \sum_{k=0}^{n-1}\left[y^{(2k)}(0)\Lambda_k(1-x) + y^{(2k)}(1)\Lambda_k(x)\right] \quad (65)$$

where $\Lambda_k(x)$ are the Lidstone polynomials of degree $2k+1$ [3], and the function $K_{2n}^x(x,t)$ is

$$K_{2n}^x(x,t) = \begin{cases} \sum_{k=0}^{n-1} \frac{t^{2n-2k-1}}{(2n-2k-1)!}\Lambda_k(1-x) & t \le x \\ \sum_{k=0}^{n-1} \frac{(1-t)^{2n-2k-1}}{(2n-2k-1)!}\Lambda_k(x) & x \le t. \end{cases} \quad (66)$$

5.2.2. Case $r = 2n+1$

If we consider the following boundary conditions

$$y(0) = \alpha_0, \quad y^{(2h-1)}(0) = \alpha_h, \quad y^{(2h-1)}(1) = \beta_h, \qquad h = 1,\dots,n \quad (67)$$

with α_0, α_h, β_h, $h = 1,\dots,n$ real constants, then P_{r-1} is the complementary Lidstone interpolating polynomial [27] of degree $2n$ [3, 24, 27, 28].

$$P_{2n}[y](x) = y(0) + \sum_{k=1}^{n}\left[y^{(2k-1)}(0)\left(v_k(1)-v_k(1-x)\right) + y^{(2k-1)}(1)\left(v_k(x)-v_k(0)\right)\right], \quad (68)$$

where $v_k(x)$ are the complementary Lidstone polynomials of degree k [27]. The kernel is

$$K_{2n-1}^x(x,t) = \begin{cases} \frac{t^{2n}}{(2n)!} + \sum_{k=1}^{n} \frac{t^{2n-2k+1}}{(2n-2k+1)!}\left(v_k(1-x)-v_k(1)\right) & t \le x \\ -\sum_{k=1}^{n} \frac{(1-t)^{2n-2k+1}}{(2n-2k+1)!}\left(v_k(x)-v_k(0)\right) & x \le t. \end{cases} \quad (69)$$

In Ref. [19], the proposed method applied to problem (2) with conditions (64) and (67), respectively, has been examined in detail.

5.2.3. Other special boundary conditions

If $r = n-1$ and $[a,b] = [0,1]$, we can consider *Bernoulli* boundary conditions [21]

$$y(0) = \beta_0, \quad y(1) = \beta_1, \quad y^{(k)}(1) - y^{(k)}(0) = \beta_{k+1}, \qquad k = 1,\dots,n-2 \quad (70)$$

with β_k, $k = 0,\dots,n-1$ real constants. The method has been examined in [14].

5.3. Multipoint boundary value problems

Let us now consider [15] the following conditions in $I = [-1,1]$

$$y^{(k)}(-1) = \alpha_k, \quad k = 0,\ldots,s-1, \qquad y^{(s)}(x_i) = \omega_i \quad i = 1,\ldots,r-s. \tag{71}$$

In this case

$$P_{r-1}[y](x) = \sum_{i=0}^{s-1} \frac{(x+1)^i}{i!}\alpha_i + \frac{1}{(s-1)!}\sum_{k=1}^{r-s}\omega_k p_{r,k}(x)\,, \tag{72}$$

with

$$p_{r,k}(x) = \int_{-1}^{x} (x-t)^{s-1} l_k(t)dt \tag{73}$$

and $l_k(t)$ are the fundamental Lagrange polynomials on the points $x_j, j = 1,\ldots,r-s$. $P_{r-1}(x)$ is the unique polynomial of degree $\le r-1$ which satisfies the Birkhoff interpolation problem

$$P_{r-1}^{(k)}(-1) = \alpha_k, \quad k = 0,\ldots,s-1, \qquad P_{r-1}^{(s)}(x_i) = \omega_i, \quad i = 1,\ldots,r-s, \quad s \le r-1 \tag{74}$$

with $-1 < x_1 < \cdots < x_{r-s} \le 1$. Hence, the solution of problem (2), with multipoint conditions (71), is

$$y(x) = P_{r-1}[y](x) + \int_{-1}^{1} K_r^x(x,t)y^{(r)}(t)dt, \tag{75}$$

with $P_{r-1}[y](x)$ given in Eq. (72) and

$$K_r^x(x,t) = \frac{1}{(r-1)!}\left[(x-t)_+^{r-1} - \binom{r-1}{s} s \sum_{i=1}^{r-s} p_{r,i,m}(x)(x_i-t)_+^{r-s-1}\right]. \tag{76}$$

Observe that Eq. (74) is a special type of Birkhoff interpolation problem with incidence matrix $E = (e_{ij})$ defined by $e_{1j} = e_{is} = 1, j = 0,\cdots,s-1, i = 2,\ldots,r-s+1$, $e_{ij} = 0$ otherwise and $r \ge 1$.

In Ref. [23], $P_{r-1}[y](x)$ is presented in a little different form:

$$P_{r-1}[y](x) = \sum_{i=0}^{s-1} \frac{(x+1)^i}{i!}\alpha_i + \sum_{k=1}^{r-s}\omega_k E_s(x, l_k(x))\,, \tag{77}$$

where $E_s(x, l_k(x)) = \underbrace{\int_{-1}^{x}\cdots\int_{-1}^{x}}_{s} l_k(t)dt\cdots dt.$

Let us now consider the following conditions [12, 20]

$$y(-1) = \omega_0, \qquad y(1) = \omega_{r-1} \qquad y''(x_i) = \omega_i \qquad i = 1,\ldots,r-2. \tag{78}$$

The solution to the Birkhoff interpolation problem

$$P_{r-1}(-1) = \omega_0, \qquad P_{r-1}(1) = \omega_{r-1}, \qquad P''_{r-1}(x_i) = \omega_i, \quad i = 1,\dots,r-2 \tag{79}$$

with $-1 < x_1 < \cdots < x_{r-2} < 1$ is [12]

$$P_{r-1}[y](x) = \frac{\omega_{r-1} + \omega_0}{2} + \frac{\omega_{r-1} - \omega_0}{2} x + \sum_{i=1}^{r-2} q_{r,i}(x)\omega_i \tag{80}$$

with

$$q_{r,i}(x) = \int_{-1}^{1} K_r^x(x,t) l_i(t) dt \tag{81}$$

and

$$K_r^x(x,t) = \begin{cases} \dfrac{(t+1)(x-1)}{2} & t \le x \\ \dfrac{(x+1)(t-1)}{2} & x < t. \end{cases} \tag{82}$$

Hence, the solution of problem (2) is

$$y(x) = P_{r-1}[y](x) + \int_{-1}^{1} K_r^x(x,t) y^{(r)}(t) dt, \tag{83}$$

with $P_{r-1}[y](x)$ given in Eq. (80) and

$$K_r^x(x,t) = \frac{1}{(r-1)!}\left[(x-t)_+^{r-1} - \frac{(1-t)^{r-1}(1+x)}{2} - (r-1)(r-2)\sum_{i=1}^{r-2} p_{r,i,m}(x)(x_i-t)_+^{r-3}\right]. \tag{84}$$

6. Numerical examples

In this section, we present some numerical results obtained by applying method (15), which we call *CGN method*, to find numerical approximations $y_{r,m}(x)$ to the solution of some test problems. In order to solve the nonlinear system (19), we use the so-called modified Newton method [29] (the same Jacobian matrix is used for more than one iteration) and we use algorithm (44) for the computation of the entries of the matrix, when polynomials $p_{r,i,m}(x)$ are not explicitly known. Since the true solutions of the analyzed problems are known, we consider the error function $e_m(x) = |y(x) - y_{r,m}(x)|$.

The maximum values of $e_m(x)$ over the interval $[a,b]$ have also been calculated by using Matlab, particularly the built-in solvers

- **ode15s**, a variable-step, variable-order multistep solver based on the numerical differentiation formulas of orders 1–5;

- **ode45**, a single-step solver, based on an explicit Runge-Kutta (4, 5) formula, the Dormand-Prince pair

for initial value problems, and the finite difference codes;

- **bvp4c** (with an optional mesh of 200 points) that implements the three-stage Lobatto IIIa formula;
- **bvp5c** that implements the four-stage Lobatto IIIa formula.

for boundary value problems.

All solvers have been used with optional parameters **RelTol=AbsTol=1e−17**.

Moreover, the powerful tool **Chebfun** [30] has been used.

Example 1 *Consider the following linear ninth-order BVP [28]*

$$\begin{cases} y^{(9)}(x) = -9e^x + y(x) & x \in [0,1] \\ y^{(j)}(0) = 1-j & j = 0,\ldots,4 \\ y^{(j)}(1) = -j\, e & j = 0,\ldots,3 \end{cases} \tag{85}$$

with exact solution $y(x) = (1-x)e^x$.

The unique polynomial $P_8(x) = P_8[y](x)$ of degree 8 satisfying the boundary conditions $P_8^{(j)}(0) = 1-j$ for $j = 0,\ldots,4$, and $P_8^{(j)}(1) = -j\, e\; j = 0,\ldots,3$ is

$$\begin{aligned} P_8(x) = \; & 1-\frac{1}{2}x^2-\frac{1}{3}x^3-\frac{1}{8}x^4+\left(\frac{31}{2}1\, e-\frac{253}{6}\right)x^5+ \\ & \left(\frac{1321}{12}-\frac{81}{2}1\, e\right)x^6+\left(\frac{71}{2}\, e-\frac{193}{2}\right)x^7+\left(\frac{685}{24}-\frac{21}{2}\, e\right)x^8. \end{aligned} \tag{86}$$

From Eq. (7), we get

$$K_9^x(x,t) = \frac{1}{8!}\cdot \begin{cases} 70t^4(x^4-4x^5+6x^6-4x^7+x^8)+56t^5(-x^3+10x^5-20x^6+15x^7-4x^8)+ \\ 28t^6(x^2-20x^5+45x^6-36x^7+10x^8)+8t^7(-x+35x^5-84x^6+70x^7-20x^8)+ \\ t^8(1-56x^5+140x^6-120x^7+35x^8) \qquad 0 \le t \le x \\ -x^8+8tx^7-28t^2x^6+56t^3x^5+70t^4(-4x^5+6x^6-4x^7+x^8)+ \\ 56t^5(10x^5-20x^6+15x^7-4x^8)+28t^6(-20x^5+45x^6-36x^7+10x^8)+ \\ 8t^7(35x^5-84x^6+70x^7-20x^8)+ \\ t^8(-56x^5+140x^6-120x^7+35x^8) \qquad x \le t \le 1. \end{cases} \tag{87}$$

Now we calculate the values of the integrals (39) by using Eq. (45), and we solve system (26). Thus, we obtain the approximate solution (15) to problem (85).

***Table** shows the numerical results. The absolute errors are compared with those obtained in Ref. [28] where a modified decomposition method is applied for the solution of problem (85). The second and third columns of **Table** show the error, respectively, in the method in Ref. [28] and in the CGN method, using in both cases polynomials of degree 12. The last column contains the error in the approximation* *by a polynomial of degree 14 using CGN method. As collocation points, equidistant nodes in $[0, 1]$ are chosen. Analogous results are obtained by using Chebyshev nodes of first and second kind, and Legendre-Gauss-Lobatto points.*

The maximum absolute error $max\{e_m(x)\}$ *on* $[0, 1]$ *has also been calculated by using Matlab* (**Table**).

x	**Method in [28]**	*CGN* $m = 4$	*CGN* $m = 6$
0.1	2.0e−10	1.45e−14	0.00
0.2	2.0e−10	3.93e−13	1.11e−16
0.3	2.0e−10	2.16e−12	9.99e−15
0.4	2.0e−10	5.70e−12	2.00e−15
0.5	2.0e−10	9.27e−12	2.55e−15
0.6	6.0e−10	1.00e−11	2.66e−15
0.7	1.0e−9	7.04e−12	2.44e−15
0.8	2.0e−9	2.70e−12	2.83e−15
0.9	3.4e−9	2.98e−13	4.91e−15

Table. Absolute error $e_m(x)$ in *MDM* and *CGN* methods for problem (85).

Chebfun	**bvp4c**	**bvp5c**
1.46	1.55e−12	4.44e−16

Table. Maximum absolute error in problem (85) using Matlab built-in functions.

x	**Cheb I**	**Cheb II**	**EqPts**
	$m = 4$	$m = 6$	$m = 9$
0.1	1.11e−16	0.00	0.00
0.2	9.54e−15	0.00	0.00
0.3	5.47e−13	3.33e−16	0.00
0.4	9.45e−12	1.11e−16	4.44e−16
0.5	8.50e−11	4.22e−15	1.11e−16
0.6	5.05e−10	3.47e−14	2.11e−15
0.7	2.25e−9	2.08e−13	1.55e−15
0.8	8.08e−9	9.68e−13	1.44e−14
0.9	2.74e−8	3.72e−12	9.18e−15
1.0	6.64e−8	1.22e−11	1.37e−14

Table. Problem (88)—example 2.

Example 2 *Consider the fifth-order initial value problem [13]*

$$\begin{cases} y^{(5)} + (32x^5 + 120x)y = 160x^3e^{-x^2} & x \in [0,1] \\ y(0) = 1, \quad y'(0) = 0, \quad y''(0) = -2 \\ y'''(0) = 0, \quad y^{(4)}(0) = 12 \end{cases} \tag{88}$$

with solution $y(x) = e^{-x^2}$.

Table *shows the absolute error in some points of the interval* $[0,1]$ *for CGN method in the case, respectively, of Chebyshev nodes of first kind* (Cheb I), *of second kind* (Cheb II) *and in the case of equidistant nodes* (EqPts).

The maximum absolute errors calculated by using Matlab are displayed in ***Table****.*

Chebfun	ode15s	ode45
2.11e–11	1.35e–13	1.33e–15

Table. Maximum absolute error in problem (88) using Matlab built-in functions.

Example 3 *Consider now the following nonlinear problem [31]*

$$\begin{cases} y^{(4)}(x) = \sin x + \sin^2 x - \left(y''(x)\right)^2 & x \in [0,1] \\ y(0) = 0 \qquad y'(0) = 1 \\ y(1) = \sin(1) \qquad y'(1) = \cos(1) \end{cases} \tag{89}$$

with exact solution $y(x) = \sin(x)$.

This kind of problems models several nonlinear phenomena such as traveling waves in suspension bridges [32] or the bending of an elastic beam [33].

Suspension bridges are generally susceptible to visible oscillations, due to the forces acting on the bridge (including the force due to the cables which are considered as a spring with a one-sided restoring, the gravitation force and the external force due to the wind or other external sources). f *represents the forcing term, while* y *represents the vertical displacement when the bridge is bending.*

In the case of elastic beam, f *represents the force exerted on the beam by the supports.* x *measures the position along the beam* ($x = 0$ *is the left-hand endpoint of the beam),* y and y' *indicate, respectively, the height and the slope of the beam at* x. y'' *measures the curvature of the graph of* y*, and, in physical terms, it measures the bending moment of the beam at* x*, that is, the torque that the load places on the beam at* x.

The considered boundary conditions state that the beam has both endpoints simply supported. Moreover, the derivative of the deflection function is not zero at those points, and it indicates that the beam at the wall is not horizontal.

Table *shows the comparison between the NMD method presented in Ref. [31] and the CGN method with* $m = 5$ and $m = 9$*, respectively. The approximating polynomial of NMD method has degree 11, while the polynomial considered in CGN method for* $m = 5$ *has degree 8.*

The maximum absolute errors calculated by using Matlab are displayed in ***Table***.

x	*NMD* [31]	*CGN* $m = 5$	*CGN* $m = 9$
0.1	7.78*e*–8	4.45*e*–10	1.53*e*–15
0.2	2.72*e*–7	5.54*e*–10	3.02*e*–15
0.3	5.24*e*–7	8.95*e*–11	7.77*e*–16
0.4	7.77*e*–7	2.03*e*–10	6.66*e*–16
0.5	9.71*e*–7	3.32*e*–11	5.55*e*–17
0.6	1.05*e*–6	1.53*e*–10	0
0.7	9.63*e*–7	9.48*e*–11	0
0.8	6.84*e*–7	5.18*e*–10	1.11*e*–16
0.9	2.71*e*–7	4.15*e*–10	0

Table. Error of *NMD* and *CGN* methods—problem (89).

Chebfun	**bvp4c**	**bvp5c**
1.67*e*–16	1.22*e*–8	8.88*e*–16

Table. Maximum absolute error in problem (89) using Matlab build-in functions.

Author details

Francesco Aldo Costabile, Maria Italia Gualtieri and Anna Napoli*

*Address all correspondence to: anna.napoli@unical.it

Department of Mathematics and Informatics, University of Calabria, Rende (Cs), Italy

References

[1] Davis P. *Interpolation and approximation*. Dover, New York. 1975.

[2] Agarwal R. *Boundary value problems for higher order differential equations*. World Scientific Publishing Co., Inc., Teaneck, NJ. 1986.

[3] Agarwal R, Wong P. *Lidstone polynomials and boundary value problems*. Computers and Mathematics with Applications. 1989; **17**(10): 1397–1421.

[4] Hairer E, Nørsett S, Wanner G. *Solving ordinary differential equations I. Nonstiff problems*. Berlin: Springer-Verlag. 1987.

[5] Henrici P. *Discrete variable methods in ordinary differential equations*. Wiley, New York. 1962.

[6] Strikwerda J. *Finite difference schemes and partial differential equations*. SIAM., Philadelphia, PA. 2004.

[7] Caglar H, Caglar N, Elfaituri K. *B-spline interpolation compared with finite difference, finite element and finite volume methods which applied to two-point boundary value problems*. Applied Mathematics and Computation. 2006; **175**(1): 72–79.

[8] Chang J, Yang Q, Zhao L. *Comparison of b-spline method and finite difference method to solve bvp of linear odes*. Journal of Computers. 2011; **6**(10): 2149–2155.

[9] Costabile F, Gualtieri MI, Serafini G. *Cubic Lidstone-Spline for numerical solution of BVPs*. submitted.

[10] Khan A. *Parametric cubic spline solution of two point boundary value problems*. Applied Mathematics and Computation. 2004; **154**(1): 175–182.

[11] Boyd J. *Chebyshev and Fourier spectral methods*. 2nd edition, Dover, Mineola, NY. 2000.

[12] Costabile F, Longo E. *A Birkhoff interpolation problem and application*. Calcolo. 2010; **47**(1): 49–63.

[13] Costabile F, Napoli A. *A class of collocation methods for numerical integration of initial value problems*. Computers and Mathematics with Applications. 2011; **62**(8): 3221–3235.

[14] Costabile F, Napoli A. *Numerical solution of high order Bernoulli boundary value problems*. Journal of Applied Mathematics. 2014, Article ID 276585. doi: 10.1155/2014/276585.

[15] Costabile F, Napoli A. *A method for high-order multipoint boundary value problems with Birkhoff-type conditions*. International Journal of Computer Mathematics. 2015; **92**(1): 192–200.

[16] Costabile F, Longo E. *A new collocation method for a BVP*. Applied and Industrial Mathematics in Italy III, 289–297. 2009; **3**: 289–297. (Ser. Adv. Math. Appl. Sci., 82, World Sci. Publ., Hackensack, NJ. 2010)

[17] Costabile F, Napoli A. *A method for global approximation of the solution of second order IVPs*. Rendiconti del Circolo Matematico, Ser. II. 2004; **24**: 239–260.

[18] Costabile F, Napoli A. *A method for polynomial approximation of the solution of general second order BVPs*. Far East Journal of Applied Mathematics. 2006; **25**(3): 289–305.

[19] Costabile F, Napoli A. *Collocation for high-order differential equations with Lidstone boundary conditions*. Journal of Applied Mathematics. 2012, Article ID 120792. doi: 10.1155/2012/120792.

[20] Costabile F, Napoli A. *A multipoint Birkhoff type boundary value problem*. Journal of Numerical Mathematics. 2015; **23**(1): 1–11.

[21] Costabile F, Serpe A, Bruzio A. *No classic boundary conditions*. In: Proceedings of World Congress on Engineering 2007; July 2–4, 2007, London 918–921.

[22] Costabile F, Napoli A. *Collocation for high order differential equations with two-points Hermite boundary conditions*. Applied Numerical Mathematics. 2015; **87**: 157–167.

[23] Dehghan M, Aryanmehr S, Eslahchi M. *A technique for the numerical solution of initial-value problems based on a class of Birkhoff-type interpolation method*. Journal of Computational and Applied Mathematics. 2013; **244**: 125–139.

[24] Wang L, Samson M, Zhao X. *A well-conditioned collocation method using a pseudospectral integration matrix*. SIAM Journal on Scientific Computing. 2014; **36**(3): 907–929.

[25] Costabile F, Napoli A. *A method for global approximation of the initial value problem*. Numerical Algorithms. 2001; **27**(2): 119–130.

[26] Coleman J, Booth A. *The Chebyshev methods of Panovsky and Richardson as Runge-Kutta-Nyström methods*. Journal of Computational and Applied Mathematics. 1995; **61**(3): 245–261.

[27] Costabile F, DellAccio F, Luceri R. *Explicit polynomial expansions of regular real functions by means of even order Bernoulli polynomials and boundary values*. Journal of Computational and Applied Mathematics. 2005; **176**(1): 77–90.

[28] Wazwaz AM. *Approximate solutions to boundary value problems of higher order by the modified decomposition method*. Computers and Mathematics with Applications. 2000; **40**(6): 679–691.

[29] Quarteroni A, Sacco R, Saleri F. *Numerical mathematics*. Second edition. Texts in Applied Mathematics, 37. Springer-Verlag: Berlin. 2007.

[30] Driscoll TA, Hale N, Trefethen NL. *Chebfun guide*. Pafnuty Publications: Oxford. 2014.

[31] Noor MA, Mohyud-Din ST. *An efficient method for fourth-order boundary value problems*. Computers and Mathematics with Applications. 2007; **54**(7): 1101–1111.

[32] Lazer A, McKenna P. *Large-amplitude periodic oscillations in suspension bridges: some new connections with nonlinear analysis*. Siam Review. 1990; **32**(4): 537–578.

[33] Yang B. *Estimates of positive solutions to a boundary value problem for the beam equation*. Communications in Mathematical Analysis. 2007; **2**(1): 13–21.

6

On Nonoscillatory Solutions of Two-Dimensional Nonlinear Dynamical Systems

Elvan Akın and Özkan Öztürk

Abstract

During the past years, there has been an increasing interest in studying oscillation and nonoscillation criteria for dynamical systems on time scales that harmonize the oscillation and nonoscillation theory for the continuous and discrete cases in order to combine them in one comprehensive theory and eliminate obscurity from both. We not only classify nonoscillatory solutions of two-dimensional systems of first-order dynamic equations on time scales but also guarantee the existence of such solutions using the Knaster, Schauder-Tychonoff and Schauder's fixed point theorems. The approach is based on the sign of components of nonoscillatory solutions. A short introduction to the time scale calculus is given as well. Examples are significant in order to see if nonoscillatory solutions exist or not. Therefore, we give several examples in order to highlight our main results for the set of real numbers $\mathbb{R}$, the set of integers $\mathbb{Z}$ and $q^{\mathbb{N}_0} = \{1, q, q^2, q^3, \ldots\}$, $q > 1$, which are the most well-known time scales.

Keywords: dynamical systems, dynamic equations, differential equations, difference equations, time scales, oscillation

1. Introduction

In this chapter, we investigate the existence and classification of nonoscillatory solutions of two-dimensional (2D) nonlinear time-scale systems of first-order dynamic equations. The method we follow is based on the sign of components of nonoscillatory solutions and the most well-known fixed point theorems. The motivation of studying dynamic equations on time scales is to unify continuous and discrete analysis and harmonize them in one comprehensive theory and eliminate obscurity from both. A *time scale* $\mathbb{T}$ is an arbitrary nonempty closed subset of the real numbers $\mathbb{R}$. The most well-known examples for time scales are $\mathbb{R}$ (which leads to differential equations), $\mathbb{Z}$ (which leads to difference equations,) and $q^{\mathbb{N}_0} := \{1, q, q^2, \cdots\}$, $q > 1$ (which leads to q-difference equations,). In 1988, the theory of time scales was initiated by Stefan Hilger in his Ph.D. thesis [5]. We assume that most readers are not familiar with the calculus of time scales and therefore we give a brief introduc-tion to time scales calculus in Section. In fact, we refer readers books [6, 7] by Bohner and Peterson for more details.

The study of 2D dynamic systems in nature and society has been motivated by their applications. Especially, a system of delay dynamic equations, considered in Section, take a lot of attention in all areas such as population dynamics, predator-prey epidemics, genomic and neuron dynamics and epidemiology in biological sciences. For instance, when the birth rate of preys is affected by the previous values rather than current values, a system of delay dynamic equations is utilized, because the rate of change at any time depends on solutions at prior times. Another novel application of delay dynamical systems is time delays that often arise in feedback loops involving actuators. A major issue faced in engineering is an unavoidable time delay between measurement and the signal received by the controller. In fact, the delay should be taken into consideration at the design stage to avoid the risk of instability.

Another special case of 2D systems of dynamic equations is the Emden-Fowler type, which is covered in Section of this chapter. The equation has several interesting applications, such as in astrophysics, gas dynamics and fluid mechanics, relativistic mechanics, nuclear physics and chemically reacting systems. For example, the fundamental problem in studying the stellar structure for gaseous dynamics in astrophysics was to look into the equilibrium formation of the mass of spherical clouds of gas for the continuous case, proposed by Kelvin and Lane. Such an equation is called Lane-Emden equation in literature. Much information about the solutions of Lane-Emden equation was provided by Ritter. In a series of 18 papers, published during 1878–1889. The mathematical foundation for the study of such an equation was made by Fowler in a series of four papers during 1914–1931.

2. Preliminaries

The set of real numbers $\mathbb{R}$, the set of integers $\mathbb{Z}$, the natural numbers $\mathbb{N}$, the nonnegative integers $\mathbb{N}_0$ and the Cantor set, $q^{\mathbb{N}_0}$, $q > 1$ and $[0,1] \cup [2,3]$ are some examples of time scales. However, the set of rational numbers $\mathbb{Q}$, the set of irrational numbers $\mathbb{R} \backslash \mathbb{Q}$, the complex numbers $\mathbb{C}$, and the open interval $(0,1)$ are not considered as time scales.

Definition 2.1. [6, Definition 1.1] Let $\mathbb{T}$ be a time scale. For $t \in \mathbb{T}$, the *forward jump operator* $\sigma : \mathbb{T} \to \mathbb{T}$ is given by

$$\sigma(t) := \inf\{s \in \mathbb{T} : \quad s > t\} \quad \text{for all} \quad t \in \mathbb{T}$$

whereas the *backward jump operator* $\rho : \mathbb{T} \to \mathbb{T}$ is defined by

$$\rho(t) := \sup\{s \in \mathbb{T} : \quad s < t\} \quad \text{for all} \quad t \in \mathbb{T}.$$

Finally, the *graininess function* $\mu : \mathbb{T} \to [0, \infty)$ is given by $\mu(t) := \sigma(t) - t \quad \text{for all} \quad t \in \mathbb{T}$.

We define $\inf\varnothing = \sup\mathbb{T}$. If $\sigma(t) > t$, then t is called *right-scattered*, whereas if $\rho(t) < t$, t is called *left-scattered*. If t is right- and left-scattered at the same time, then we say that t is *isolated*. If $t < \sup\mathbb{T}$ and $\sigma(t) = t$, then t is called *right-dense*, while if $t > \inf \mathbb{T}$ and $\rho(t) = t$, we say that t is *left-dense*. Also, if t is right- and left-dense at the same time, then we say that t is *dense*.

Table shows some examples of the forward and backward jump operators and the graininess function for most known time scales.

$\mathbb{T}$	$\sigma(t)$	$\rho(t)$	$\mu(t)$
$\mathbb{R}$	t	t	0
$\mathbb{Z}$	$t+1$	$t-1$	1
$q^{\mathbb{N}_0}$	tq	$\frac{t}{q}$	$(q-1)t$

Table. Examples of most known time scales.

If $\sup\mathbb{T} < \infty$, then $\mathbb{T}^\kappa = \mathbb{T}\backslash(\rho(\sup\mathbb{T}), \sup\mathbb{T}]$ and $\mathbb{T}^\kappa = \mathbb{T}$ if $\sup\mathbb{T} = \infty$. Suppose that $f : \mathbb{T} \to \mathbb{R}$ is a function. Then $f^\sigma : \mathbb{T} \to \mathbb{R}$ is defined by $f^\sigma(t) = f(\sigma(t)) \quad \text{for all} \quad t \in \mathbb{T}$.

Definition 2.2. [6, Definition 1.10] For any ε, if there exists a $\delta > 0$ such that

$$|f(\sigma(t)) - f(s) - f^\Delta(t)(\sigma(t) - s)| \leq \varepsilon|\sigma(t) - s| \quad \text{for all} \quad s \in (t-\delta, t+\delta) \cap \mathbb{T},$$

then f is called *delta (or Hilger) differentiable* on $\mathbb{T}^\kappa$ and f^Δ is called *delta derivative* of f.

Theorem 2.3 [6, Theorem 1.16] *Let $f : \mathbb{T} \to \mathbb{R}$ be a function with $t \in \mathbb{T}^\kappa$. Then*

a. *If f is differentiable at t, f is continuous at t.*

b. *If f is continuous at t and t is right-scattered, then f is differentiable at t and*

$$f^\Delta(t) = \frac{f(\sigma(t)) - f(t)}{\mu(t)}.$$

c. *If t is right dense, then f is differentiable at t if and only if*

$$f^\Delta(t) = \lim_{s \to t} \frac{f(t) - f(s)}{t - s}$$

exists as a finite number.

d. *If f is differentiable at t, then $f(\sigma(t)) = f(t) + \mu(t) f^\Delta(t)$.*

If $\mathbb{T} = \mathbb{R}$, then f^Δ turns out to be the usual derivative f' while f^Δ is reduced to forward difference operator Δf if $\mathbb{T} = \mathbb{Z}$. Finally, if $\mathbb{T} = q^{\mathbb{N}_0}$, then the delta derivative turns out to be *q-difference* operator Δ_q. The following theorem presents the sum, product and quotient rules on time scales.

Theorem 2.4 [6, Theorem 1.20] *Let $f, g : \mathbb{T} \to \mathbb{R}$ be differentiable at $t \in \mathbb{T}^{\kappa}$. Then*

a. *The sum $f + g : \mathbb{T} \to \mathbb{R}$ is differentiable at t with*

$$(f+g)^{\Delta}(t) = f^{\Delta}(t) + g^{\Delta}(t).$$

b. *If $fg : \mathbb{T} \to \mathbb{R}$ is differentiable at t, then*

$$(fg)^{\Delta}(t) = f^{\Delta}(t)g(t) + f(\sigma(t))g^{\Delta}(t) = f(t)g^{\Delta}(t) + f^{\Delta}(t)g(\sigma(t)).$$

c. *If $g(t)g(\sigma(t)) \neq 0$, then $\frac{f}{g}$ is differentiable at t with*

$$\left(\frac{f}{g}\right)^{\Delta}(t) = \frac{f^{\Delta}(t)g(t) - f(t)g^{\Delta}(t)}{g(t)g(\sigma(t))}.$$

The following concepts must be introduced in order to define delta-integrable functions.

Definition 2.5. [6, Definition 1.58] $f : \mathbb{T} \to \mathbb{R}$ is called *right dense continuous* (rd-continuous), denoted by $C_{rd}, C_{rd}(\mathbb{T})$, or $C_{rd}(\mathbb{T}, \mathbb{R})$, if it is continuous at right dense points in $\mathbb{T}$ and its left-sided limits exist as a finite number at left dense points in $\mathbb{T}$. We denote continuous functions by C throughout this chapter.

Theorem 2.6 [6, Theorem 1.60] *Let $f : \mathbb{T} \to \mathbb{R}$.*

a. *If f is continuous, then f is rd-continuous.*

b. *The jump operator σ is rd-continuous.*

Also, the Cauchy integral is defined by

$$\int_a^b f(t)\Delta t = F(b) - F(a) \quad \text{for all} \quad a, b \in \mathbb{T}.$$

The following theorem presents the existence of antiderivatives.

Theorem 2.7 [6, Theorem 1.74] *Every rd-continuous function has an antiderivative. Moreover, F given by*

$$F(t) = \int_{t_0}^{t} f(s)\Delta s \quad for \quad t \in \mathbb{T}$$

is an antiderivative of f.

Theorem 2.8 [6, Theorems 1.76–1.77] *Let $a, b, c \in \mathbb{T}, \alpha \in \mathbb{R}$, and $f, g \in C_{rd}$. Then we have:*

1. *If $f^{\Delta} \geq 0$, then f is nondecreasing.*

2. *If $f(t) \geq 0$ for all $a \leq t \leq b$, then $\int_a^b f(t)\Delta t \geq 0$.*

3. $\int_a^b [(\alpha f(t)) + (\alpha g(t))]\Delta t = \alpha \int_a^b f(t)\Delta t + \alpha \int_a^b g(t)\Delta t.$

4. $\int_a^b f(t)\Delta t = -\int_b^a f(t)\Delta t.$

5. $\int_a^b f(t)\Delta t = \int_a^C f(t)\Delta t + \int_C^b f(t)\Delta t.$

6. $\int_a^b f(t)g^{\Delta}(t)\Delta t = (fg)(b) - (fg)(a) - \int_a^b f^{\Delta}(t)g(\sigma(t))\Delta t$

7. $\int_a^b f(\sigma(t))g^{\Delta}(t)\Delta t = (fg)(b) - (fg)(a) - \int_a^b f^{\Delta}(t)g(t)\Delta t$

8. $\int_a^a f(t)\Delta t = 0.$

Table shows the derivative and integral definitions for the most known time scales for $a, b \in \mathbb{T}$.

$\mathbb{T}$	$f^{\Delta}(t)$	$\int_a^b f(t)\Delta t$
$\mathbb{R}$	$f'(t)$	$\int_a^b f(t)dt$
$\mathbb{Z}$	$\Delta f(t)$	$\sum_{t=a}^{b-1} f(t)$
$q^{\mathbb{N}_0}$	$\Delta_q f(t)$	$\sum_{t \in [a,b)_{q^{\mathbb{N}_0}}} f(t)\mu(t)$

Table. Derivatives and integrals for most common time scales.

Finally, we finish the section by the following fixed point theorems.

Theorem 2.9 (Schauder's Fixed Point Theorem) [23, Theorem 2.A] *Let S be a nonempty, closed, bounded, convex subset of a Banach space X and suppose that $T : S \to S$ is a compact operator. Then, T has a fixed point.*

The Schauder fixed point theorem was proved by Juliusz Schauder in 1930. In 1934, Tychonoff proved the same theorem for the case when S is a compact convex subset of a locally convex space X. In the literature, this version is known as the Schauder-Tychonoff fixed point theorem.

Theorem 2.10 (Schauder-Tychonoff Fixed Point Theorem). *Let S be a compact convex subset of a locally convex (linear topological) space X and T a continuous map of S into itself. Then, T has a fixed point.*

Finally, we provide the Knaster fixed point theorem.

Theorem 2.11 (Knaster Fixed Point Theorem) *If $(S, \leq)$ is a complete lattice and $T : S \to S$ is order-preserving (also called monotone or isotone), then T has a fixed point. In fact, the set of fixed points of T is a complete lattice.*

3. Dynamical Systems on Time Scales

In this section, we consider the following system

$$\begin{cases} x^{\Delta}(t) = a(t)f(y(t)) \\ y^{\Delta}(t) = -b(t)g(x(t)), \end{cases} \tag{1}$$

where $f, g \in C(\mathbb{R}, \mathbb{R})$ are nondecreasing such that $uf(u) > 0$, $ug(u) > 0$ for $u \neq 0$ and $a, b \in C_{rd}([t_0, \infty)_{\mathbb{T}}, \mathbb{R}^+)$. The main results in this section come from Ref. [26]. If $\mathbb{T} = \mathbb{R}$ and $\mathbb{T} = \mathbb{Z}$, Eq. (1) turns out to be a system of first-order differential equations and difference equations, Refs. [27] and [28], respectively. Recent advances in oscillation and nonoscillation criteria for two-dimensional time scale systems have been studied in Refs. [29–31].

Throughout this chapter, we assume that $\mathbb{T}$ is unbounded above. Whenever we write $t \geq t_1$, we mean $t \in [t_1, \infty)_{\mathbb{T}} := [t_1, \infty) \cap \mathbb{T}$. We call (x, y) a *proper solution* if it is defined on $[t_0, \infty)_{\mathbb{T}}$ and $\sup\{|x(s)|, |y(s)| : s \in [t, \infty)_{\mathbb{T}}\} > 0$ for $t \geq t_0$. A solution (x, y) of Eq. (1) is said to be *nonoscillatory* if the component functions x and y are both nonoscillatory, i.e., either eventually positive or eventually negative. Otherwise, it is said to be *oscillatory.* The definitions above are also valid for systems considered in the next sections.

Assume that (x, y) is a nonoscillatory solution of system (1) such that x oscillates but y is eventually positive. Then the first equation of system (1) yields $x^{\Delta}(t) = a(t)f(y(t)) > 0$ eventually one sign for all large $t \geq t_0$, a contradiction. The case where y is eventually negative is similar. Therefore, we have that the component functions x and y are themselves nonoscillatory. In other words, any nonoscillatory solution (x, y) of system (1) belongs to one of the following classes:

$$M^{+} := \{(x, y) \in M : \ xy > 0 \quad \text{eventually}\}$$

$$M^{-} := \{(x, y) \in M : \ xy < 0 \quad \text{eventually}\},$$

where M is the set of all nonoscillatory solutions of system (1).

In this section, we only focus on the existence of nonoscillatory solutions of system (1) in M^{-}, whereas M^{+} is considered together with delay system (12) in the following section.

For convenience, let us set

$$Y(t) = \int_t^{\infty} a(s)\Delta s \qquad \text{and} \qquad Z(t) = \int_t^{\infty} b(s)\Delta s. \tag{2}$$

We begin with the following results playing an important role in this chapter.

Lemma 3.1 *Let (x, y) be a nonoscillatory solution of system (1) and $t_0 \in \mathbb{T}$. Then we have the followings:*

a. [29, Lemma 2.3] *If $Y(t_0) < \infty$ and $Z(t_0) < \infty$, then system (1) is nonoscillatory.*

b. [29, Lemma 2.2] *If* $Y(t_0) = \infty$ *and* $Z(t_0) = \infty$, *then system* (1) *is oscillatory.*

c. *If* $Y(t_0) < \infty$ *and* $Z(t_0) = \infty$, *then* $M^+ = \varnothing$.

d. *If* $Y(t_0) = \infty$ *and* $Z(t_0) < \infty$, *then* $M^- = \varnothing$.

e. *Let* $Y(t_0) < \infty$. *Then x has a finite limit.*

f. *If* $Y(t_0) = \infty$ *or* $Z(t_0) < \infty$, *then y has a finite limit.*

Proof. Here, we only prove (a), (c) and (e) and the reader is asked to finish the proof in Exercise 3.2. To prove (a), choose $t_1 \in [t_0, \infty)_{\mathbb{T}}$ such that

$$\int_{t_1}^{\infty} a(t)f(1+g(2)\int_t^{\infty} b(s)\Delta s)\Delta t < 1.$$

Let X be the space of all continuous functions on $\mathbb{T}$ with the norm $\|x\| = \sup_{t\in[t_1,\infty)_{\mathbb{T}}} |x(t)|$ and with the usual point-wise ordering $\leq$. Define a subset Ω of X as

$$\Omega := \{x \in X: \quad 1 \leq x(t) \leq 2, \quad t \geq t_1\}.$$

For any subset S of Ω, we have $\inf S \in \Omega$ and $\sup S \in \Omega$. Define an operator $F: \Omega \to X$ such that

$$(Fx)(t) = 1 + \int_{t_1}^{t} a(s)f\Big(1+\int_s^{\infty} b(u)g(x(u))\Delta u\Big)\Delta s, \quad t \geq t_1.$$

By using the monotonicity and the fact that $x \in \Omega$, we have

$$1 \leq (Fx)(t) \leq 1 + \int_{t_1}^{t} a(s)f\Big(1+g(2)\int_s^{\infty} b(u)\Delta u\Big)\Delta s \leq 2, \quad t \geq t_1.$$

It is also easy to show that F is an increasing mapping. So by Theorem 2.11, there exists $\overline{x} \in \Omega$ such that $F\overline{x} = \overline{x}$. Then we have

$$\overline{x}^{\Delta}(t) = a(t)f\Big(1+\int_t^{\infty} b(u)g(\overline{x}(u))\Delta u\Big).$$

Setting

$$\overline{y}(t) = 1 + \int_t^{\infty} b(u)g(\overline{x}(u))\Delta u > 0, \quad t \geq t_1$$

gives us

$$\overline{y}^{\Delta}(t) = -b(t)g(\overline{x}(t)) \quad \text{and} \quad \overline{x}^{\Delta}(t) = a(t)f(\overline{y}(t)),$$

that is, $(\overline{x}, \overline{y})$ is a nonoscillatory solution of Eq. (1). In order to prove part (c), assume that there exists a nonoscillatory solution (x, y) of system (1) in M^+ such that $x(t) > 0$ for $t \geq t_1$. Then by monotonicity of x and g, there exists a number $k > 0$ such that $g(x(t)) \geq k$ for $t \geq t_1$. Integrating the second equation of system from t_1 to t gives us

$$y(t) \leq y(t_1) - k\int_{t_1}^{t} b(s)\Delta s.$$

As $t \to \infty$, it follows $y(t) \to -\infty$. But this contradicts that y is eventually positive. Finally for part (e), without loss of generality, we assume that there exists $t_1 \geq t_0$ such that $x(t) > 0$ for $t \geq t_1$. If $(x, y) \in M^-$, then by the first equation of system (1), $x^{\Delta}(t) < 0$ for $t \geq t_1$. Hence, the limit of x exists. So let us show that the assertion follows if $(x, y) \in M^+$. Suppose $(x, y) \in M^+$. Then from the first equation of system (1), we have $x^{\Delta}(t) > 0$ for $t \geq t_1$. Now let us show that $\lim_{t\to\infty} x(t) = \infty$ cannot happen. Integrating the first equation of system (1) from t_1 to t and using the monotonicity of y and f yield

$$x(t) \leq x(t_1) + f(y(t_1))\int_{t_1}^{t} a(s)\Delta s.$$

Taking the limit as $t \to \infty$, it follows that x has a finite limit. This completes the proof.

Exercise 3.2. Prove the remainder of Lemma 3.1.

Throughout this section, we assume $Y(t_0) < \infty$ and $Z(t_0) = \infty$. Note that Lemma 3.1 (c) indicates $M^+ = \emptyset$. Therefore, every nonoscillatory solution of system (1) belongs to M^-. Let (x, y) be a nonoscillatory solution of system (1) such that the component function x of solution (x, y) is eventually positive. Then, the second equation of system (1) yields $y < 0$ and eventually decreasing. Then for $k < 0$, we have that y approaches k or $-\infty$. In view of Lemma 3.1 (e), x has a finite limit. So in light of this information, any nonoscillatory solution of system (1) in M^- belongs to one of the following subclasses for $0 < c < \infty$ and $0 < d < \infty$:

$$M^-_{0,B} = \{(x, y) \in M^- : \lim_{t\to\infty} |x(t)| = 0, \quad \lim_{t\to\infty} |y(t)| = d\},$$

$$M^-_{B,B} = \{(x, y) \in M^- : \lim_{t\to\infty} |x(t)| = c, \quad \lim_{t\to\infty} |y(t)| = d\},$$

$$M^-_{0,\infty} = \{(x, y) \in M^- : \lim_{t\to\infty} |x(t)| = 0, \; \lim_{t\to\infty} |y(t)| = \infty\},$$

$$M^-_{B,\infty} = \{(x, y) \in M^- : \lim_{t\to\infty} |x(t)| = c, \; \lim_{t\to\infty} |y(t)| = \infty\}.$$

Nonoscillatory solutions in $M^-_{0,\infty}$ is called *slowly decaying solutions* in literature. The following theorems show the existence of nonoscillatory solutions in subclasses of M^- given above. Our approach for the next two theorems is based on the Schauder fixed point theorem.

Theorem 3.3 *$M^-_{0,B} \neq \emptyset$ if and only if*

$$\int_{t_0}^{\infty} b(t) g\left(c_1 \int_{t}^{\infty} a(s)\Delta s\right)\Delta t < \infty, \quad c_1 \neq 0. \tag{3}$$

Proof. Suppose that there exists a solution $(x, y) \in M^-_{0,B}$ such that $x(t) > 0$ for $t \geq t_0$, $x(t) \to 0$ and

$y(t) \to -d$ as $t \to \infty$, where $d > 0$. Integrating the first equation of system (1) from t to ∞ and the monotonicity of f yield that there exists $c > 0$ such that

$$x(t) \geq c \int_t^{\infty} a(s)\Delta s, \quad t \geq t_0. \tag{4}$$

By integrating the second equation from t_0 to t, using inequality (4) with $c = c_1$ and the monotonicity of g, we have

$$y(t) = y(t_0) - \int_{t_0}^{t} b(s)g(x(s))\Delta s \leq -\int_{t_0}^{t} b(s)g\Big(c_1 \int_s^{\infty} a(u)\Delta u\Big)\Delta s.$$

So as $t \to \infty$, the assertion follows since y has a finite limit. (For the case $x < 0$ eventually, the proof can be shown similarly with $c_1 < 0$.)

Conversely, suppose that Eq. (3) holds for some $c_1 > 0$. (For the case $c_1 < 0$ can be shown similarly.) Then there exist $t_1 \geq t_0$ and $d > 0$ such that

$$\int_{t_1}^{\infty} b(t)g\Big(c_1 \int_t^{\infty} a(s)\Delta s\Big)\Delta t < d, \quad t \geq t_1, \tag{5}$$

where $c_1 = -f(-3d)$. Let X be the space of all continuous and bounded functions on $[t_1, \infty)_{\mathbb{T}}$ with the norm $\|y\| = \sup_{t \in [t_1, \infty)_{\mathbb{T}}} |y(t)|$. Then X is a Banach space, Let Ω be the subset of X such that

$$\Omega := \{y \in X : \quad -3d \leq y(t) \leq -2d, \quad t \geq t_1\}$$

and define an operator $T : \Omega \to X$ such that

$$(Ty)(t) = -3d + \int_t^{\infty} b(s)g\Big(-\int_s^{\infty} a(u)f(y(u))\Delta u\Big)\Delta s.$$

It is easy to see that T maps into itself. Indeed, we have

$$-3d \leq (Ty)(t) \leq -3d + \int_t^{\infty} b(s)g\Big(-\int_s^{\infty} a(u)f(-3d)\Delta u\Big)\Delta s \leq -2d$$

by Eq. (5). Let us show that T is continuous on Ω. To accomplish this, let y_n be a sequence in Ω such that $y_n \to y \in \Omega = \overline{\Omega}$. Then

$$\begin{aligned} &|(Ty_n)(t) - (Ty)(t)| \\ &\leq \int_{t_1}^{\infty} b(s)|[g\Big(-\int_s^{\infty} a(u)f(y_n(u))\Delta u\Big) - g\Big(-\int_s^{\infty} a(u)f(y(u))\Delta u\Big)]|\Delta s. \end{aligned}$$

Then the Lebesque dominated convergence theorem and the continuity of g give $\|(Ty_n) - (Ty)\| \to 0$ as $n \to \infty$, i.e., T is continuous. Also, since

$$0 < -(Ty)^{\Delta}(t) = b(t)g\left(-\int_t^{\infty} a(u)f(y(u))\Delta u\right) < \infty,$$

it follows that $T(\Omega)$ is relatively compact. Then by Theorem 2.9, we have that there exists $\overline{y} \in \Omega$ such that $\overline{y} = T\overline{y}$. So as $t \to \infty$, we have $\overline{y}(t) \to -3d < 0$. Setting

$$\overline{x}(t) = -\int_t^{\infty} a(u)f(\overline{y}(u))\Delta u > 0, \quad t \geq t_1$$

gives that $\overline{x}(t) \to 0$ as $t \to \infty$ and implies $\overline{x}^{\Delta} = af(\overline{y})$, i.e., $(\overline{x}, \overline{y})$ is a nonoscillatory solution in $M_{0,B}^-$.

In the following example, we apply Theorem 3.3 to show the nonemptiness of $M_{0,B}^-$.

Example 3.4 *Let $\mathbb{T} = q^{\mathbb{N}_0}, q > 1$ and consider the system*

$$\begin{cases} \Delta_q x(t) = \dfrac{t^{\frac{1}{3}}}{(t+1)(tq+1)(2t-1)^{\frac{1}{3}}} y^{\frac{1}{3}}(t) \\ \Delta_q y(t) = -\dfrac{(t+1)^{\frac{5}{3}}}{qt^2} x^{\frac{5}{3}}(t). \end{cases} \tag{6}$$

Since

$$\int_1^T a(s)\Delta s = (q-1) \sum_{s \in [1,T)_{q^{\mathbb{N}_0}}} \frac{s^{\frac{4}{3}}}{(s+1)(sq+1)(2s-1)^{\frac{1}{3}}} \leq (q-1) \sum_{s \in [1,T)_{q^{\mathbb{N}_0}}} \frac{1}{s^{\frac{2}{3}}},$$

where $t = q^n$ and $s = tq^m$, $n, m \in \mathbb{N}_0$, we obtain

$$Y(1) \leq (q-1) \sum_{n=0}^{\infty} \left(\frac{1}{q^{\frac{2}{3}}}\right)^n < \infty.$$

Also,

$\int_1^T b(s)\Delta s = \sum_{s \in [1,T)_{q^{\mathbb{N}_0}}} \frac{(s+1)^{\frac{5}{3}}}{qs^2}(q-1)s \geq \frac{q-1}{q} \sum_{s \in [1,T)_{q^{\mathbb{N}_0}}} s^{\frac{2}{3}}$ *implies* $Z(1) \geq \frac{q-1}{q} \sum_{m=0}^{\infty} (q^{\frac{2}{3}})^m = \infty$. *Now let us show that Eq. (3) holds. First,*

$$\int_t^T a(s)\Delta s \leq (q-1) \sum_{s \in [t,T)_{q^{\mathbb{N}_0}}} \frac{1}{s^{\frac{2}{3}}} \quad \textit{implies} \quad \int_t^{\infty} a(s)\Delta s \leq (q-1) \sum_{s \in [t,\infty)_{q^{\mathbb{N}_0}}} \frac{1}{s^{\frac{2}{3}}} = \frac{q^{\frac{2}{3}}(q-1)}{(q^{\frac{2}{3}}-1)t^{\frac{2}{3}}}.$$

Therefore,

$$\int_1^T b(t)g\left(c_1 \int_t^{\infty} a(s)\Delta s\right)\Delta t \leq \alpha \sum_{t \in [1,T)_{q^{\mathbb{N}_0}}} \frac{(t+1)^{\frac{5}{3}}}{t^{\frac{19}{10}}},$$

where $\alpha = \frac{(q-1)^2 q^{\frac{1}{9}}}{(q^{\frac{2}{3}}-1)^{\frac{5}{3}}}$. So as $T \to \infty$, we have that Eq. (3) holds by the Ratio test. One can also show that $\left(\frac{1}{t+1}, -2 + \frac{1}{t}\right)$ of system (6) such that $x(t) \to 0$ and $y(t) \to -2$ as $t \to \infty$, i.e., $M_{0,B}^- \neq \emptyset$.

The proof of the following theorem is similar to the proof of Theorem 3.3.

Theorem 3.5 *$M^{-}_{B,B}\neq\emptyset$ if and only if*

$$\int_{t_0}^{\infty} b(t)g\left(d_1-c_1\int_t^{\infty} a(s)\Delta s\right)\Delta t < \infty$$

for some $c_1 < 0$ and $d_1 > 0$. (Or $c_1 > 0$ and $d_1 < 0$.)

Exercise 3.6. Prove Theorem 3.5 by means of Theorem 2.9.

The following theorem follows from the Knaster fixed point theorem.

Theorem 3.7 *$M^{-}_{B,\infty}\neq\emptyset$ if and only if*

$$\int_{t_0}^{\infty} a(s)f\left(c_1\int_{t_0}^{s} b(u)\Delta u\right)\Delta s < \infty \tag{7}$$

for some $c_1\neq 0$, where f is an odd function.

Proof. Suppose that there exists a nonoscillatory solution $(x,y)\in M^{-}_{B,\infty}$ such that $x > 0$ eventually, $x(t)\to c_2$ and $y(t)\to -\infty$ as $t\to\infty$, where $0 < c_2 < \infty$. Because of the monotonicity of x and the fact that x has a finite limit, there exist $t_1\geq t_0$ and $c_3 > 0$ such that

$$c_2\leq x(t)\leq c_3 \quad \text{for} \quad t\geq t_1. \tag{8}$$

Integrating the first equation from t_1 to t gives us

$$c_2\leq x(t) = x(t_1) + \int_{t_1}^{t} a(s)f(y(s))\Delta s\leq c_3, \quad t\geq t_1.$$

So by taking the limit as $t\to\infty$, we have

$$\int_{t_1}^{\infty} a(s)|f(y(s))|\Delta s < \infty. \tag{9}$$

The monotonicity of g, Eq. (8) and integrating the second equation from t_1 to t yield

$$y(t)\leq y(t_1)-g(c_2)\int_{t_1}^{t} b(s)\Delta s\leq -g(c_2)\int_{t_1}^{t} b(s)\Delta s.$$

Since $f(-u) = -f(u)$ for $u\neq 0$ and by the monotonicity of f, we have

$$|f(y(t))|\geq f\left(g(c_2)\int_{t_1}^{t} b(s)\Delta s\right), \quad t\geq t_1. \tag{10}$$

By Eqs. (9) and (10), we have

$$\int_{t_1}^{t} a(s)|f(y(s))|\Delta s\geq\int_{t_1}^{t} a(s)f\left(g(c_2)\int_{t_1}^{s} b(u)\Delta u\right)\Delta s, \quad \text{where} \quad g(c_2) = c_1.$$

As $t\to\infty$, the proof is finished. (The case $x < 0$ eventually can be proved similarly with $c_1 < 0$.)

Conversely, suppose $\int_{t_0}^{\infty} a(s)f\left(c_1\int_{t_0}^{s} b(u)\Delta u\right)\Delta s < \infty$ for some $c_1 \neq 0$. Without loss of generality, assume $c_1 > 0$. (The case $c_1 < 0$ can be done similarly.) Then, we can choose $t_1 \geq t_0$ and $d > 0$ such that

$$\int_{t_1}^{\infty} a(s)f\left(c_1\int_{t_1}^{s} b(u)\Delta u\right)\Delta s < d, \quad t \geq t_1,$$

where $c_1 = g(2d) > 0$. Let X be the partially ordered Banach space of all real-valued continuous functions endowed with supremum norm $\|x\| = \sup_{t\in[t_1,\infty)_{\mathbb{T}}} |x(t)|$ and with the usual pointwise ordering $\leq$. Define a subset Ω of X such that

$$\Omega =: \{x \in X : \quad d \leq x(t) \leq 2d, \quad t \geq t_1\}.$$

For any subset B of Ω, $\inf B \in \Omega$ and $\sup B \in \Omega$, i.e., $(\Omega, \leq)$ is complete. Define an operator $F : \Omega \to X$ as

$$(Fx)(t) = d + \int_{t}^{\infty} a(s)f\left(\int_{t_1}^{s} b(u)g(x(u))\Delta u\right)\Delta s, \quad t \geq t_1.$$

The rest of the proof can be completed similar to the proof of Lemma 3.1(a). So, it is omitted.

Exercise 3.8 Let $\mathbb{T} = \mathbb{Z}$. Use Theorem 3.7 to justify that $(x_n, y_n) = (1 + 2^{-n}, -2^n)$ is a nonoscillatory solution in $M_{B,\infty}^-$ of

$$\begin{cases} \Delta x_n = 2^{\frac{-6n}{5}-1}(y_n)^{\frac{1}{5}} \\ \Delta y_n = -\dfrac{4^n}{1+2^n}(x_n). \end{cases}$$

For convenience, set

$$I = \int_{t_0}^{\infty} a(t)f\left(k\int_{t}^{\infty} b(s)\Delta s\right)\Delta t, \quad k \neq 0. \tag{11}$$

In order to obtain the nonemptiness of $M_{0,\infty}^-$, we apply Theorem 2.11 and use the similar discussion as in Lemma 3.1(a).

Theorem 3.9 *$M_{0,\infty}^- \neq \emptyset$ if for some $k > 0$ and any $d_1 > 0$ ($k < 0$ and $d_1 < 0$)*

$$I < \infty \quad \text{and} \quad \int_{t_0}^{\infty} b(t)g\left(d_1\int_{t}^{\infty} a(s)\Delta s\right)\Delta t = \infty,$$

where I is defined as in Eq. (11) *and f is an odd function.*

Exercise 3.10. Prove Theorem 3.9.

We reconsider system (1) in the next section to emphasize the existence of nonoscillatory solutions in M^+.

4. Delay Dynamical Systems on Time Scales

This section is concerned with the delay system

$$\begin{cases} x^{\Delta}(t) = a(t)f(y(t)) \\ y^{\Delta}(t) = -b(t)g(x(\tau(t))) \end{cases} \tag{12}$$

with $a,b \in C_{rd}([t_0,\infty)_{\mathbb{T}},\mathbb{R}^+)$, $\tau \in C_{rd}([t_0,\infty)_{\mathbb{T}},[t_0,\infty)_{\mathbb{T}})$, $\tau(t) \leq t$ and $\tau(t) \to \infty$ as $t \to \infty$, $f,g \in C(\mathbb{R},\mathbb{R})$ are nondecreasing functions such that $uf(u) > 0$ and $ug(u) > 0$ for $u \neq 0$. Motivated by Ref. [34] in which $\tau(t) = t-\eta$, $\eta > 0$, our purpose in this section is to obtain the criteria for the existence of nonoscillatory solutions of Eq. (12) based on $Y(t_0)$ and $Z(t_0)$. However, note that the results in Ref. [34] do not hold for any time scale, e.g., $\mathbb{T} = q^{\mathbb{N}_0}$, $q > 1$, because $t-\eta$ is not necessarily in $\mathbb{T}$. In fact, theoretical claims in this section follow from Ref. [35].

Since system (12) is oscillatory for the case $Y(t_0) = \infty$ and $Z(t_0) = \infty$, the existence results on any time scale are obtained in the next subsections based on the other three cases of $Y(t_0)$ and $Z(t_0)$. Let (x,y) be a nonoscillatory solution of system (12) in M^+ such that the component function x is eventually positive. Then by the second equation of system (12), y is eventually decreasing. In addition, using the first equation of system (12), we have that $x(t) \to c$ or ∞ and $y(t) \to d$ or 0 as $t \to \infty$ for $0 < c < \infty$ and $0 < d < \infty$. Therefore, we have the following subclasses of M^+:

$$M^+_{B,B} = \{(x,y) \in M^+ : \lim_{t\to\infty}|x(t)| = c, \quad \lim_{t\to\infty}|y(t)| = d\},$$

$$M^+_{B,0} = \{(x,y) \in M^+ : \lim_{t\to\infty}|x(t)| = c, \quad \lim_{t\to\infty}|y(t)| = 0\},$$

$$M^+_{\infty,B} = \{(x,y) \in M^+ : \lim_{t\to\infty}|x(t)| = \infty, \quad \lim_{t\to\infty}|y(t)| = d\},$$

$$M^+_{\infty,0} = \{(x,y) \in M^+ : \lim_{t\to\infty}|x(t)| = \infty, \quad \lim_{t\to\infty}|y(t)| = 0\}.$$

In the literature, solutions in $M^+_{B,0}$, $M^+_{\infty,B}$ and $M^+_{\infty,0}$ are called *subdominant, dominant* and *intermediate solutions*, respectively. Any nonoscillatory solution of system (12) belongs to M^+ or M^- given in Section. Also, it is important to emphasize that Lemma 3.1 holds for system (12) as well.

4.1. The case $Y(t_0) = \infty$ and $Z(t_0) < \infty$

We restrict our attention to M^+ in this subsection because $M^- = \emptyset$ when $Y(t_0) = \infty$ and $Z(t_0) < \infty$. The following lemma specifies the limit behavior of the component functions of nonoscillatory solutions (x,y) under the case $Y(t_0) = \infty$ and $Z(t_0) < \infty$.

Lemma 4.1 *If* $|x(t)| \to c$, *then* $y(t) \to 0$ *as* $t \to \infty$ *for* $0 < c < \infty$.

Proof. Assume to the contrary. So $y(t) \to d$ for $0 < d < \infty$ as $t \to \infty$. Then since $y(t) > 0$ and decreasing eventually, there exists $t_1 \geq t_0$ such that $f(y(\tau(t))) \geq f(d) = k$ for $t \geq t_1$. By the same discussion as in the proof of Theorem 3.3, we obtain

$$x(t) \geq k \int_{t_1}^{t} a(s)\Delta s, \quad t \geq t_1.$$

However, this gives us a contradiction to the fact that $x(t) \to c$ as $t \to \infty$. So the assertion follows.

Remark 4.2. The discussion above and Lemma 4.1 yield us $M_{B,B}^{+} = \emptyset$.

Theorem 4.3. $M_{B,0}^{+} \neq \emptyset$ *if and only if* $I < \infty$.

Proof. Suppose that there exists a solution $(x,y) \in M_{B,0}^{+}$ such that $x(t) > 0$, $x(\tau(t)) > 0$ for $t \geq t_0$, $x(t) \to c_1$ and $y(t) \to 0$ as $t \to \infty$. Because x is eventually increasing, there exist $t_1 \geq t_0$ and $c_2 > 0$ such that $c_2 \leq g(x(\tau(t)))$ for $t \geq t_1$. Integrating the second equation from t to ∞ gives

$$y(t) = \int_{t}^{\infty} b(s) g(x(\tau(s)))\Delta s, \quad t \geq t_1. \tag{13}$$

Also, integrating the first equation from t_1 to t, Eq. (13) and the monotonicity of g result in

$$x(t) \geq \int_{t_1}^{t} a(s) f\Big(\int_{s}^{\infty} b(u) g(x(\tau(u)))\Delta u\Big)\Delta s \geq \int_{t_1}^{t} a(s) f\Big(c_2 \int_{s}^{\infty} b(u)\Delta u\Big)\Delta s.$$

Setting $c_2 = k$ and taking the limit as $t \to \infty$ prove the assertion. (For the case $x < 0$ eventually, the proof can be shown similarly with $k < 0$.)

Conversely, suppose $I < \infty$ for some $k > 0$. (For the case $k < 0$ can be shown similarly.) Then, choose $t_1 \geq t_0$ so large that

$$\int_{t_1}^{\infty} a(t) f\Big(k \int_{t}^{\infty} b(s)\Delta s\Big)\Delta t < \frac{c_1}{2}, \quad t \geq t_1,$$

where $k = g(c_1)$. Let X be the space of all continuous and bounded functions on $[t_1, \infty)_{\mathbb{T}}$ with the norm $\|y\| = \sup_{t \in [t_1,\infty)_{\mathbb{T}}} |y(t)|$. Then, X is a Banach space. Let Ω be the subset of X such that

$$\Omega := \{x \in X : \quad \frac{c_1}{2} \leq x(\tau(t)) \leq c_1, \quad \tau(t) \geq t_1\},$$

and define an operator $F : \Omega \to X$ such that

$$(Fx)(t) = c_1 - \int_{t}^{\infty} a(s) f\Big(\int_{s}^{\infty} b(u) g(x(\tau(u)))\Delta u\Big)\Delta s, \quad \tau(t) \geq t_1.$$

It is easy to see that Ω is bounded, convex and a closed subset of X. It can also be shown that F maps into itself, relatively compact and continuous on Ω by the Lebesques dominated convergence theorem. Then, Theorem 2.9 gives that there exists $\bar{x} \in \Omega$ such that $\bar{x} = F\bar{x}$. As $t \to \infty$, we get $\bar{x}(t) \to c_1 > 0$. Setting

$$\overline{y}(t)=\int_t^{\infty} b(u)g(\overline{x}(\tau(u)))\Delta u>0,\quad \tau(t)\geq t_1$$

shows $\overline{y}(t)\to 0$ as $t\to\infty$. Taking the derivatives of $\overline{x}$ and $\overline{y}$ yield that $(\overline{x},\overline{y})$ is a solution of system (12). Hence, $M_{B,0}^{+}\neq\emptyset$.

We demonstrate the following example to highlight Theorem 4.3.

Example 4.4 *Let* $\mathbb{T}=2^{\mathbb{N}_0}$ *and consider the system*

$$\begin{cases}\Delta_2 x(t)=\dfrac{1}{2t^{\frac{4}{5}}}\left(y(t)\right)^{\frac{3}{5}}\\ \Delta_2 y(t)=-\dfrac{3}{4t^2(8t-4)}x(\dfrac{t}{4}).\end{cases}\tag{14}$$

First, it must be shown $Y(t_0)=\infty$ *and* $Z(t_0)<\infty$. *Indeed,*

$$\int_{t_0}^{t} a(s)\Delta s=\frac{1}{2}\sum_{s\in[4,t)_{2^{\mathbb{N}_0}}} s^{\frac{1}{5}}\quad \textit{implies}\quad Y(t_0)=\frac{1}{2}\lim_{n\to\infty}\sum_{m=2}^{n-1}(2^m)^{\frac{1}{5}}=\infty$$

and

$$\int_{t_0}^{t} b(s)\Delta s\leq\frac{3}{16}\sum_{s\in[4,t)_{2^{\mathbb{N}_0}}}\frac{1}{s}\quad \textit{implies}\quad Z(t_0)\leq\frac{3}{16}\lim_{n\to\infty}\sum_{m=2}^{n-1}\frac{1}{2^m}<\infty$$

by the geometric series, where $t=2^n$, $s=2^m$, $m,n\geq 2$. *Note that*

$$\int_{t}^{T} b(s)\Delta s\leq\frac{3}{16}\sum_{s\in[t,T)_{2^{\mathbb{N}_0}}}\frac{1}{s}\quad \textit{implies}\quad Z(t)\leq\frac{3}{16}\lim_{n\to\infty}\sum_{m=2}^{n-1}\frac{1}{2^m}=\frac{3}{8}\lim_{n\to\infty}\left(\frac{1}{t}-\frac{1}{t2^n}\right)=\frac{3}{8t}.$$

Letting $k=1$ *and using the last inequality gives*

$$\int_{t_0}^{T} a(t)f\left(k\int_t^{\infty} b(s)\Delta s\right)\Delta t\leq\int_{t_0}^{T}\frac{1}{2t^{\frac{4}{5}}}\left(\frac{3}{8t}\right)^{\frac{3}{5}}\Delta t=\left(\frac{3}{8}\right)^{\frac{3}{5}}\frac{1}{2}\sum_{t\in[1,T)_{2^{\mathbb{N}_0}}}\frac{1}{t^{\frac{2}{5}}}.$$

Therefore, we have

$$\int_{t_0}^{\infty} a(t)f\left(k\int_t^{\infty} b(s)\Delta s\right)\Delta t\leq\left(\frac{3}{8}\right)^{\frac{3}{5}}\frac{1}{2}\sum_{n=0}^{\infty}\frac{1}{2^{\frac{2n}{5}}}<\infty$$

by the geometric series. It can be seen that $(x,y)=\left(8-\frac{1}{t},\frac{1}{t^2}\right)$ *is a nonoscillatory solution of Eq.* (14) *such that* $x(t)\to 8$ *and* $y(t)\to 0$ *as* $t\to\infty$, *i.e.,* $M_{B,0}^{+}\neq\emptyset$.

The existence in subclasses $M_{\infty,B}^{+}$ and $M_{\infty,0}^{+}$ is not obtained on general time scales. The main reason is that setting an operator including a delay function gives a struggle when the fixed points theorems are applied. In fact, when we restrict the delay function to $\tau(t) = t - \eta$ for $\eta \geq 0$, it was shown $M_{\infty,B}^{+} \neq \emptyset$, Nevertheless, the existence in $M_{\infty,B}^{+}$ and $M_{\infty,0}^{+}$ for system (1) is shown in Subsection.

4.2. The case $Y(t_0) < \infty$ and $Z(t_0) < \infty$

Because the component functions x and y have finite limits by Lemma 3.1(e) and (f), the subclasses $M_{\infty,B}^{+}$ and $M_{\infty,0}^{+}$ are empty. Since the existence of nonoscillatory solutions in $M_{B,0}^{+}$ is shown in Theorem 4.3, we only focus on $M_{B,B}^{+}$ in this subsection.

The Knaster fixed point theorem is utilized in order to prove the following theorem.

Theorem 4.5 $M_{B,B}^{+} \neq \emptyset$ *if and only if*

$$\int_{t_0}^{\infty} a(s) f\left(d_1 + k \int_s^{\infty} b(u) \Delta u\right) \Delta s < \infty, \quad k, d_1 \neq 0. \tag{15}$$

Proof. The proof of the necessity part is very similar to those of previous theorems. So for sufficiency, suppose Eq. (15) holds. Choose $t_1 \geq t_0$, $k > 0$ and $d_1 > 0$ such that

$$\int_{t_1}^{\infty} a(s) f\left(d_1 + k \int_s^{\infty} b(u) \Delta u\right) \Delta s < d_1,$$

where $k = g(2d_1)$. (The case $k, d_1 < 0$ can be done similarly.) Let X be the Banach space of all continuous real-valued functions endowed with the norm $\|x\| = \sup_{t \in [t_1,\infty)_{\mathbb{T}}} |x(t)|$ and with usual point-wise ordering $\leq$. Define a subset Ω of X as

$$\Omega := \{x \in X : \quad d_1 \leq x(\tau(t)) \leq 2d_1, \quad \tau(t) \geq t_1\}.$$

For any subset B of Ω, it is clear that $\inf B \in \Omega$ and $\sup B \in \Omega$. An operator $F : \Omega \to X$ is defined as

$$(Fx)(t) = d_1 + \int_{t_1}^{t} a(s) f\left(d_1 + \int_s^{\infty} b(u) g(x(\tau(u))) \Delta u\right) \Delta s, \quad \tau(t) \geq t_1.$$

It is obvious that F is an increasing mapping into itself. Therefore,

$$d_1 \leq (Fx)(t) \leq d_1 + \int_{t_1}^{t} a(s) f\left(d_1 + g(2d_1) \int_s^{\infty} b(u) \Delta u\right) \Delta s \leq 2d_1, \quad \tau(t) \geq t_1.$$

Then, by Theorem 2.11, there exists $\bar{x} \in \Omega$ such that $\bar{x} = F\bar{x}$. By setting

$$\bar{y}(t) = d_1 + \int_t^{\infty} b(u) g(\bar{x}(\tau(u))), \quad \tau(t) \geq t_1,$$

we get that

$$\overline{y}^{\Delta}(t) = -b(t)g(\overline{x}(\tau(t))). \tag{16}$$

Also taking the derivative of $\overline{x}$ and Eq. (16) give that $(\overline{x}, \overline{y})$ is a solution of system (12). Hence, we conclude that $\overline{x}(t) \to \alpha$ and $\overline{y}(t) \to d_1$ as $t \to \infty$, where $0 < \alpha < \infty$, i.e., $M^+_{B,B} \neq \emptyset$. Note that a similar proof can be done for the case $k < 0$ and $d_1 < 0$ with $x < 0$.

Example 4.6 *Let* $\mathbb{T} = 2^{\mathbb{N}_0}$ *and consider the system*

$$\begin{cases} \Delta_2 x(t) = \dfrac{1}{2t^{\frac{5}{3}}(3t+1)^{\frac{1}{3}}} y^{\frac{1}{3}}(t) \\ \Delta_2 y(t) = -\dfrac{1}{2t(6t-4)} x\left(\dfrac{t}{4}\right). \end{cases} \tag{17}$$

We first demonstrate $Y(t_0) < \infty$ *and* $Z(t_0) < \infty$.

$$\int_{t_0}^{t} a(s)\Delta s = \frac{1}{2} \sum_{s \in [4,t)_{2^{\mathbb{N}_0}}} \frac{1}{s^{\frac{2}{3}}(3s+1)^{\frac{1}{3}}} \quad \textit{implies} \quad Y(t_0) = \frac{1}{2} \lim_{n\to\infty} \sum_{m=2}^{n-1} \frac{1}{(2^m)^{\frac{2}{3}}(3 \cdot 2^m + 1)^{\frac{1}{3}}} < \infty$$

by the Ratio test for $t = 2^n$, $s = 2^m$, $n \geq 2$. *Similarly,*

$$\int_{t_0}^{t} b(s)\Delta s = \frac{1}{2} \sum_{s \in [4,t)_{2^{\mathbb{N}_0}}} \frac{1}{6s-4} \quad \textit{implies} \quad Z(t_0) = \frac{1}{2} \lim_{n\to\infty} \sum_{m=2}^{n-1} \frac{1}{6.2^m - 4} < \infty.$$

Because $Y(t_0) < \infty$ *and* $Z(t_0) < \infty$, *it is easy to show that Eq. (15) holds. One can also verify that* $(6 - \frac{1}{t}, 3 + \frac{1}{t})$ *is a nonoscillatory solution of system (17) such that* $x(t) \to 6$ *and* $y(t) \to 3$ *as* $t \to \infty$, *i.e.,* $M^+_{B,B} \neq \emptyset$ *by Theorem 4.5.*

4.3. The case $Y(t_0) < \infty$ and $Z(t_0) = \infty$

Lemma 3.1(c) yields $M^+ = \emptyset$ for the case $Y(t_0) < \infty$ and $Z(t_0) = \infty$. Thus, we pay our attention to M^- in this subsection. The proof of the following remark is similar to that of Theorem 3.7.

Remark 4.7 $M^-_{B,\infty} \neq \emptyset$ *if and only if integral condition* (7) *holds.*

Exercise 4.8 Prove Remark 4.7 and also show that $(3 + \frac{1}{t}, -t - \frac{1}{t})$ is a nonoscillatory solution of

$$\begin{cases} \Delta_2 x(t) = \dfrac{1}{2t^{\frac{7}{5}}(t^2+1)^{\frac{3}{5}}} (y(t))^{\frac{3}{5}} \\ \Delta_2 y(t) = -\dfrac{2t^2 - 1}{2t^{\frac{9}{5}}(3t+4)^{\frac{1}{5}}} \left(x\left(\dfrac{t}{4}\right)\right)^{\frac{1}{5}} \end{cases}$$

in $M^-_{B,\infty} \neq \emptyset$ when $\mathbb{T} = 2^{\mathbb{N}_0}$.

4.4. Dominant and intermediate solutions of Eq. (1)

Note that the existence of nonoscillatory solutions of system (1) in $M^-_{0,\infty}$, $M^-_{B,B}$ and $M^-_{0,B}$ is not shown on a general time scale. In fact, the existence in these subclasses is obtained for system

(1) in Section. Since system (12) is reduced to system (1) when $\tau(t) = t$, notice that the results obtained for system (12) in Section also hold for system (1). Therefore, we only need to show the existence of nonoscillatory solutions for Eq. (1) in $M^{+}_{\infty,B}$ and $M^{+}_{\infty,0}$, which are not acquired for Eq. (12) on a general time scale. To achieve the goal, we assume $Y(t_0) = \infty$ and $Z(t_0) < \infty$.

Theorem 4.9 *$M^{+}_{\infty,B} \neq \emptyset$ if and only if*

$$\int_{t_0}^{\infty} b(s)g\Big(c_1\int_{t_0}^{s} a(u)\Delta u\Big)\Delta s < \infty, \quad c_1 \neq 0. \tag{18}$$

Proof. The necessity part is left to readers as an exercise. Therefore, for sufficiency, suppose that Eq. (18) holds. Choose $t_1 \geq t_0$, $c_1 > 0$ and $d_1 > 0$ such that

$$\int_{t_1}^{\infty} b(s)g\Big(c_1\int_{t_1}^{s} b(u)\Delta u\Big)\Delta s < d_1, \quad t \geq t_1, \tag{19}$$

where $c_1 = f(2d_1) > 0$. (The case $c_1 < 0$ can be done similarly.) Let X be the partially ordered Banach space of all real-valued continuous functions endowed with supremum norm $\|x\| = \sup_{t\in[t_1,\infty)_{\mathbb{T}}} \frac{|x(t)|}{\int_{t_1}^{t} a(s)\Delta s}$ and with the usual point-wise ordering $\leq$. Define a subset Ω of X such that

$$\Omega =: \{x \in X : \ f(d_1)\int_{t_1}^{t} a(s)\Delta s \leq x(t) \leq f(2d_1)\int_{t_1}^{t} a(s)\Delta s, \quad t \geq t_1\}.$$

For any subset B of Ω, $\inf B \in \Omega$ and $\sup B \in \Omega$, i.e., $(\Omega, \leq)$ is complete. Define an operator $F : \Omega \to X$ as

$$(Fx)(t) = \int_{t_1}^{t} a(s)f\Big(d_1 + \int_{t}^{\infty} b(u)g(x(u))\Delta u\Big)\Delta s, \quad t \geq t_1.$$

It is obvious that it is an increasing mapping, so let us show $F := \Omega \to \Omega$.

$$\begin{aligned} f(d_1)\int_{t_1}^{t} a(s)\Delta s &\leq (Fx)(t) \\ &\leq \int_{t_1}^{t} a(s)f\Big(d_1 + \int_{s}^{\infty} b(u)g\Big(f(2d_1)\int_{t_1}^{u} a(\lambda)\Delta\lambda\Big)\Delta u\Big)\Delta s \\ &\leq f(2d_1)\int_{t_1}^{t} a(s)\Delta s \end{aligned}$$

by Eq. (19). Then, by Theorem 2.11, there exists $\overline{x} \in \Omega$ such that $\overline{x} = F\overline{x}$ and so

$$\overline{x}^{\Delta}(t) = a(t)f\Big(d_1 + \int_{t}^{\infty} b(u)g(\overline{x}(u))\Delta u\Big), \quad t \geq t_1.$$

Setting $\overline{y}(t) = d_1 + \int_{t}^{\infty} b(u)g(\overline{x}(u))\Delta u$ leads us $\overline{y}^{\Delta} = -bg(\overline{x})$ and so, $(\overline{x}, \overline{y})$ is a solution of system (1) such that $\overline{x}(t) > 0$ and $\overline{y}(t) > 0$ for $t \geq t_1$ and $\overline{x}(t) \to \infty$ and $\overline{y}(t) \to d_1 > 0$ as $t \to \infty$, i.e., $M^{+}_{\infty,B} \neq \emptyset$.

Theorem 4.10 $M^+_{\infty,0}{\neq}\emptyset$ *if*

$$I = \infty \quad and \quad \int_{t_0}^{\infty} b(t)g\left(l\int_{t_0}^{\infty} a(s)\Delta s\right)\Delta t < \infty,$$

where I is defined as in Eq. (11), *for any* $k > 0$ *and some* $l > 0$ ($k < 0$ *and* $l < 0$).

Exercise 4.11 Prove Theorem 4.10 using Theorem 2.11.

5. Emden-Fowler Dynamical Systems on Time Scales

Motivated by the papers [28, 36, 37], we deal with the classification and existence of nonoscillatory solutions of the Emden-Fowler dynamical system

$$\begin{cases} x^{\Delta}(t) = a(t)|y(t)|^{\frac{1}{\alpha}}\text{sgn } y(t) \\ y^{\Delta}(t) = -b(t)|x^{\sigma}(t)|^{\beta}\text{sgn } x^{\sigma}(t), \end{cases} \tag{20}$$

where $\alpha, \beta > 0$ $a, b \in C_{rd}([t_0, \infty)_{\mathbb{T}}, \mathbb{R}^+)$ and $x^{\sigma}(t) = x(\sigma(t))$. The main results of this section follow from Ref. [38]. If $\mathbb{T} = \mathbb{Z}$, system (20) is reduced to a Emden-Fowler system of difference equations while it is reduced to a Emden-Fowler system of differential equations when $\mathbb{T} = \mathbb{R}$,. We also refer readers to Refs. [41 46] for quasilinear and Emden-Fowler dynamic equations on time scales .

Note that any nonoscillatory solution of system (20) belongs to M^+ or M^- given in Section. Also, it could be shown that Lemma 3.1 holds for system (20) as well.

5.1. The case $Y(t_0) = \infty$ and $Z(t_0) < \infty$

In this case, we have $M^- = \emptyset$, see Lemma 3.1(d). By a similar discussion as in Subsection. 1, solutions in M^+ belongs to one of the subclasses $M^+_{B,0}$, $M^+_{\infty,B}$ and $M^+_{\infty,0}$.

Let us set

$$J_{\alpha} = \int_{t_0}^{\infty} a(t)\left(\int_{t}^{\infty} b(s)\Delta s\right)^{\frac{1}{\alpha}}\Delta t$$
$$K_{\beta} = \int_{t_0}^{\infty} b(t)\left(\int_{t_0}^{\sigma(t)} a(s)\Delta s\right)^{\beta}\Delta t.$$

Note that integral I, defined as in Eq. (11), is reduced to J_{α} by replacing $f(z) = z^{\frac{1}{\alpha}}$ and $g(z) = z^{\beta}$. The following theorem can be proven similar to Theorem 4.3.

Theorem 5.1 $M^+_{B,0}{\neq}\emptyset$ *if and only if* $J_{\alpha} < \infty$.

Exercise 5.2 Prove Theorem 5.1.

Next, we provide the existence of dominant and intermediate solutions of system (20) along with examples.

Theorem 5.3 $M^+_{\infty,B} \neq \varnothing$ *if and only if* $K_\beta < \infty$.

Proof. Suppose that there exists $(x,y) \in M^+$ such that $x > 0$ eventually, $x(t) \to \infty$ and $y(t) \to d$ as $t \to \infty$ for $0 < d < \infty$. Integrating the first equation from t_1 to $\sigma(t)$, using the monotonicity of y and integrating the second equation from t_1 to t of system (20) give us

$$x^\sigma(t) = x^\sigma(t_1) + \int_{t_1}^{\sigma(t)} a(s) y^{\frac{1}{\alpha}}(s) \Delta s > d^{\frac{1}{\alpha}} \int_{t_1}^{\sigma(t)} a(s) \Delta s. \tag{21}$$

and

$$y(t_1) - y(t) = \int_{t_1}^{t} b(s) \left(x^\sigma(s)\right)^\beta \Delta s, \tag{22}$$

respectively. Then, by Eqs. (21) and (22), we have

$$\int_{t_1}^{t} b(s) \left(\int_{t_1}^{\sigma(s)} a(u) \Delta u \right)^\beta \Delta s < d^{\frac{-\beta}{\alpha}} \int_{t_1}^{t} b(s) \left(x^\sigma(s)\right)^\beta \Delta s = d^{\frac{-\beta}{\alpha}} \left(y(t_1) - y(t) \right)$$

So as $t \to \infty$, it follows $K_\beta < \infty$.

Conversely, suppose $K_\beta < \infty$. Choose $t_1 \geq t_0$ so large that

$$\int_{t_1}^{\infty} b(s) \left(\int_{t_1}^{\sigma(s)} a(u) \Delta u \right)^\beta \Delta s < \frac{d^{1-\beta}}{2^\beta}$$

for arbitrarily given $d > 0$. Let X be the partially ordered Banach Space of all real valued continuous functions with the norm $\|x\| = \sup_{t > t_1} \frac{|x(t)|}{\int_{t_1}^{t} a(s) \Delta s}$ and the usual point-wise ordering $\leq$.

Define a subset Ω of X as follows:

$$\Omega : \{x \in X : \quad d^{\frac{1}{\alpha}} \int_{t_1}^{t} a(s) \Delta s \leq x(t) \leq (2d)^{\frac{1}{\alpha}} \int_{t_1}^{t} a(s) \Delta s \quad \text{for} \quad t > t_1\}.$$

First, since every subset of Ω has a supremum and infimum in Ω, $(\Omega, \leq)$ is a complete lattice. Define an operator $F : \Omega \to X$ as

$$(Fx)(t) = \int_{t_1}^{t} a(s) \left(d + \int_{s}^{\infty} b(u) \left(x^\sigma(u) \right)^\beta \Delta \tau \right)^{\frac{1}{\alpha}} \Delta s.$$

The rest of the proof can be finished via the Knaster fixed point theorem, and thus is left to readers.

Example 5.4 *Let $\mathbb{T} = q^{\mathbb{N}_0}$, $q > 1$ and consider the system*

$$\begin{cases} x^{\Delta} = \dfrac{t}{1+2t}|y|\text{sgn}\, y \\ y^{\Delta} = -\dfrac{1}{q^{1+\beta}t^{\beta+2}}|x^{\sigma}|^{\beta}\text{sgn}\, x. \end{cases} \tag{23}$$

It is left to readers to show $Y(t_0) = \infty$ and $Z(t_0) < \infty$. In order to show $K_\beta < \infty$, we first calculate

$$\int_{t_0}^{T} b(t)\Big(\int_{t_0}^{\sigma(t)} a(s)\Delta s\Big)^{\beta}\Delta t = \sum_{t\in[1,T)_{q^{\mathbb{N}_0}}} \frac{1}{q^{1+\beta}t^{\beta+2}}\left(\sum_{s\in[1,\sigma(t))_{q^{\mathbb{N}_0}}} \frac{s^2(q-1)}{1+2s}\right)^{\beta}(q-1)t$$

$$< \frac{(q-1)^{\beta+1}}{q^{1+\beta}} \sum_{t\in[1,T)_{q^{\mathbb{N}_0}}} \frac{1}{t^{1+\beta}}\Big(\sum_{s\in[1,\sigma(t))_{q^{\mathbb{N}_0}}} s\Big)^{\beta} < \frac{q-1}{q}\sum_{t\in[1,T)_{q^{\mathbb{N}_0}}}\frac{1}{t},$$

where $s = q^m$ and $t = q^n$ for $m, n \in \mathbb{N}_0$. Since

$$\lim_{T\to\infty} \sum_{t\in[1,T)_{q^{\mathbb{N}_0}}} \frac{1}{t} = \sum_{n=0}^{\infty}\frac{1}{q^n} < \infty$$

by the geometric series, we have $K_\beta < \infty$. It can be verified that $(t, \frac{1}{t}+2)$ is a nonoscillatory solution of system (23) in $M^{+}_{\infty,B}$.

Theorem 5.5 *$M^{+}_{\infty,0} \neq \emptyset$ if $J_\alpha = \infty$ and $K_\beta < \infty$.*

Proof. Suppose that $J_\alpha = \infty$ and $K_\beta < \infty$ hold. Since $Y(t_0) = \infty$, we can choose t_1 and t_2 so large that

$$\int_{t_2}^{\infty} b(t)\Big(\int_{t_0}^{\sigma(t)} a(s)\Delta s\Big)^{\beta}\Delta t \leq 1 \quad \text{and} \quad \int_{t_1}^{t_2} a(s)\Delta s \geq 1, \quad t \geq t_2 \geq t_1.$$

Let X be the Fréchet Space of all continuous functions on $[t_1,\infty)_{\mathbb{T}}$ endowed with the topology of uniform convergence on compact subintervals of $[t_1,\infty)_{\mathbb{T}}$. Set

$$\Omega := \{x \in X : \quad 1 \leq x(t) \leq \int_{t_1}^{t} a(s)\Delta s \quad \text{for} \quad t \geq t_1\}$$

and define an operator $T : \Omega \to X$ by

$$(Tx)(t) = 1 + \int_{t_2}^{t} a(s)\Big(\int_{s}^{\infty} b(u)\big(x^{\sigma}(u)\big)^{\beta}\Delta u\Big)^{\frac{1}{\alpha}}. \tag{24}$$

We can show that $T : \Omega \to \Omega$ is continuous on $\Omega \subset X$ by the Lebesque dominated convergence theorem. Since

$$0 \leq [(Tx)(t)]^{\Delta} = a(t)\Big(\int_t^{\infty} b(u)\big(x^{\sigma}(u)\big)^{\beta}\Delta u\Big)^{\frac{1}{\alpha}}$$
$$\leq a(t)\Big(\int_t^{\infty} b(u)\Big(\int_{t_1}^{\sigma(u)} a(\lambda)\Delta\lambda\Big)^{\beta}\Delta u\Big)^{\frac{1}{\alpha}} < \infty,$$

it follows that T is equibounded and equicontinuous. Then by Theorem 2.10, there exists $\overline{x} \in \Omega$ such that $\overline{x} = T\overline{x}$. Thus, it follows that $\overline{x}$ is eventually positive, i.e nonoscillatory. Then differentiating $\overline{x}$ and the first equation of system (20) give us

$$\overline{y}(t) = \Big(\frac{1}{a(t)}\Big)^{\alpha}\big(\overline{x}^{\Delta}(t)\big)^{\alpha} = \int_t^{\infty} b(u)\big(\overline{x}^{\sigma}(u)\big)^{\beta}\Delta u > 0, \quad t \geq t_1. \tag{25}$$

This results in that $\overline{y}$ is eventually positive and hence $(\overline{x}, \overline{y})$ is a nonoscillatory solution of system (20) in M^+. Also by monotonicity of $\overline{x}$, we have

$$\overline{x}(t) = 1 + \int_{t_2}^{t} a(s)\Big(\int_s^{\infty} b(u)\big(\overline{x}^{\sigma}(u)\big)^{\beta}\Delta u\Big)^{\frac{1}{\alpha}} \geq \big(\overline{x}(t_2)\big)^{\beta}\int_{t_2}^{t} a(s)\Big(\int_s^{\infty} b(u)\Delta u\Big)^{\frac{1}{\alpha}}.$$

Hence as $t \to \infty$, it follows $\overline{x}(t) \to \infty$. And by Eq. (25), we have $\overline{y}(t) \to 0$ as $t \to \infty$. Therefore $M^+_{\infty,0} \neq \emptyset$.

Example 5.6 *Let $\mathbb{T} = q^{\mathbb{N}_0}$, $q > 1$ and $\beta < 1$. Consider the system*

$$\begin{cases} x^{\Delta} = (1+t)|y|^{\frac{1}{\alpha}}\operatorname{sgn} y \\ y^{\Delta} = -\dfrac{1}{(1+t)(1+tq)^{\beta+1}}|x^{\sigma}|^{\beta}\operatorname{sgn} x. \end{cases} \tag{26}$$

It is easy to verify $Y(t_0) = \infty$ and $Z(t_0) < \infty$. Letting $s = q^m$ and $t = q^n$, where $m, n \in \mathbb{N}_0$ gives

$$\int_{t_0}^{T} a(t)\Big(\int_t^{T} b(s)\Delta s\Big)^{\frac{1}{\alpha}}\Delta t = \sum_{t \in [1,T)_{q^{\mathbb{N}_0}}} (1+t)\left(\sum_{s \in [t,T)_{q^{\mathbb{N}_0}}} \frac{(q-1)s}{(1+s)(1+sq)^{\beta+1}}\right)(q-1)t$$
$$\geq (q-1)^2 \sum_{t \in [1,T)_{q^{\mathbb{N}_0}}} (1+t)\left(\frac{t}{(1+t)(1+tq)^{\beta+1}}\right)t = (q-1)^2 \sum_{t \in [1,T)_{q^{\mathbb{N}_0}}} \frac{t^2}{(1+tq)^{\beta+1}}.$$

So we have

$$\lim_{T\to\infty} \sum_{t \in [1,T)_{q^{\mathbb{N}_0}}} \frac{t^2}{(1+tq)^{\beta+1}} = \sum_{n=0}^{\infty} \frac{q^{2n}}{(1+q^{n+1})^{\beta+1}} = \infty$$

by the Test for Divergence and $\beta < 1$. Now let us show that $K_{\beta} < \infty$. Since

$$\int_{t_0}^{\sigma(t)} a(s)\Delta s = \sum_{s \in [1,t)_{q^{\mathbb{N}_0}}} (1+s)(q-1)s \leq tq(1+tq),$$

we have

$$\int_{t_0}^{T} b(t)\Big(\int_{t_0}^{\sigma(t)} a(s)\Delta s\Big)^{\beta}\Delta t \leq \sum_{t\in[1,T)_{q^{\mathbb{N}_0}}} \frac{1}{(1+t)(1+tq)^{\beta+1}}\Big(tq(1+tq)\Big)^{\beta} t(q-1)$$

$$\leq q^{\beta}(q-1)\sum_{t\in[1,T)_{q^{\mathbb{N}_0}}} \frac{t^{\beta}}{1+t}.$$

Therefore by the Ratio test,

$$\lim_{T\to\infty} q^{\beta}(q-1)\sum_{t\in[1,T)_{q^{\mathbb{N}_0}}} \frac{t^{\beta}}{1+t} = q^{\beta}(q-1)\sum_{n=0}^{\infty}\frac{(q^n)^{\beta}}{(1+q^n)} < \infty$$

gives $K_\beta < \infty$. *It can also be verified that* $\left(1+t, \frac{1}{t+1}\right)$ *is a nonoscillatory solution of Eq. (26) in* $M^{+}_{\infty,0}$.

Exercise 5.7 Show that the following system

$$\begin{cases} x' = e^{2t}|y|^{\frac{1}{\alpha}}\operatorname{sgn} y \\ y' = -\alpha e^{-t(\alpha+\beta)}|x|^{\beta}\operatorname{sgn} x \end{cases}$$

has a nonoscillatory solution $(e^t, e^{-\alpha t})$ in $M^{+}_{\infty,0}$.

Next, we intend to derive a conclusion for the existence of nonoscillatory solutions of system (20) based on α and β. The proof of the following lemma is similar to the proofs of Lemmas 1.1, 3.2, 3.3, 3.6 and 3.7 in [47].

Lemma 5.8

a. *If* $J_\alpha < \infty$, *or* $K_\beta < \infty$ *then* $Z_b < \infty$.

b. *If* $K_\beta = \infty$, *then* $Y(t_0) = \infty$ *or* $Z(t_0) = \infty$.

c. *If* $J_\alpha = \infty$, *then* $Y(t_0) = \infty$ *or* $Z(t_0) = \infty$.

d. *Let* $\alpha \geq 1$. *If* $J_\alpha < \infty$, *then* $K_\alpha < \infty$.

e. *Let* $\beta \leq 1$. *If* $K_\beta < \infty$, *then* $J_\beta < \infty$.

f. *Let* $\alpha < \beta$. *If* $K_\beta < \infty$, *then* $J_\alpha < \infty$ *and* $K_\alpha < \infty$.

g. *Let* $\alpha > \beta$. *If* $J_\alpha < \infty$, *then* $K_\beta < \infty$ *and* $J_\beta < \infty$.

Exercise 5.9 Prove Lemma 5.8.

The following corollary summarizes the existence of subdominant and dominant solutions of system (20) in this subsection by means of Lemma 5.8.

Corollary 5.10 *Suppose that* $Y(t_0) = \infty$ *and* $Z(t_0) < \infty$. *Then*

a. $M^{+}_{B,0} \neq \emptyset$ *if any of the followings hold:*

(i) $J_\alpha < \infty$, **(ii)** $\alpha < \beta$, $\beta \geq 1$ *and* $J_\beta < \infty$,

(iii) $\alpha < \beta$ *and* $K_\beta < \infty$, **(iv)** $\alpha \leq 1$ *and* $K_\alpha < \infty$.

b. $M^{+}_{\infty,B} \neq \emptyset$ *if any of the followings hold:*

(i) $K_\beta < \infty$, **(ii)** $\alpha \geq 1$ *and* $J_\beta < \infty$,

(iii) $\alpha > \beta$ *and* $J_\alpha < \infty$.

5.2. The Case $Y(t_0) < \infty$ and $Z(t_0) < \infty$

With the similar discussion as in Subsection. we concentrate on $M^{+}_{B,B}$ and $M^{+}_{B,0}$. Actually, the existence in $M^{+}_{B,0}$ is shown in Subsection. Also, we use the same argument of the proof of Lemma 3.1(a) so that the criteria for the existence of nonoscillatory solutions of system (20) in $M^{+}_{B,B}$ is $Y(t_0) < \infty$ and $Z(t_0) < \infty$.

The most important question that arose in this section is about the existence of nonoscillatory solutions of the Emden-Fowler system in M^{-}. The existence of such solutions in $M^{-}_{B,\infty}$, $M^{-}_{0,\infty}$ can similarly be shown as in Theorems 3.7 and 3.9. When concerns about and $M^{-}_{0,B}$ come to our attention, we need to assume that σ must be differentiable, which is not necessarily true on arbitrary time scales. The following exercise is a great observation about the discussion mentioned above.

Exercise 5.11 Consider the system

$$\begin{cases} x^{\Delta}(t) = \dfrac{t^{\frac{1}{2}}}{2(t+1)(t+2)(3t-1)^{\frac{1}{2}}} |y(t)|^{\frac{1}{2}} \operatorname{sgn} y(t) \\ y^{\Delta}(t) = -\dfrac{(t+1)^{\frac{1}{3}}}{2^{\frac{2}{3}} t^2 (4t+5)^{\frac{1}{3}}} |x^{\sigma}(t)|^{\frac{1}{3}} \operatorname{sgn} x^{\sigma}(t) \end{cases} \tag{27}$$

in $\mathbb{T} = 2^{\mathbb{N}_0}$ and show that $(2 + \frac{1}{t+2}, -3 + \frac{1}{t})$ is a nonoscillatory solution of system (27) in $M^{-}_{B,B}$. Note that $\sigma(t) = 2t$ is differentiable on $\mathbb{T} = 2^{\mathbb{N}_0}$.

Author details

Elvan Akın* and Özkan Öztürk

*Address all correspondence to: akine@mst.edu

Missouri University of Science and Technology, Missouri, USA

References

[1] Kelley W. G., Peterson A. C. The Theory of Differential Equations: Classical and Qualitative, Springer, 2010. 413 p. DOI: 10.1007/978-1-4419-5783-2

[2] Kelley W. G., Peterson A. C. Difference Equations, Second Edition: An Introduction with Applications, Academic Press, 2001. 403 p.

[3] Agarwal R. P. Difference Equations and Inequalities: Theory, Methods and Applications, Marcel Dekker, 2000. 971 p.

[4] Kac V., Cheung P. Quantum Calculus. Universitext, Springer-Verlag, New York, 2002. 112 p. DOI: 10.1007/978-1-4613-0071-7

[5] Hilger S. Ein Maß kettenkalkül mit Anwendung auf Zentrumsmannigfaltigkeiten [thesis]. Universität Würzburg, 1988.

[6] Bohner M., Peterson A. Dynamic Equations on Time Scales: An Introduction with Applications, Birkhäuser, Boston, 2001. 358 p. DOI: 10.1007/978-1-4612-0201-1

[7] Bohner M., Peterson A. Advances in Dynamic Equations on Time Scales. Birkhäuser, Boston, 2003. 348 p. DOI: 10.1007/978-0-8176-8230-9

[8] Agarwal R. P., O'Regan D., Saker S. H. Oscillation and Stability of Delay Models in Biology, Springer, 2014. 340 p. DOI: 10.1007/978-3-319-06557-1

[9] Ruan S., Wolkowicz G. S. K., Wu J. Differential Equations with Applications to Biology, AMS, 1999. 509 p.

[10] Kyrychko Y. N., Hogan S. J. On the Use of Delay Equations in Engineering Applications. J. Vib. Control., 2010;**16(7-8)**:943–960. DOI: 10.1177/1077546309341100

[11] Wu L., Lam H., Zhao Y., Shu Z. Time-Delay Systems and Their Applications in Engineering. Math. Prob. Eng., 2015. 1–3 p. Article ID: 246351

[12] Arthur A. M., Robinson P. D. Complementary variational principle for $\nabla^2 = f(\Phi)$ with applications to the Thomas – Fermaind Liouville equations, Proc. Cambridge Philos. Soc. 1969;**65**:535–542.

[13] Davis H. T. Introduction to Nonlinear Differential Integral Equations, U.S. Atomic Energy Commission, Washington, D.C., 1960. 592 p. (Reprint: Dover, NewYork, 1962).

[14] Nehari Z. On a nonlinear differential equatioan rising in nuclear physics. Proc. Roy. Irish Academy Sect. A, 1963;**62**:117–135.

[15] Shevyelo V. N. Problems methods and fundamental results in the theory of oscillation of solutions of nonlinear nonautonomous ordinary differential equations, Proc. 2nd All-Union Conf. on Theoretical and Appl. Mech., Moscow, 1965. pp. 142–157.

[16] Homerlane I. J. On the Theoretical Temperature of the Sun under the Hypothesis of a Gaseous Mass Maintaining Its Volume by Its Internal Heat and Depending on the Laws of Gases Known to Terrestial Experiment. Am. J. Sci. Arts, **4**(1869–70):57–74.

[17] Thompson W. (Lord Kelvin). On the Convective Equilibrium of Temperature in the Atmosphere. Manchester Philos. Soc. Proc., **2**(1860–62):170–176; reprint, Math and Phys., Papers by Lord Kelvin, **3**(1890):255–260.

[18] Ritter A. Untersuchungen über die Höhe der Atmosphäre und die Konstitution gasförmiger Weltkörper, 18 articles, Wiedemann Annalender Physik, 5–20 (1878–1883).

[19] Fowler R. H. The Form Near Infinity of Real, Continuous Solutions of a Certain Differential Equation of the Second Order. Quart. J. Math., 1914;**45**:289–350.

[20] Fowler R. H. The Solution of Emden's and Similar Differential Equations. Monthly Notices Roy. Astro. Soc., 1930;**91**:63–91.

[21] Fowler R. H. Some Results on the Form Near Infinity of Real Continuous Solutions of a Certain Type of Second Order Differential Equations. Proc. London Math. Soc., 1914;**13**:341–371. DOI:10.1112/plms/s2-13.1.341

[22] Fowler R. H. Further Studies of Emden's and Similar Differential Equations. Quart. J. Math., 1931; **2**:259–288.

[23] Zeidler E. Nonlinear Functional Analysis and its Applications - I: Fixed Point Theorems, Springer Verlag, New York, 1986. 909 p. DOI: 10.1007/978-1-4612-4838-5

[24] Sidney M. A., Noussair E. S. The Schauder-Tychonoff Fixed Point Theorem and Applications. Mat. Časopis Sloven. Akad. Vied., 1975; **25(2)**:165–172.

[25] Knaster B. Un théorème sur les fonctions d'ensembles. Ann. Soc. Polon. Math., 1928; **6**:133–134.

[26] Öztürk Ö., Akın E. Nonoscillation Criteria for Two Dimensional Timeâ€"-Scale Systems. Nonauton. Dyn. Syst. 2016; **3**:1–13. DOI:10.1515/msds-2016-0001

[27] Li W. T., Cheng S. Limiting Behaviors of Non-oscillatory Solutions of a Pair of Coupled Nonlinear Differential Equations. Proc. Edinb. Math. Soc., 2000; **43**:457–473.

[28] Li W. T. Classification Schemes for Nonoscillatory Solutions of Two-Dimensional Nonlinear Difference Systems. Comput. Math. Appl., 2001; **42**:341–355. DOI:10.1016/S0898-1221(01)00159-6

[29] Anderson D. R. Oscillation and Nonoscillation Criteria for Two-dimensional Time-Scale Systems of First Order Nonlinear Dynamic Equations. Electron. J. Differential Equations, 2009; **2009(24)**:1–13.

[30] Hassan T. S. Oscillation Criterion for Two-Dimensional Dynamic Systems on Time Scales. Tamkang J. Math., 2013; **44(3)**:227–232. DOI: 10.5556/j.tkjm.44.2013.1189

[31] Zhu S., Sheng C. Oscillation and Nonoscillation Criteria for Nonlinear Dynamic Systems on Time Scales. Discrete Dyn. Nature Soc., 2012; **2012**:1–14. Article ID 137471.

[32] Agarwal R. P., Manojlović J. V. On the Existence and the Asymptotic Behavior of Nonoscillatory Solutions of Second Order Quasilinear Difference Equations. Funkcialaj Ekvacioj, 2013; **56**:81–109. DOI: 10.1619/fesi.56.81

[33] Ciarlet P. G. Linear and Nonlinear Functional Analysis with Applications, Siam, 2013. 832 p.

[34] Zhang X. Nonoscillation Criteria for Nonlinear Delay Dynamic Systems on Time Scales. Int. J. Math. Comput. Natural Phys. Eng., 2014; **8(1)**:222–226.

[35] Öztürk Ö., Akın E. On Nonoscillatory Solutions of Two Dimensional Nonlinear Delay Dynamical Systems. Opuscula Math., 2016; **36(5)**:651–669.

[36] Došlá Z., Marini M. On Super-linear Emden-Fowler Type Differential Equations. Elsevier, J. Math. Anal. Appl., 2014; **416**:497–510. DOI: 10.1016/j.jmaa.2014.02.052

[37] Cecchi M., Došlá Z., Marini M. Unbounded Solutions of Quasilinear Difference Equations. Comp. Math. Appl., 2003; **45(6-9)**:1113–1123.

[38] Öztürk Ö., Akın E., Tiryaki I. U. On Nonoscillatory Solutions of Emden-Fowler Dynamic Systems on Time Scales. Filomat, Accepted, 2015.

[39] Cecchi M., Došlá Z., Marini M. On Oscillation and Nonoscillation Properties of Emden-Fowler Difference Equations. Cent. Eur. J. Math., 2009; **7(2)**:322–334. DOI: 10.2478/s11533-009-0014-7

[40] Kusano T., Naito Y. Oscillation and Nonoscillation Criteria for Second Order Quasilinear Differential Equations. Acta Math. Hungar., 1997; **76(1-2)**:81–99. DOI: 10.1007/BF02907054

[41] Erbe L., Baoguo J., Peterson A. On the Asymptotic Behavior of Solutions of Emden â€" Fowler Equations on Time Scales. Annali di Mathematica, 2012; **191**:205–217. DOI:10.1007/s10231-010-0179-5

[42] Tanigawa T. Existence and Asymptotic Behavior of Positive solutions of Second Order Quasilinear Differential Equations. Adv. Math. Sci. Appl., Gakkotosho, Tokyo, 1999; **9 (2)**:907–938.

[43] Akın-Bohner E. Positive Decreasing Solutions of Quasilinear Dynamic Equations. Math. Comput. Model., 2006; **43(3-4)**:283–293. DOI: 10.1016/j.mcm.2005.03.006

[44] Akın-Bohner E. Positive Increasing Solutions of Quasilinear Dynamic Equations. Math. Inequal. Appl., 2007; **10(1)**:99–110. DOI: 10.7153/mia-10-10

[45] Akın-Bohner E. Regularly and Strongly Decaying Solutions for Quasilinear Dynamic Equations. Adv. Dyn. Syst. Appl., 2008; **3(1)**:15–24.

[46] Akın-Bohner E., Bohner M., Saker S. H. Oscillation Criteria for a Certain Class of Second Order Emden Fowler Dynamic Equations. Electron. Trans. Numer. Anal., 2007; **27**:1–12.

[47] Öztürk Ö., Akın E. Classification of Nonoscillatory Solutions of Emden-Fowler Dynamic Equations on Time Scales. Dynam. Syst. Appl., 2016; **25**:219–236.

7

Generalized Ratio Control of Discrete-Time Systems

Dušan Krokavec and Anna Filasová

Abstract

This chapter exposes the important connection between ratio control and the state control reflecting equality constraint for linear discrete-time systems, which allows significant reduction in computational complexity and efforts. Based on an enhanced bounded real lemma form, to outperform known approaches, the existence of the state feedback for such defined singular task is proven, and the design procedure based on the linear matrix inequalities is provided. The proposed principle, guaranteeing feasibility of the set of inequalities, improves steady-state accuracy of the ratio control and essentially reduces the design effort. The approach is illustrated on simulation examples, where the validity of the proposed method is demonstrated.

Keywords: discrete-time systems, ratio control, state feedback, equality constraint, singular systems, linear matrix inequalities

1. Introduction

The problem of the ratio feedback control is one of the specific topics in the theory of control synthesis. It is well practically motivated by applied realizations but not favorable developed in a state control technique or in combination with the state estimation theory. However, a considerable number of problems in the ratio control design have to deal with systems subjected to constraint conditions, which are other than linear, or directly formulated as singular constrained tasks. In the typical case [1, 2] where the system state reflects certain physical entities, constraints usually prescribe the system state, the region of technological conditions. If the ratio control is not formulated as a task with the equality constraints, the application requires further procedures of controlling the evolution of the set-valued ratio. Notably, a special form of the problems can be defined while the system state variables satisfy constraints and interpreted as descriptor systems [3–6]; but, the system with state equality constraints generally does not satisfy the conditions under which the results of descriptor systems can be used. Moreover, if the design task is interpreted as a singular problem, constraint associated methods have to be developed to design the controller.

In principle, it is possible to design the controller that stabilizes a system and simultaneously forces its closed-loop properties to satisfy given constraints [7, 8]. Following the idea of linear quadratic (LQ) control application, these approaches heavily rely on set-valued calculus as well as on min-max theory [9, 10], which are not simple and lead to rather cumbersome technical and numerical procedures. A more simple technique, using equality constraints formulation for discrete-time multiinput/multioutput (MIMO) systems, is introduced in Refs. [11, 12]. Based on the eigenstructure assignment principle, a slight modification of equality constraint technique is presented in Ref. [13].

Many tasks that arise in state-feedback control formulation can be reduced to standard convex problems that involve matrix inequalities. Generally, optimal solutions of such problems can be computed by using the interior point method [14], which converges in polynomial time with respect to the problem size. A review of the progress made in this field can be found in Refs. [15–17] and the references therein. In the given sense, the stability conditions are expressed in terms of linear matrix inequalities (LMI), which have a notable practical interest due to the existence of numerical LMI solvers [18, 19].

The chapter devotes the design conditions to obtain a closed-loop system in which minimally two state variables are rebind by the prescribed ratio. The generalized ratio control principle is reformulated as the full-state feedback control with one equality constraint. Solving this problem, the technique for an enhanced BRL representation [20, 21] is exploited, to circumvent potentially ill-conditioned singular task concerning the discrete-time systems control design with state equality constraints [22]. Due to application of the enhanced BRL, which decouple the Lyapunov matrix and the system matrices, the design task stays well-conditioned. These conditions impose such control that assures asymptotic stability for time-invariant discrete control under defined equality constraints. The presented way, based on projecting the target state variables into a subset of the system state space, adapts the idea of performing the LQ control principle in the fault tolerant control and the constraint control of discrete-time stochastic systems [23, 24].

The outline of this chapter is as follows. Continuing the introduction outlines in Section, the problem formulation is principally presented in Section. Section is dedicated to the mathematical backgrounds supporting the problem solution and the exploited discrete-time LMI modifications are given in Section. These results are used in Section to examine the linearization problems in bilinear matrix inequalities, so that in Section, these results can be given with convex formulation of control design condition, guaranteeing a feasible solution of the generally singular design task. Subsequently, numerical examples to illustrate basic properties of the proposed method are presented in Section, and Section is finally devoted to a brief concluding remarks.

Throughout the chapter, the following notations are used: x^T and X^T denote the transpose of the vector x and matrix X, respectively, for a square matrix $X < 0$ that X is a symmetric negative definite matrix, the symbol $\boldsymbol{I}_n$ represents the nth order unit matrix, $\boldsymbol{Y}^{\ominus 1}$ denotes the Moore-Penrose pseudoinverse of a nonsquare $\boldsymbol{Y}$, $\|\cdot\|$ represents the Euclidean norm for vectors and the spectral norm for matrices, $\boldsymbol{R}$ denotes the set of real numbers and $\boldsymbol{R}^{n \times r}$ the set of all $n \times r$ real matrices.

2. Problem formulation

Through this chapter, the task is concerned with design of the full-state feedback control to discrete-time linear dynamic systems in such a way that the closed-loop system state variables are constrained in the prescribed ratio. The systems are defined by the set of state equations

$$\boldsymbol{q}(i+1) = \boldsymbol{F}\boldsymbol{q}(i) + \boldsymbol{G}\boldsymbol{u}(i), \tag{1}$$

$$\boldsymbol{y}(i) = \boldsymbol{C}\boldsymbol{q}(i), \tag{2}$$

where $\boldsymbol{q}(i) \in \boldsymbol{R}^n$ is the vector of the state variables, $\boldsymbol{u}(i) \in \boldsymbol{R}^r$ is the vector of the input variables, $\boldsymbol{y}(i) \in \boldsymbol{R}^m$ is the vector of the output variables, and nominal system matrices $\boldsymbol{F} \in \boldsymbol{R}^{n \times n}$, $\boldsymbol{G} \in \boldsymbol{R}^{n \times r}$, and $\boldsymbol{C} \in \boldsymbol{R}^{m \times n}$ are real matrices, and $i \in Z_+$.

The discrete transfer function matrix of dimension $m \times r$, associated with the system Eqs. (1) and (2) is defined as

$$\boldsymbol{H}(z) = \boldsymbol{C}(z\boldsymbol{I} - \boldsymbol{F})^{-1}\boldsymbol{G} = \frac{\tilde{\boldsymbol{y}}(z)}{\tilde{\boldsymbol{u}}(z)} \tag{3}$$

where $\boldsymbol{I}_n \in \boldsymbol{R}^{n \times n}$ is the identity matrix, $\tilde{\boldsymbol{y}}(z)$ and $\tilde{\boldsymbol{u}}(z)$ stand for the Z transform of m dimensional output vector and r dimensional input vector, respectively, and a complex number z is the transform variable of the Z transform [25].

In practice, the ratio control maintains the relationship between two state variables [26, 27] and is defined for all $i \in Z$ as

$$\frac{q_h(i+1)}{q_k(i+1)} = a_h \Rightarrow q_h(i+1) - a_h q_k(i+1) = 0. \tag{4}$$

Assuming the parameter vector $\boldsymbol{e}_h$, the task can be expressed by using the system state vector $\boldsymbol{q}(i+1)$ as

$$\boldsymbol{e}_h^T \boldsymbol{q}(i+1) = 0, \tag{5}$$

where

$$\boldsymbol{e}_h^T = [0_1 \quad \cdots \quad 1_h \quad \cdots \quad -a_h \quad \cdots \quad 0_n], \tag{6}$$

$$\boldsymbol{q}^T(i+1) = \begin{bmatrix} q_1(i+1) & \cdots & q_h(i+1) & \cdots & q_k(i+1) & \cdots & q_n(i+1) \end{bmatrix}. \tag{7}$$

It is evident that the generalized ratio control can be defined by a composed structure of **e**, as well as by a structured matrix $\boldsymbol{E}$ [28].

The task formulated above means the design problem that can be generally defined as the stable closed-loop system synthesis using the linear full-state feedback controller of the form

$$\boldsymbol{u}(i) = -\boldsymbol{K}\boldsymbol{q}(i), \tag{8}$$

where $\boldsymbol{K} \in \mathbb{R}^{r \times n}$ is the controller feedback gain matrix, and the design constraint is considered in the general matrix equality form

$$\boldsymbol{E}\boldsymbol{q}(i+1) = 0, \tag{9}$$

with $\boldsymbol{E} \in \mathbb{R}^{p \times n}$, rank $\boldsymbol{E} = p \leq r$. In general, the matrix $\boldsymbol{E}$ reflects prescribed fixed ratio of two or more state variables. The equality Eq. (9) evidently implies $\boldsymbol{\Lambda}\boldsymbol{E}\boldsymbol{q}(i+1) = 0$, where $\boldsymbol{\Lambda} \in \mathbb{R}^{p \times p}$ is an arbitrary matrix.

It is considered in the following the discrete-time system is controllable and observable that is, $\text{rank}(z\boldsymbol{I} - \boldsymbol{F}, \boldsymbol{G}) = n \ \forall z \in \mathcal{C}$ and $\text{rank}(z\boldsymbol{I} - \boldsymbol{F}^T, \boldsymbol{C}^T) = n \ \forall z \in \mathcal{C}$, respectively [29], and that all state variables are measurable.

3. Basic preliminaries

Proposition 1. *(Matrix pseudoinverse) Let* $\boldsymbol{\Theta}$ *is a matrix variable and* $\boldsymbol{A}$, $\boldsymbol{B}$, *and* $\boldsymbol{\Pi}$ *are known nonsquare matrices of appropriate dimensions such that*

$$\boldsymbol{A}\boldsymbol{\Theta}\boldsymbol{B} = \boldsymbol{\Pi}. \tag{10}$$

Then all solution to $\boldsymbol{\Theta}$ *means that*

$$\boldsymbol{\Theta} = \boldsymbol{A}^{\ominus 1}\boldsymbol{\Lambda}\boldsymbol{B}^{\ominus 1} + \boldsymbol{\Theta}^{\circ} - \boldsymbol{A}^{\ominus 1}\boldsymbol{A}\boldsymbol{\Theta}^{\circ}\boldsymbol{B}\boldsymbol{B}^{\ominus 1}, \tag{11}$$

where

$$\boldsymbol{A}^{\ominus 1} = \boldsymbol{A}^T\left(\boldsymbol{A}\boldsymbol{A}^T\right)^{-1}, \quad \boldsymbol{B}^{\ominus 1} = \left(\boldsymbol{B}^T\boldsymbol{B}\right)^{-1}\boldsymbol{B}^T, \tag{12}$$

while $\boldsymbol{A}^{\ominus 1}$ *is the left Moore-Penrose pseudoinverse of* $\boldsymbol{A}$, $\boldsymbol{B}^{\ominus 1}$ *is the right Moore-Penrose pseudoinverse of* $\boldsymbol{B}$ *and* $\boldsymbol{\Theta}^{\circ}$ *is an arbitrary matrix of appropriate dimension.*

Proof.

Proposition 2. *Let* $\boldsymbol{\Xi} \in \mathbb{R}^{n \times n}$ *is a real square matrix with nonrepeated eigenvalues, satisfying the equality constraint*

$$\boldsymbol{e}^T\boldsymbol{\Xi} = \boldsymbol{0}, \tag{13}$$

then one from its eigenvalues is zero, and the (normalized) vector $\boldsymbol{e}^T$ *is the left raw eigenvector of* $\boldsymbol{\Xi}$ *associated with the zero eigenvalue.*

Proof. If $\Xi \in R^{n \times n}$ is a real square matrix satisfying the above given eigenvalues limitation, then the eigenvalue decomposition of Ξ takes the following form

$$\Xi = N\Sigma M^T, \tag{14}$$

$$N = [\, n_1 \quad \cdots \quad n_n \,], \quad M = [\, m_1 \quad \cdots \quad m_n \,], \quad M^T N = I, \quad \Sigma = \mathrm{diag}[\, z_1 \quad \cdots \quad z_n \,], \tag{15}$$

where n_l is the right eigenvector and m_l^T is the left eigenvector associated with the eigenvalue z_l of Ξ, and $\{z_l,\ l = 1, 2, \ldots n\}$ is the set of the eigenvalues of Ξ. Then Eq. (13) can be rewritten as follows:

$$0 = e^T [\, n_1 \quad \cdots \quad n_h \quad \cdots \quad n_n \,] \mathrm{diag}[\, z_1 \quad \cdots \quad z_h \quad \cdots \quad z_n \,] M^T. \tag{16}$$

If $e^T = m_h^T$, then orthogonal property Eq. (15) implies

$$0 = [\, 0_1 \quad \cdots \quad 1_h \quad \cdots \quad 0_n \,] \mathrm{diag}[\, z_1 \quad \cdots \quad z_h \quad \cdots \quad z_n \,] M^T \tag{17}$$

and it is evident that Eq. (17) can be satisfied only if $z_h = 0$. This concludes the proof. □

Proposition 3. *(Quadratic performance) Given a stable system of the structure* Eqs. (1) and (2), *then it yields*

$$\sum_{l=0}^{\infty} \left(y^T(l) y(l) - \gamma_\infty^2 u^T(l) u(l) \right) > 0, \tag{18}$$

where $\gamma_\infty \in R$ is the H_∞ norm of the transfer function matrix of the system Eq. (3).

Proof. Since Eq. (3) implies

$$\tilde{y}(z) = H(z) \tilde{u}(z), \tag{19}$$

then, evidently,

$$\| \tilde{y}(z) \| \le \| H(z) \|_2 \| \tilde{u}(z) \|, \tag{20}$$

where $\| H(z) \|$ is H_2 norm of the discrete transfer function matrix $H(z)$.

Since the H_∞ norm property states

$$\frac{1}{\sqrt{m}} \| H(z) \|_\infty \le \| H(z) \|_2 \le \sqrt{r} \| H(z) \|_\infty, \tag{21}$$

using the notation $\| H(z) \|_\infty = \gamma_\infty$, then Eq. (21) can be naturally rewritten as

$$\frac{1}{\sqrt{m}} \le 1 < \frac{1}{\gamma_\infty} \frac{\| \tilde{y}(z) \|}{\| \tilde{u}(z) \|} \le \frac{1}{\gamma_\infty} \| H(z) \|_2 \le \sqrt{r}. \tag{22}$$

Thus, based on the Parseval's theorem, Eq. (22) gives

$$1 < \frac{\|\tilde{\boldsymbol{y}}(z)\|}{\gamma_{\infty}\|\tilde{\boldsymbol{u}}(z)\|} = \frac{\sqrt{\sum_{i=0}^{\infty}\boldsymbol{y}^T(i)\boldsymbol{y}(i)}}{\gamma_{\infty}\sqrt{\sum_{i=0}^{\infty}\boldsymbol{u}^T(i)\boldsymbol{u}(i)}} \tag{23}$$

and using squares of the elements, the inequality Eq. (23) subsequently results in

$$\sum_{i=0}^{\infty}\boldsymbol{y}^T(i)\boldsymbol{y}(i) - \gamma_{\infty}^2\sum_{i=0}^{\infty}\boldsymbol{u}^T(i)\boldsymbol{u}(i) > 0. \tag{24}$$

Thus, Eq. (24) implies Eq. (18). This concludes the proof. □

If it is not in contradiction with other design constraints, Eq. (18) can be used as the extension to a Lyapunov function candidate for linear discrete-time systems, since it is positive.

4. Quadratic performances

The above presented assumptions are imposed to obtain LMI structures exploiting H_{∞} norm, known as the bounded real lemma LMIs. To simplify proofs of theorems in following, proof sketches of the BRL are presented, since more versions of BRL can be constructed.

Proposition 4. *(Bounded real lemma) The autonomous system* Eqs. (1) and (2) *is stable with the quadratic performance γ_{∞}, if there exist a symmetric positive definite matrix $\boldsymbol{P} \in \mathbb{R}^{n \times n}$ and a positive scalar $\gamma_{\infty} \in \mathbb{R}$ such that*

$$\boldsymbol{P} = \boldsymbol{P}^T > 0, \qquad \gamma_{\infty} > 0, \tag{25}$$

$$\begin{bmatrix} -\boldsymbol{P} & * & * & * \\ \boldsymbol{F}^T\boldsymbol{P} & -\boldsymbol{P} & * & * \\ \boldsymbol{G}^T\boldsymbol{P} & \boldsymbol{0} & -\gamma_{\infty}\boldsymbol{I}_r & * \\ \boldsymbol{0} & \boldsymbol{C} & \boldsymbol{0} & -\gamma_{\infty}\boldsymbol{I}_m \end{bmatrix} < 0, \tag{26}$$

where $\boldsymbol{I}_r \in \mathbb{R}^{r \times r}$ and $\boldsymbol{I}_m \in \mathbb{R}^{m \times m}$ are identity matrices, respectively.

Hereafter, $*$ denotes the symmetric item in a symmetric matrix.

Proof. (compare, e.g., Refs. [16] and [23]) Defining the Lyapunov function candidate as follows:

$$v(\boldsymbol{q}(i)) = \boldsymbol{q}^T(i)\boldsymbol{P}\boldsymbol{q}(i) + \gamma_{\infty}^{-1}\sum_{l=0}^{i-1}\left(\boldsymbol{y}^T(l)\boldsymbol{y}(l) - \gamma_{\infty}^2\boldsymbol{u}^T(l)\boldsymbol{u}(l)\right) > 0, \tag{27}$$

then Eq. (18) implies that with the H_{∞} norm γ_{∞} of the transform function matrix Eq. (3), the inequality Eq. (27) is positive. The forward difference of Eq. (27) along a solution of the autonomous system Eq. (1) can be written as

$$\begin{aligned}\Delta v(\boldsymbol{q}(i)) &= v(\boldsymbol{q}(i+1)) - v(\boldsymbol{q}(i)) \\ &= \boldsymbol{q}^T(i+1)\boldsymbol{P}\boldsymbol{q}(i+1) - \boldsymbol{q}^T(i)\boldsymbol{P}\boldsymbol{q}(i) + \gamma_\infty^{-1}\boldsymbol{y}^T(i)\boldsymbol{y}(i) - \gamma_\infty\boldsymbol{u}^T(i)\boldsymbol{u}(i) < 0\end{aligned} \tag{28}$$

and, using the description of the state system Eqs. (1) and (2), the inequality Eq. (28) becomes

$$\begin{aligned}\Delta v(\boldsymbol{q}(i)) = \boldsymbol{q}^T(i)(\gamma_\infty^{-1}\boldsymbol{C}^T\boldsymbol{C} - \boldsymbol{P} + \boldsymbol{F}^T\boldsymbol{P}\boldsymbol{F})\boldsymbol{q}(i) + \boldsymbol{u}^T(i)\boldsymbol{G}^T\boldsymbol{P}\boldsymbol{F}\boldsymbol{q}(i) \\ +\boldsymbol{q}^T(i)\boldsymbol{F}^T\boldsymbol{P}\boldsymbol{G}\boldsymbol{u}(i) + \boldsymbol{u}^T(i)(\boldsymbol{G}^T\boldsymbol{P}\boldsymbol{G} - \gamma_\infty\boldsymbol{I}_r)\boldsymbol{u}(i) < 0.\end{aligned} \tag{29}$$

Thus, introducing the notation

$$\boldsymbol{q}_c^T(i) = \begin{bmatrix}\boldsymbol{q}^T(i) & \boldsymbol{u}^T(i)\end{bmatrix}, \tag{30}$$

it is obtained

$$\Delta v(\boldsymbol{q}_c(i)) = \boldsymbol{q}_c^T(i)\boldsymbol{P}_c\boldsymbol{q}_c(i) < 0, \tag{31}$$

where

$$\boldsymbol{P}_c = \begin{bmatrix}\boldsymbol{F}^T\boldsymbol{P}\boldsymbol{F} + \gamma_\infty^{-1}\boldsymbol{C}^T\boldsymbol{C} - \boldsymbol{P} & \boldsymbol{F}^T\boldsymbol{P}\boldsymbol{G} \\ \boldsymbol{G}^T\boldsymbol{P}\boldsymbol{F} & \boldsymbol{G}^T\boldsymbol{P}\boldsymbol{G} - \gamma_\infty\boldsymbol{I}_r\end{bmatrix} < 0. \tag{32}$$

Since, using the Schur complement property with respect to the matrix element $\gamma_\infty^{-1}\boldsymbol{C}^T\boldsymbol{C}$, Eq. (32) can be rewritten as

$$\boldsymbol{P}_c = \begin{bmatrix}-\boldsymbol{P} & \boldsymbol{0} & \boldsymbol{C}^T \\ \boldsymbol{0} & -\gamma_\infty\boldsymbol{I}_r & \boldsymbol{0} \\ \boldsymbol{C} & \boldsymbol{0} & -\gamma_\infty\boldsymbol{I}_m\end{bmatrix} + \begin{bmatrix}\boldsymbol{F}^T\boldsymbol{P} \\ \boldsymbol{G}^T\boldsymbol{P} \\ \boldsymbol{0}\end{bmatrix}\boldsymbol{P}^{-1}[\boldsymbol{P}\boldsymbol{F} \quad \boldsymbol{P}\boldsymbol{G} \quad \boldsymbol{0}] < 0, \tag{33}$$

then, applying the dual Schur complement property, Eq. (33) implies Eq. (26). This concludes the proof. □

Direct application of the second Lyapunov method [30, 31] and BRL in the structure given by Eqs. (25) and (26) for affine uncertain systems as well as in constrained control design is in general ill-conditioned owing to singular design conditions [13]. To circumvent this problem, an enhanced LMI representation of BRL is proposed, where design condition proof is based on another form of LMIs.

Proposition 5. *(Enhanced LMI representation of BRL) The autonomous system* Eqs. (1) and (2) *is stable with the quadratic performance γ_∞, if there exist a symmetric positive definite matrix $\boldsymbol{P} \in \mathbb{R}^{n \times n}$, a regular square matrix $\boldsymbol{Q} \in \mathbb{R}^{n \times n}$, and a positive scalar $\gamma_\infty \in \mathbb{R}$ such that*

$$\boldsymbol{P} = \boldsymbol{P}^T > 0, \quad \gamma_\infty > 0, \tag{34}$$

$$\boldsymbol{Y} = \begin{bmatrix}\boldsymbol{P} - \boldsymbol{Q} - \boldsymbol{Q}^T & * & * & * \\ \boldsymbol{F}^T\boldsymbol{Q}^T & -\boldsymbol{P} & * & * \\ \boldsymbol{G}^T\boldsymbol{Q}^T & \boldsymbol{0} & -\gamma_\infty\boldsymbol{I}_r & * \\ \boldsymbol{0} & \boldsymbol{C} & \boldsymbol{0} & -\gamma_\infty\boldsymbol{I}_m\end{bmatrix} < 0, \tag{35}$$

where $\boldsymbol{I}_r \in \mathbb{R}^{r \times r}$ and $\boldsymbol{I}_m \in \mathbb{R}^{m \times m}$ are identity matrices.

Proof. Since, Eq. (1) can be rewritten as

$$\boldsymbol{F}\boldsymbol{q}(i) + \boldsymbol{G}\boldsymbol{u}(i) - \boldsymbol{q}(i+1) = \boldsymbol{0}, \tag{36}$$

with an arbitrary square matrix $\boldsymbol{Q} \in \mathbb{R}^{n \times n}$, it yields

$$\boldsymbol{q}^T(i+1)\boldsymbol{Q}(\boldsymbol{F}\boldsymbol{q}(i) + \boldsymbol{G}\boldsymbol{u}(i) - \boldsymbol{q}(i+1)) = \boldsymbol{0}. \tag{37}$$

Now, not substituting Eq. (1) into Eq. (28), but adding Eq. (37) and its transposition to Eq. (28), it can be obtained that

$$\begin{aligned} \Delta v(\boldsymbol{q}(i)) = {} & \boldsymbol{q}^T(i+1)\boldsymbol{P}\boldsymbol{q}(i+1) - \boldsymbol{q}^T(i)\boldsymbol{P}\boldsymbol{q}(i) + \gamma_\infty^{-1}\boldsymbol{y}^T(i)\boldsymbol{y}(i) - \gamma_\infty \boldsymbol{u}^T(i)\boldsymbol{u}(i) \\ & + (\boldsymbol{F}\boldsymbol{q}(i) + \boldsymbol{G}\boldsymbol{u}(i) - \boldsymbol{q}(i+1))^T\boldsymbol{Q}^T\boldsymbol{q}(i+1) \\ & + \boldsymbol{q}^T(i+1)\boldsymbol{Q}(\boldsymbol{F}\boldsymbol{q}(i) + \boldsymbol{G}\boldsymbol{u}(i) - \boldsymbol{q}(i+1)) < 0. \end{aligned} \tag{38}$$

Thus, considering Eq. (2), then Eq. (38) can be rewritten as

$$\boldsymbol{q}^{\circ T}(i)\boldsymbol{P}^{\circ}\boldsymbol{q}^{\circ}(i) < 0, \tag{39}$$

where

$$\boldsymbol{q}^{\circ T}(i) = \begin{bmatrix} \boldsymbol{q}^T(i) & \boldsymbol{q}^T(i+1) & \boldsymbol{u}^T(i) \end{bmatrix} \tag{40}$$

and

$$\boldsymbol{P}^{\circ} = \begin{bmatrix} -\boldsymbol{P} + \gamma_\infty^{-1}\boldsymbol{C}^T\boldsymbol{C} & \boldsymbol{F}^T\boldsymbol{Q}^T & \boldsymbol{0} \\ \boldsymbol{Q}\boldsymbol{F} & \boldsymbol{P} - \boldsymbol{Q} - \boldsymbol{Q}^T & \boldsymbol{Q}\boldsymbol{G} \\ \boldsymbol{0} & \boldsymbol{G}^T\boldsymbol{Q}^T & -\gamma_\infty \boldsymbol{I}_r \end{bmatrix} < 0. \tag{41}$$

Since Eq. (41) can be written as

$$\boldsymbol{P}^{\circ} = \begin{bmatrix} -\boldsymbol{P} & \boldsymbol{F}^T\boldsymbol{Q}^T & \boldsymbol{0} \\ \boldsymbol{Q}\boldsymbol{F} & \boldsymbol{P} - \boldsymbol{Q} - \boldsymbol{Q}^T & \boldsymbol{Q}\boldsymbol{G} \\ \boldsymbol{0} & \boldsymbol{G}^T\boldsymbol{Q}^T & -\gamma_\infty \boldsymbol{I}_r \end{bmatrix} + \gamma_\infty^{-1} \begin{bmatrix} \boldsymbol{C}^T \\ \boldsymbol{0} \\ \boldsymbol{0} \end{bmatrix} \begin{bmatrix} \boldsymbol{C} & \boldsymbol{0} & \boldsymbol{0} \end{bmatrix} < 0, \tag{42}$$

then, using the dual Schur complement property, Eq. (43) can be transformed in the form

$$\begin{bmatrix} -\gamma_\infty \boldsymbol{I}_m & \boldsymbol{C} & \boldsymbol{0} & \boldsymbol{0} \\ \boldsymbol{C}^T & -\boldsymbol{P} & \boldsymbol{F}^T\boldsymbol{Q}^T & \boldsymbol{0} \\ \boldsymbol{0} & \boldsymbol{Q}\boldsymbol{F} & \boldsymbol{P} - \boldsymbol{Q} - \boldsymbol{Q}^T & \boldsymbol{Q}\boldsymbol{G} \\ \boldsymbol{0} & \boldsymbol{0} & \boldsymbol{G}^T\boldsymbol{Q}^T & -\gamma_\infty \boldsymbol{I}_r \end{bmatrix} < 0. \tag{43}$$

To obtain a LMI structure visually comparable with Eq. (26), the following block permutation matrix is defined

$$T_a{}^{\circ} = \begin{bmatrix} 0 & 0 & I_n & 0 \\ 0 & I_n & 0 & 0 \\ 0 & 0 & 0 & I_r \\ I_m & 0 & 0 & 0 \end{bmatrix}. \tag{44}$$

Then, premultiplying the left side of Eq. (43) by $T_a{}^{\circ}$ and postmultiplying the right side of Eq. (43) by the transposition of $T_a{}^{\circ}$ lead to the inequality in Eq. (35). This concludes the proof.□

It is evident that Lyapunov matrix $\boldsymbol{P}$ is separated from the matrix parameters of the system $\boldsymbol{F}$, $\boldsymbol{G}$, and $\boldsymbol{C}$, i.e., there are no terms containing the product of $\boldsymbol{P}$ and any of them. By introducing the slack variable matrix $\boldsymbol{Q}$, the product forms are relaxed to new products $\boldsymbol{QF}$ and $\boldsymbol{QG}$, where $\boldsymbol{Q}$ needs not be symmetric and positive definite. This enables a robust BRL, which can be obtained to deal with linear systems with parametric uncertainties, as well as with singular system matrices.

Considering a symmetric positive definite matrix $\boldsymbol{Q} \in \boldsymbol{R}^{n \times n}$, the following symmetric enhanced LMI representation of BRL is evidently obtained.

Corollary 1. *(Enhanced symmetric LMI representation of BRL) The autonomous system* Eqs. (1) and (2) *is stable with the quadratic performance γ_∞, if there exist symmetric positive definite matrices $\boldsymbol{P}, \boldsymbol{Q} \in \boldsymbol{R}^{n \times n}$ and a positive scalar $\gamma_\infty \in \boldsymbol{R}$ such that*

$$\boldsymbol{P} = \boldsymbol{P}^T > 0, \quad \boldsymbol{Q} = \boldsymbol{Q}^T > 0, \quad \gamma_\infty > 0, \tag{45}$$

$$\begin{bmatrix} \boldsymbol{P} - 2\boldsymbol{Q} & * & * & * \\ \boldsymbol{F}^T\boldsymbol{Q} & -\boldsymbol{P} & * & * \\ \boldsymbol{G}^T\boldsymbol{Q} & 0 & -\gamma_\infty \boldsymbol{I}_r & * \\ 0 & \boldsymbol{C} & 0 & -\gamma_\infty \boldsymbol{I}_m \end{bmatrix} < 0, \tag{46}$$

where $\boldsymbol{I}_r \in \boldsymbol{R}^{r \times r}$, $\boldsymbol{I}_m \in \boldsymbol{R}^{m \times m}$ are identity matrices.

Note, Corollary 1 provides the identical condition of existence to Proposition 4, if the equality $\boldsymbol{P} = \boldsymbol{Q}$ is set.

5. Control law parameter design

The state-feedback control problem is finding, for an optimized (or prescribed) scalar $\gamma > 0$, the state-feedback gain $\boldsymbol{K}$ such that the control law guarantees an upper bound of γ_∞ of the closed-loop transfer function, while the closed-loop is stable. Note, all the above presented BRL structures applied in the control law synthesis lead to bilinear matrix inequalities and have to be linearized.

Theorem 1. *System* Eqs. (1) and (2) *under control* Eq. (3) *is stable with quadratic performance γ_∞, if there exist a positive definite symmetric matrix $\boldsymbol{R} \in \boldsymbol{R}^{n \times n}$, a matrix $\boldsymbol{Y} \in \boldsymbol{R}^{r \times n}$, and a positive scalar $\gamma_\infty \in \boldsymbol{R}$ such that*

$$R = R^T > 0, \qquad \gamma_\infty > 0, \tag{47}$$

$$\begin{bmatrix} -R & * & * & * \\ RF^T - Y^T G^T & -R & * & * \\ G^T & 0 & -\gamma_\infty I_r & * \\ 0 & CR & 0 & -\gamma_\infty I_m \end{bmatrix} < 0. \tag{48}$$

When these inequalities are satisfied, the control law gain matrix is given as

$$K = YR^{-1}. \tag{49}$$

Proof. Since P is positive definite, the transform matrix T_∞ can be defined as follows:

$$T_\infty = \text{diag}[R \quad R \quad I_r \quad I_m], \quad R = P^{-1}. \tag{50}$$

Then, premultiplying the left side of Eq. (35) and postmultiplying the right side of Eq. (35) by T_∞ gives

$$\begin{bmatrix} -R & FR & G & 0 \\ RF^T & -R & 0 & RC^T \\ G^T & 0 & -\gamma_\infty I_r & 0 \\ 0 & CR & 0 & -\gamma_\infty I_m \end{bmatrix} < 0. \tag{51}$$

Inserting $F \leftarrow F_c = (F - GK)$ into Eq. (51) gives

$$\begin{bmatrix} -R & (F-GK)R & G & 0 \\ R(F-GK)^T & -R & 0 & RC^T \\ G^T & 0 & -\gamma_\infty I_r & 0 \\ 0 & CR & 0 & -\gamma_\infty I_m \end{bmatrix} < 0 \tag{52}$$

and with

$$Y = KR \tag{53}$$

Eq. (53) implies Eq. (48). This concludes the proof. □

Theorem 2. *System* Eqs. (1) and (2) *under control* Eq. (3) *is stable with quadratic performance γ_∞, if there exist positive definite symmetric matrices S, $O \in \mathbb{R}^{n \times n}$, a matrix $Y \in \mathbb{R}^{r \times n}$, and a positive scalar $\gamma_\infty \in \mathbb{R}$ such that*

$$S = S^T > 0, \quad O = O^T > 0, \quad \gamma_\infty > 0, \tag{54}$$

$$\begin{bmatrix} O - 2S & * & * & * \\ SF^T - Y^T G^T & -O & * & * \\ G^T & 0 & -\gamma_\infty I_r & asl \\ 0 & CS & 0 & -\gamma_\infty I_m \end{bmatrix} < 0. \tag{55}$$

When these inequalities are satisfied, the control law gain matrix is given as

$$K = YS^{-1}. \tag{56}$$

Proof. Considering that Q is positive definite, the transform matrix T°_{∞} can be defined as follows:

$$T^{\circ}_{\infty} = \mathrm{diag}[\,S \quad S \quad I_r \quad I_m\,], \quad S = Q^{-1}. \tag{57}$$

Therefore, premultiplying the left side of Eq. (46) and postmultiplying the right side of Eq. (46) by the matrix T°_{∞} gives

$$\begin{bmatrix} SPS - 2S & FS & G & 0 \\ SF^T & -SPS & 0 & SC^T \\ G^T & 0 & -\gamma_{\infty} I_r & 0 \\ 0 & CS & 0 & -\gamma_{\infty} I_m \end{bmatrix} < 0. \tag{58}$$

Substituting $F \leftarrow F_c = (F - GK)$ into Eq. (58) gives

$$\begin{bmatrix} SPS - 2S & (F - GK)S & G & 0 \\ S(F - GK)^T & -SPS & 0 & SC^T \\ G^T & 0 & -\gamma_{\infty} I_r & 0 \\ 0 & CS & 0 & -\gamma_{\infty} I_m \end{bmatrix} < 0. \tag{59}$$

and with

$$Y = KQ, \quad O = SPS, \tag{60}$$

Eq. (59) implies Eq. (55). This concludes the proof. □

6. Ratio control design

Using the control law Eq. (3), the closed-loop system equations take the form

$$q(i+1) = (F - GK)q(i), \tag{61}$$

$$y(i) = Cq(i). \tag{62}$$

Prescribed by a matrix $E \in R^{p \times n}$, rank $E = p \le r$, it is considered the design constraint Eq. (9) for all nonzero natural numbers i. From Proposition 2, it is clear that such kind of design is a singular task, where Eq. (9) gives

$$Eq(i+1) = E(F - GK)q(i) = 0, \tag{63}$$

which evidently implies

$$E(F - GK) = 0. \tag{64}$$

Evidently, the equality

$$EF = EGK \tag{65}$$

can be satisfied, as well as the closed-loop system matrix $F_c = F - GK$ has to stable (all its eigenvalues are from the unit circle in the complex plane Z).

Lemma 1. *The equivalent state-space description of the system* Eqs. (1) and (2) *under control* Eq. (3), *in which closed-loop state variables satisfying the condition* Eq. (9) *is*

$$q(i+1) = (F - GK)q(i), \tag{66}$$

$$y(i) = Cq(i), \tag{67}$$

where

$$K = J + LK^{\circ}, \quad J = (EG)^{\ominus 1}EF, \quad L = I_r - (EG)^T\left(EG(EG)^T\right)^{-1}EG \tag{68}$$

while $L \in R^{r \times r}$ *is the projection matrix (the orthogonal projector of* EG *onto the null space* $\mathcal{N}_{EG}$ [23]) *and* $K^{\circ} \in R^{r \times n}$ *is the ratio control gain matrix.*

Proof. Premultiplying the left side of Eq. (65) by the identity matrix, it yields

$$EG(EG)^T\left(EG(EG)^T\right)^{-1}EF = EGK, \tag{69}$$

which implies the particular solution

$$K = (EG)^{\ominus 1}EF, \tag{70}$$

where

$$(EG)^{\ominus 1} = (EG)^T\left(EG(EG)^T\right)^{-1} \tag{71}$$

is the left Moore-Penrose pseudoinverse of EG.

Using the equality Eq. (65), then Eq. (69) can be also written as

$$EG(EG)^T\left(EG(EG)^T\right)^{-1}EGK = EGK, \tag{72}$$

which implies

$$EG\left(I_r - (EG)^T\left(EG(EG)^T\right)^{-1}EG\right)K = 0, \tag{73}$$

$$EG\left(I_r - (EG)^{\ominus 1}EG\right)K = 0, \tag{74}$$

respectively, where $I_r \in R^{p \times p}$ is the identity matrix. It is evident that Eq. (74) can be satisfied only if

$$I_r - (EG)^{\ominus 1} EG = 0. \tag{75}$$

Thus, Eq. (11) implies all solutions of K as follows

$$K = (EG)^{\ominus 1} EF + \left(I_r - (EG)^{\ominus 1} EG \right) K^{\circ}, \tag{76}$$

where K° is an arbitrary matrix with appropriate dimension, and evidently Eq. (76) gives Eq. (68). This concludes the proof. □

Considering the model involving the given ratio constraint on the closed-loop system state variables Eqs. (66)–(68), the design conditions are presented in the following theorems.

Theorem 3. *System* Eqs. (1) and (2) *under the control* (3), *and satisfying the constraint* Eq. (4) *is stable with the quadratic performance* γ_{∞}, *if there exist positive definite matrices* $S, O \in R^{n \times n}$, *a matrix* $Y^{\circ} \in R^{r \times n}$, *and a positive scalar* $\gamma_{\infty} \in R$ *such that*

$$S = S^T > 0, \quad O = O^T > 0, \quad \gamma_{\infty} > 0, \tag{77}$$

$$\begin{bmatrix} O - 2S & * & * & * \\ S(F - GJ)^T - Y^{\circ T} L^T G^T & -O & * & * \\ G^T & 0 & -\gamma_{\infty} I_r & * \\ 0 & CS & 0 & -\gamma_{\infty} I_m \end{bmatrix} < 0. \tag{78}$$

When these inequalities are satisfied, the control law gain matrices are given as

$$K^{\circ} = Y^{\circ} S^{-1}, \quad K = J + LK^{\circ}, \tag{79}$$

where J, L *are defined in* Eq. (68).

Proof. Substituting Eq. (68) into Eq. (59) gives

$$\begin{bmatrix} O - 2S & (F - GL - GLK^{\circ})S & G & 0 \\ S(F - GJ - GLK^{\circ})^T & -O & 0 & SC^T \\ G^T & 0 & -\gamma_{\infty} I_r & 0 \\ 0 & CS & 0 & -\gamma_{\infty} I_m \end{bmatrix} < 0. \tag{80}$$

Using the notation

$$Y^{\circ} = K^{\circ} S \tag{81}$$

Eq. (80) implies Eq. (78). This concludes the proof. □

The ratio control does not exclude a forced regime given by the control law

$$u(i) = -Kq(i) + Ww(i), \tag{82}$$

where $w(i) \in R^m$ is desired output signal vector and $W \in R^{m \times m}$ is the signal gain matrix. Using the static decoupling principle, the conditions to design the signal gain matrix W can be proven.

Lemma 2. *If the system* Eqs. (1) and (2) *is square, which is stabilizable by the control policy* Eq. (82) *and* Ref. [32]

$$rank\begin{bmatrix} F & G \\ C & 0 \end{bmatrix} = n + m, \tag{83}$$

then the matrix W takes the form

$$W = \left(C(I_n - (F - GK))^{-1}G\right)^{-1}, \tag{84}$$

where $I_n \in R^{n \times n}$ is the identity matrix.

Proof. In a steady state, the system equations Eqs. (1) and (2), and the control law Eq. (82) imply

$$q_o = (F - GK)q_o + GWw_o, \tag{85}$$

where q_o, w_o are the steady-state values of the vectors $q(i)$, $w(i)$, respectively. Since from Eq. (85), it can be derived that

$$q_o = (I_n - (F - GK))^{-1}GWw_o \tag{86}$$

and

$$y_o = C(I_n - (F - GK))^{-1}GWw_o, \tag{87}$$

considering $y_o = w_o$, Eq. (87) implies Eq. (84). This concludes the proof. □

Theorem 4. *If the closed-loop system state variables satisfy the state constraint* Eq. (63), *then the common state variable vector $q_d(i) = Eq(i)$, $q_d(i) \in R^k$ attains the steady-state value*

$$q_{dw} = EGWw_o. \tag{88}$$

Proof. Using the control policy Eq. (82), then

$$Eq(i+1) = E(F - GK)q(i) + EGWw(i). \tag{89}$$

Since K satisfies Eq. (65), then Eq. (89) implies

$$Eq(i+1) = EGWw(i) \tag{90}$$

and it is evident that the tied state variable $q_d(i)$ of the closed-loop system in a steady state is proportional to the steady state of the desired signal w_o and takes the value Eq. (88). This concludes the proof. □

7. Illustrative examples

To demonstrate properties of proposed approach, the classical example for a helicopter control [33] is taken, where the discrete-time state-space representation Eqs. (1) and (2) for the sampling period $\Delta t = 0.05s$ consists of the following parameters

$$F = \begin{bmatrix} 0.9982 & 0.0013 & 0.0004 & -0.0229 \\ 0.0023 & 0.9507 & -0.0048 & -0.1962 \\ 0.0049 & 0.0176 & 0.9670 & 0.0679 \\ 0.0001 & 0.0004 & 0.0492 & 1.0017 \end{bmatrix}, \quad G = \begin{bmatrix} 0.0221 & 0.0086 \\ 0.1733 & -0.3705 \\ -0.2697 & 0.2173 \\ -0.0068 & 0.0055 \end{bmatrix},$$

$$C = \begin{bmatrix} 0 & 1 & 0 & 0 \\ 1 & 0 & 0 & 0 \end{bmatrix}. \tag{91}$$

The state constraint, defining the ratio control of two state system variables, is specified as

$$\frac{q_4(t)}{q_1(t)} = 1.5 \quad \Rightarrow E = \begin{bmatrix} -1.5 & 0 & 0 & 1 \end{bmatrix} \tag{92}$$

and subsequently it yields

$$(EG)^{\ominus 1} = \begin{bmatrix} -24.1737 \\ -4.4828 \end{bmatrix}, \quad L = \begin{bmatrix} 0.0332 & -0.1793 \\ -0.1793 & 0.9668 \end{bmatrix}, \tag{93}$$

$$J = \begin{bmatrix} 36.1914 & 0.0372 & -1.1753 & -25.0447 \\ 6.7113 & 0.0069 & -0.2179 & -4.6443 \end{bmatrix}. \tag{94}$$

Solving Eqs. (77) and (78) using self-dual-minimization (SeDuMi) package for Matlab [19], the feedback gain matrix design problem in the constrained control is feasible with the results

$$O = \begin{bmatrix} 2.9027 & 0.2117 & 0.1103 & -1.7595 \\ 0.2117 & 1.3174 & -0.1751 & -0.1245 \\ 0.1103 & -0.1751 & 0.4162 & 0.0060 \\ -1.7595 & -0.1245 & 0.0060 & 3.2464 \end{bmatrix},$$

$$S = \begin{bmatrix} 2.4910 & 0.1375 & 0.0792 & -1.4957 \\ 0.1375 & 1.0779 & -0.0910 & -0.0030 \\ 0.0792 & -0.0910 & 0.3735 & -0.0348 \\ -1.4957 & -0.0030 & -0.0348 & 3.0926 \end{bmatrix}, \tag{95}$$

$$Y^{\circ} = \begin{bmatrix} -2.2113 & 0.2435 & -0.0819 & 1.4281 \\ 11.9245 & -1.3129 & 0.4416 & -7.7011 \end{bmatrix}, \quad \gamma_{\infty} = 8.5565. \tag{96}$$

Inserting Y° and S into Eq. (79), the gain matrix K° is computed as

$$K^{\circ} = \begin{bmatrix} -0.8887 & 0.3441 & 0.0562 & 0.0329 \\ 4.7926 & -1.8555 & -0.3028 & -0.1775 \end{bmatrix} \tag{97}$$

and Eq. (79) implies the full-state feedback gain matrix values

$$K = \begin{bmatrix} 35.3027 & 0.3813 & -1.1191 & -25.0117 \\ 11.5040 & -1.8486 & -0.5208 & -4.8217 \end{bmatrix}. \tag{98}$$

It can be easily verified that the closed-loop system matrix takes the format

$$F_c = F - GK = \begin{bmatrix} 0.1179 & 0.0088 & 0.0296 & 0.5722 \\ -1.8528 & 0.1997 & -0.0038 & 2.3515 \\ 7.0258 & 0.5223 & 0.7783 & -5.6297 \\ 0.1768 & 0.0132 & 0.0444 & 0.8583 \end{bmatrix}, \tag{99}$$

while the ratio control law rises up the stable closed-loop system with the closed-loop system matrix eigenvalues spectrum

$$\rho(F_c) = \{\, 0.9527, \quad 0.7566, \quad 0.0000, \quad 0.2449 \,\}. \tag{100}$$

Note that one from the resulting eigenvalue of F_c is zero (rank(E) = 1)), because Proposition 2 prescribes this constrained design task as a singular problem. Using the connection between the eigenvector matrix N and M as given by Eq. (17), it is possible to show that this instance is documented also by the structure of M, while

$$N = \begin{bmatrix} -0.3109 & -0.1105 & -0.0800 & -0.0184 \\ -0.6937 & -0.3384 & -0.4690 & -0.7382 \\ 0.4522 & 0.9197 & 0.8793 & 0.6738 \\ -0.4664 & -0.1657 & -0.0218 & -0.0276 \end{bmatrix},$$

$$M = \begin{bmatrix} -3.4197 & -0.3938 & -0.5157 & 0.2213 \\ 10.2685 & 1.3777 & 1.4844 & -7.4555 \\ -15.2705 & 0.0000 & 0.0000 & 10.1803 \\ 8.2076 & -1.6162 & -0.1958 & -3.2577 \end{bmatrix}, \tag{101}$$

where the structure of the third row of M correspondents to the structure of the constraint vector E, while $a_4 = m_3^T(1)/m_3^T(4) = -1.5$.

To illustrate the closed-loop system property in the forced mode, the signal gain matrix W is computed by using Eq. (84) as follows

$$W = \begin{bmatrix} 1.4575 & 35.9137 \\ -1.7651 & 11.6521 \end{bmatrix}. \tag{102}$$

Therefore, according to Theorem 4, the constraint given on the states of the system under study is satisfied with zero offset in the autonomous regime and with offset value equal q_{dw} in the forced mode, i.e.,

$$q_d = 0, \qquad q_{dw} = \boldsymbol{EGW}\boldsymbol{w}_o = 3.0001, \tag{103}$$

while

$$\boldsymbol{w}(i) = \begin{bmatrix} 1 \\ -2 \end{bmatrix} \text{ for all } i. \tag{104}$$

The simulation results of the closed-loop system response in the autonomous and forced mode are presented, where **Figure** is concerned with the system state variables response in the autonomous regime and **Figure** with the system state variables response in the forced mode. It is evident that the condition Eq. (9) is satisfied at all time instant, except initial time instant in the above given way (see the time response of the additive of variable, which is included as $q_d(i)$ in the figures).

For comparison, an example is given for default design of state feedback gain matrix using BRL structure of LMIs. Solving Eqs. (54) and (55), the task is feasible with the Lyapunov matrix variables

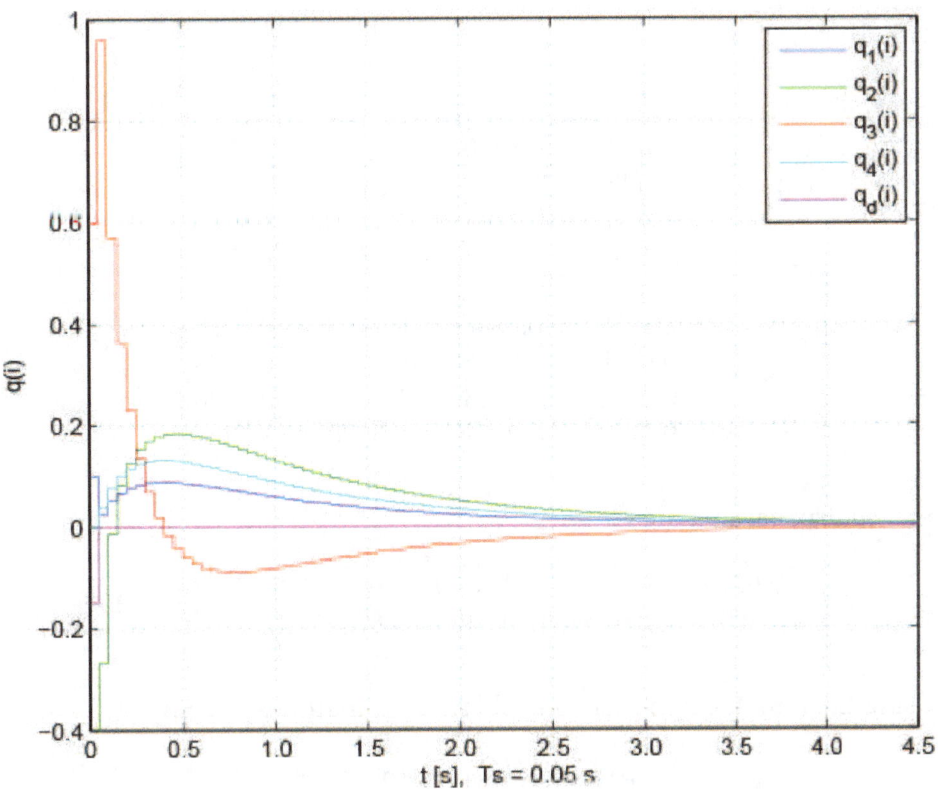

Figure. State response in autonomous regime.

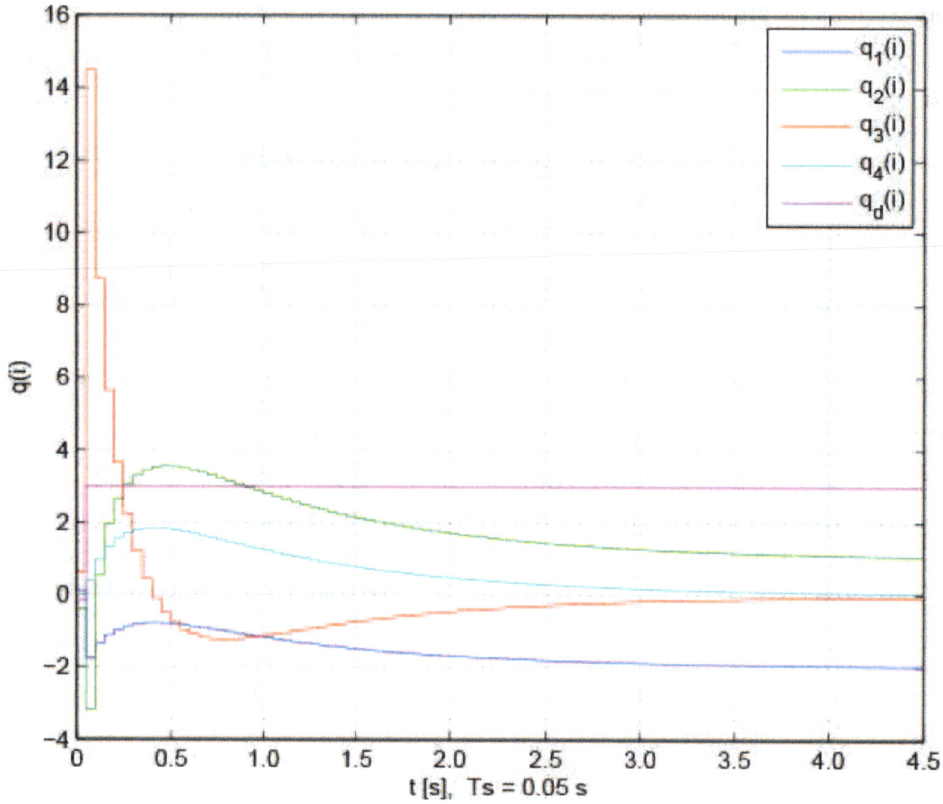

Figure. State response in forced mode.

$$
O = \begin{bmatrix} 0.1438 & -0.1090 & -0.1619 & -0.2191 \\ -0.1090 & 1.5603 & -0.2198 & 0.2945 \\ -0.1619 & -0.2198 & 1.6006 & -0.4711 \\ -0.2191 & 0.2945 & -0.4711 & 1.8586 \end{bmatrix},
$$

$$
S = \begin{bmatrix} 0.1338 & -0.0840 & -0.1490 & -0.1928 \\ -0.0840 & 1.2736 & -0.2314 & 0.2439 \\ -0.1490 & -0.2314 & 1.6729 & -0.5520 \\ -0.1928 & 0.2439 & -0.5520 & 1.8296 \end{bmatrix}, \tag{105}
$$

and parameter matrix variable

$$
Y = \begin{bmatrix} 0.6210 & -0.8607 & -2.6800 & -0.7582 \\ 0.4017 & -2.6793 & -0.3804 & 0.1788 \end{bmatrix}, \quad \gamma_\infty = 3.1301. \tag{106}
$$

Therefore, using Eq. (56), the nominal control law gain matrix K is computed as

$$
K = \begin{bmatrix} 0.8951 & 0.8107 & -1.8928 & -0.7830 \\ 2.4671 & -2.0742 & -0.0947 & 0.6056 \end{bmatrix}, \tag{107}
$$

the closed-loop system matrix takes the form

$$\boldsymbol{F}_c = \boldsymbol{F} - \boldsymbol{G}\boldsymbol{K} = \begin{bmatrix} 0.9571 & 0.0371 & 0.0431 & -0.0108 \\ 0.7613 & 0.3227 & 0.2881 & 0.1639 \\ -0.2898 & 0.2498 & 0.4771 & -0.2749 \\ -0.0073 & 0.0063 & 0.0368 & 0.9931 \end{bmatrix}, \tag{108}$$

while the closed-loop system matrix eigenvalues spectrum is

$$\rho(\boldsymbol{F}_c) = \{0.1207, \quad 0.6570, \quad 0.9733, \quad 0.9990\}. \tag{109}$$

To apply in the forced mode, the signal gain matrix $\boldsymbol{W}$ is now computed by using Eq. (84) as follows:

$$\boldsymbol{W} = \begin{bmatrix} -0.8296 & 0.9567 \\ -2.2360 & 2.4922 \end{bmatrix}. \tag{110}$$

The simulation results of the nominal closed-loop system response are illustrated in **Figures** where **Figure** is concerned with the system state variables response in the autonomous regime and **Figure** with the system state variables response in the forced mode.

Since these two control structures are of interest in the context of full-state control design, matching the presented results, it is evident that the system dynamics in both cases are comparable.

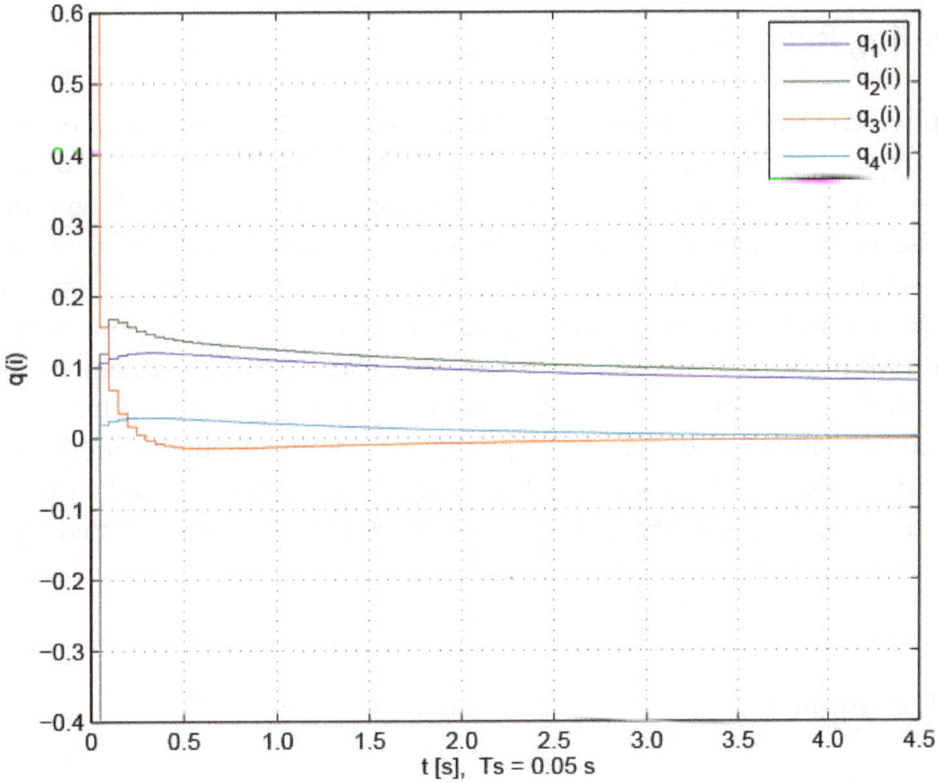

Figure. State response in autonomous regime.

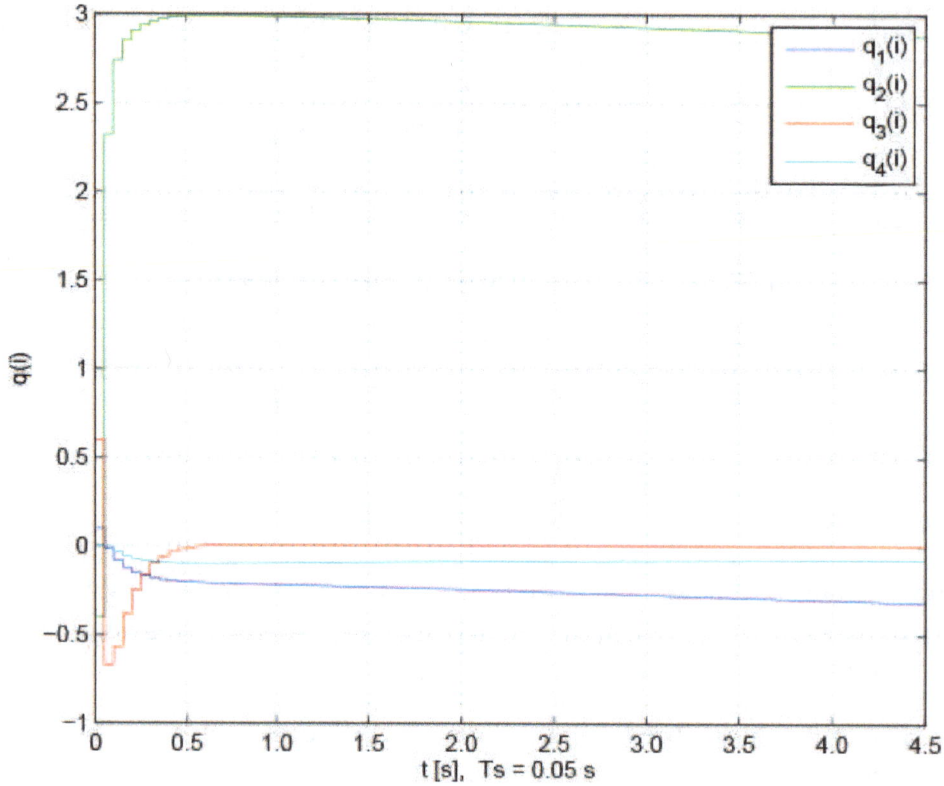

Figure. State response in forced mode.

8. Concluding Remarks

In this chapter, an extended method is presented, based on the classical memoryless feedback H_∞ control principle of discrete-time systems, if the ratio control is reformulated by an equality constraint setting on associated state variables. The asymptotic stability of the control scheme is guaranteed in the sense of the enhanced representation of BRL, while resulting LMIs are linear with respect to the system state variables, and does not involve products of the Lyapunov matrix and the system matrix parameters, which provides one way of solving the singular LMI problem. Moreover, formulated as a stabilization problem with the full-state feedback controller, the control gain matrix takes no special structure. The formulation allows to find a solution without restrictive assumptions and additional specifications on the design parameters. It is clear from Theorem 4 that the control law strictly solves the problem even in the unforced mode. The validity of the proposed method is demonstrated by numerical examples.

Acknowledgements

The work presented in this chapter was supported by VEGA, the Grant Agency of the Ministry of Education and the Academy of Science of Slovak Republic, under Grant No. 1/0608/17. These supports are very gratefully acknowledged.

Author details

Dušan Krokavec* and Anna Filasová

*Address all correspondence to: dusan.krokavec@tuke.sk

Department of Cybernetics Artificial Intelligence, Faculty of Electrical Engineering Informatics, Technical University of Košice, Košice, Slovakia

References

[1] Benzaouia A., Gurgat C. Regulator problem for linear discrete-time systems with nonsymmetrical constrained control. International Journal of Control. 1988. **48**(6):2441–2451. DOI: 10.1109/CDC.1991.261705.

[2] Castelan E.B., Hennet J.C. Eigenstructure assignment for state constrained linear continuous time systems. Automatica. 1992. **28**(3):605–611. DOI: 10.1016/0005-1098(92) 90185-I.

[3] Hahn H. Linear systems controlled by stabilized constraint relations. In: Proceedings of the 31st IEEE Conference on Decision and Control; 16–18 December 1992; Tucson, USA. pp. 840–848.

[4] Tarbouriech S., Castelan E.B. An eigenstructure assignment approach for constrained linear continuous-time singular systems. Systems & Control Letters. 1995. **24**(5):333–343. DOI: 10.1016/0167-6911(94)00046-X.

[5] Kaczorek T. Externally and internally positive singular discrete-time linear systems. International Journal of Applied Mathematics and Computer Science. 2002. **12**(2):197–202.

[6] Filasová A., and Krokavec D. Enhanced approach to PD control design for linear time-invariant descriptor systems. Journal of Physics: Conference Series. 2017. **783**. 12p. (13th European Workshop on Advanced Control and Diagnosis (ACD 2016)). DOI:10.1088/1742-6596/783/1/012037

[7] Yu T.J., Lin C.F., Müller P.C. Design of LQ regulator for linear systems with algebraic-equation constraints. In: Proceedings of the 35th IEEE Conference on Decision and Control; 13 December 1996; Kobe, Japan. pp. 4146–4151.

[8] Oloomi H., Shafai B. Constrained stabilization problem and transient mismatch phenomenon in singularity perturbed systems. International Journal of Control. 1997. **67**(2):435–454. DOI: 10.1080/002071797224199.

[9] Petersen I.R. Minimax LQG control. International Journal of Applied Mathematics and Computer Science. 2006. **16**(3):309–323. http://eudml.org/doc/207795.

[10] Xue Y., Wei Y., Duan G. Eigenstructure assignment for linear systems with constrained input via state feedback. A parametric approach. In: Proceedings of the 25th Chinese Control Conference; 7–11 August 2006; Harbin, China. pp. 108–113.

[11] Ko S., Bitmead R.R. State estimation for linear systems with state equality constraints. Automatica. 2007. **43**(9):1363–1368. DOI: 10.1016/j.automatica.2007.01.017.

[12] Ko S., Bitmead R.R. Optimal control for linear systems with state equality constraints. Automatica. 2007. **43**(9):1573–1582. DOI: 10.1016/j.automatica.2007.01.024.

[13] Filasová A., Krokavec D. Observer state feedback control of discrete-time systems with state equality constraints. Archives of Control Sciences. 2010. **10**(3):253–266. DOI: 10.2478/v10170-010-0016-5.

[14] Nesterov Y., Nemirovsky A. Interior Point Polynomial Methods in Convex Programming. Theory and Applications. Philadelphia: SIAM; 1994. 407 p. DOI: 10.1137/1.9781611970791.fm

[15] Boyd D., El Ghaoui L., Peron E., Balakrishnan V. Linear Matrix Inequalities in System and Control Theory. Philadelphia: SIAM; 1994. 205 p. DOI: 10. 1137/1.9781611970777.

[16] Skelton R.E., Iwasaki T., Grigoriadis K. A Unified Algebraic Approach to Linear Control Design. London: Taylor & Francis; 1998. 285 p. DOI: 10.1002/rnc.694.

[17] Herrmann G., Turner M.C., Postlethwaite I. Linear matrix inequalities in control. In: Turner M.C., Bates D.G., editors. Mathematical Methods for Robust and Nonlinear Control. Berlin: Springer-Verlag; 2007. pp. 123–142. DOI: 10.1007/978-1-84800-025-4-4.

[18] Gahinet P., Nemirovski A., Laub A.J., Chilali M. LMI Control Toolbox User's Guide. Natick: The MathWorks; 1995. 356 p.

[19] Peaucelle D., Henrion D., Labit Y., Taitz K. User's Guide for SeDuMi Interface 1.04. Toulouse: LAAS-CNRS; 2002. 36 p.

[20] Oliveira de M.C., Bernussou J., Geromel J.C. A new discrete-time robust stability condition. Systems & Control Letters. 1999. **37**(4):261–265. DOI: 10.1016/S0167-6911(99)00035-3.

[21] Wu A.I., Duan G.R. Enhanced LMI representations for H_2 performance of polytopic uncertain systems. Continuous-time case. International Journal of Automation and Computing. 2006. **3**(3):304–308. http://www.ijac.net/EN/Y2006/V3/I3/304.

[22] Filasová A., Krokavec D. $H\infty$ control of discrete-time linear systems constrained in state by equality constraints. International Journal of Applied Mathematics and Computer Science. 2012. **22**(3):551–560. DOI: 10.2478/v10006-012-0042-5.

[23] Krokavec D., Filasová A. Constrained control of discrete-time stochastic systems. IFAC Proceedings Volumes. 2008. **41**(2):15315–15320. DOI: 10.3182/20080706-5-KR-1001.02590.

[24] Krokavec D., Filasová A. Control reconfiguration based on the constrained LQ control algorithms. IFAC Proceedings Volumes. 2009. **42**(8):143–148. DOI: 10.3182/20090630 -4-ES-2003.00024.

[25] Ogata K. Discrete-Time Control Systems. Upper Saddle River: Prentice-Hall; 1995. 760 p.

[26] Debiane L., Ivorra B., Mohammadi B., Nicoud F., Ernz A., Poinsot T., Pitsch H. Temperature and pollution control in flames. In: Proceedings of the Summer Program 2004; 2004; University of Montpellier, France, pp. 1–9.

[27] Cakmakci M., Ulsoy A.G. Modular discrete optimal MIMO controller for a VCT engine. In: Proceedings of the 2009 American Control Conference; 10–12 June 2009; St. Louis, USA, pp. 1359–1364.

[28] Krokavec D., Filasová A. Performance of reconfiguration structures based on the constrained control. IFAC Proceedings Volumes. 2008. **41**(2):1243–1248.

[29] Heij C., Ran A., van Schagen F. Introduction to Mathematical Systems Theory. Linear Systems, Identification and Control. Basel: Birkhäuser Verlag; 2007. 168 p. DOI: 10.1007/978-3-7643-7549-2.

[30] Gajic Z., Qureshi M.T.J. Lyapunov Matrix Equation in System Stability and Control. San Diego: Academic Press; 1995. 271 p. DOI: 10.1137/1038139.

[31] Mason O., Shorten R. On common quadratic Lyapunov functions for stable discrete-time LTI systems. IMA Journal of Applied Mathematics. 2004. **69**(3):271–283. DOI: 10.1093/imamat/69.3.271.

[32] Wang Q.G. Decoupling Control. Berlin: Springer-Verlag; 2003. 369 p. DOI: 10.1007/3-540-46151-5.

[33] Wen C.C., Cheng C.C. Design of sliding surface for mismatched uncertain systems to achieve asymptotical stability. Journal of the Franklin Institute. 2008. **345**(8):926–941. DOI: 10.1016/j.jfranklin.2008.06.003.

8

Solution of Differential Equations with Applications to Engineering Problems

Cheng Yung Ming

Abstract

Over the last hundred years, many techniques have been developed for the solution of ordinary differential equations and partial differential equations. While quite a major portion of the techniques is only useful for academic purposes, there are some which are important in the solution of real problems arising from science and engineering. In this chapter, only very limited techniques for solving ordinary differential and partial differential equations are discussed, as it is impossible to cover all the available techniques even in a book form. The readers are then suggested to pursue further studies on this issue if necessary. After that, the readers are introduced to two major numerical methods commonly used by the engineers for the solution of real engineering problems.

Keywords: differential equations, analytical solution, numerical solution

1. Introduction

1.1. Classification of ordinary and partial equations

To begin with, a differential equation can be classified as an ordinary or partial differential equation which depends on whether only ordinary derivatives are involved or partial derivatives are involved. The differential equation can also be classified as linear or nonlinear. A differential equation is termed as linear if it exclusively involves linear terms (that is, terms to the power 1) of y, y', y'' or higher order, and all the coefficients depend on only one variable x as shown in Eq. (1). In Eq. (1), if $f(x)$ is 0, then we term this equation as homogeneous. The general solution of non-homogeneous ordinary differential equation (ODE) or partial differential equation (PDE) equals to the sum of the *fundamental solution* of the corresponding homogenous equation (i.e. with $f(x) = 0$) plus the *particular solution* of the non-homogeneous ODE or PDE. On the other hand, nonlinear differential equations involve nonlinear terms in any of y, y', y'', or higher order term. A nonlinear differential equation is generally more difficult to solve than linear equations. It is common that nonlinear equation is approximated as linear equation (over acceptable solution domain) for many practical problems, either in an analytical or numerical form. The nonlinear nature of the problem is then approximated as series of linear differential equation by simple increment or with correction/deviation from the nonlinear behaviour. This approach is adopted for the solution of many non-linear engineering problems. Without such procedure, most of the non-linear differential equations cannot be solved. Differential equation can further be classified by the order of differential. In general, higher-order differential equations are difficult to solve, and analytical solutions are not available for many higher differential equations. A linear differential equation is generally governed by an equation form as Eq. (1).

$$\frac{d^n y}{dx^n} + a_1(x)\frac{d^{n-1}y}{dx^{n-1}} + \ldots + a_n(x)y = f(x) \tag{1}$$

"Non-linear" differential equation can generally be further classified as

1. *Truly nonlinear* in the sense that F is nonlinear in the derivative terms.

$$F\left(x_1, x_1, x_n, u, \frac{\partial u}{\partial x_1}, \frac{\partial u}{\partial x_2}, \frac{\partial^2 u}{\partial x_1 \partial x_2}\right) = 0 \tag{2}$$

2. *Quasi-linear 1st* PDE if nonlinearity in F only involves u but not its derivatives

$$A_1(x_1, x_2, u)\frac{\partial u}{\partial x_1} + A_2(x_1, x_2, u)\frac{\partial u}{\partial x_2} = B(x_1, x_2, u) \tag{3}$$

3. *Quasi-linear 2nd* PDE if nonlinearity in F only involves u and its first derivative but not its second-order derivatives

$$A_{11}\left(x_1, x_2, u, \frac{\partial u}{\partial x_1}, \frac{\partial u}{\partial x_2}\right)\frac{\partial^2 u}{\partial x_1^2} + A_{12}\left(x_1, x_2, u, \frac{\partial u}{\partial x_1}, \frac{\partial u}{\partial x_2}\right)\frac{\partial^2 u}{\partial x_1 \partial x_2} + A_{22}\left(x_1, x_2, u, \frac{\partial u}{\partial x_1}, \frac{\partial u}{\partial x_2}\right)\frac{\partial^2 u}{\partial x_2^2}$$
$$= F\left(x_1, x_2, u\frac{\partial u}{\partial x_1}, \frac{\partial u}{\partial x_2}\right) \tag{4}$$

Examples of differential equations:

1. $\frac{dy}{dx} = 3x + 2$; first-order ODE (linear)/nonhomogeneous
2. $(y - 2x)dy - 3ydx = 0$; first-order ODE (nonlinear)/homogenous
3. $\frac{d^2y}{dt^2} + t^2 y\left(\frac{dy}{dt}\right)^3 + y = 0$; second-order ODE (nonlinear)/homogenous
4. $\frac{d^4x}{dt^4} + 5\frac{d^2x}{dt^2} + 7x = sint$; fourth-order ODE (linear)/nonhomogeneous
5. $\frac{\partial z}{\partial x} + \frac{\partial z}{\partial y} = 2z$; first-order PDE (linear)/homogeneous

6. $\frac{\partial^2 u}{\partial x^2}+\frac{\partial^2 u}{\partial y^2}+4x+3y-uz=0$; second-order PDE (linear)/nonhomogeneous

7. $x\frac{\partial^2 u}{\partial x^2}+2u\frac{\partial^2 u}{\partial y^2}+3u^2=0$; second-order PDE (linear)/homogeneous

8. $\frac{du}{dx}-\frac{dv}{dx}=6x$; 1st ODE (linear) for two unknowns/nonhomogeneous

1.2. Typical differential equations in engineering problems

Many engineering problems are governed by different types of partial differential equations, and some of the more important types are given below.

Tricomi equation: $y\frac{\partial^2 u}{\partial x^2}+\frac{\partial^2 u}{\partial y^2}=0\begin{cases} y>0: \textit{elliptic} \\ y<0: \textit{hyperbolic}\end{cases}$

Laplace equation (or variants): $\frac{\partial^2\varphi}{\partial x^2}+\frac{\partial^2\varphi}{\partial y^2}=\nabla^2\varphi=0$

Poisson's equation: $\frac{\partial^2\varphi}{\partial x^2}+\frac{\partial^2\varphi}{\partial y^2}=f(x,y)$

Helmholtz equation: $\frac{\partial^2\varphi}{\partial x^2}+\frac{\partial^2\varphi}{\partial y^2}+c^2\varphi=0$

Plate bending: $\nabla^2\nabla^2 w=\nabla^4 w=\frac{q}{D}$

Wave equation (1D-3D): $\frac{\partial^2 u}{\partial t^2}-c^2\left(\frac{\partial^2 u}{\partial x^2}+\frac{\partial^2 u}{\partial y^2}\right)=0$

Fourier equation: $\frac{\partial T}{\partial t}=\alpha\left(\frac{\partial^2 T}{\partial x^2}\right)$

There are many methods of solutions for different types of differential equations, but most of these methods are not commonly used for practical problems. In this chapter, the most important and basic methods for solving ordinary and partial differential equations will be discussed, which will then be followed by numerical methods such as finite difference and finite element methods (FEMs). For other numerical methods such as boundary element method, they are less commonly adopted by the engineers; hence, these methods will not be discussed in this chapter.

1.3. Separable differential equations

For equations which can be expressed in separable form as shown below, the solution can be obtained easily as

$$\frac{dy}{dx}=F(x,y)\frac{dy}{\Phi(y)}=f(x)dx\int\frac{dy}{\Phi(y)}=\int f(x)dx+c \tag{5}$$

$$M(x,y)dx+N(x,y)dy=0\; M(x)dx=-N(y)dy \tag{6}$$

$$\text{then}\int M(x)dx=-\int N(y)dy+c \tag{7}$$

Example:

$$\frac{dy}{dx} = x^3(y^2+1) \Rightarrow \frac{dy}{y^2+1} = x^3 dx$$

$$\int \frac{dy}{y^2+1} = \int x^3 dx + c \Rightarrow tan^{-1}y = \frac{1}{4}x^4 + C \Rightarrow y = tan\left(\frac{1}{4}x^4 + c\right)$$

Example:

$\frac{dy}{dx} = \frac{3x^2+4x+2}{2(y-1)}$ subject to $y(0) = -1$

Since this is a separable function, the problem can be solved as

$$2(y-1)dy = (3x^2+4x+2)dx$$
$$y^2 - 2y = x^3 + 2x^2 + 2x + c$$

Based on the boundary condition, $c = 3$, hence $y^2 - 2y = x^3 + 2x^2 + 2x + 3$.

This quadratic equation in y^2 can be solved with two solutions by the quadratic equation as

$$y = 1 - \sqrt{x^3 + 2x^2 + 2x + 4} \text{ and } y = 1 + \sqrt{x^3 + 2x^2 + 2x + 4}.$$

Since the second solution does not satisfy the boundary condition, it will not be accepted; hence, the solution to this differential equation is obtained.

1.4. Variation of parameters

For the following equation form, it is possible to solve it by variations of parameters.

$$\text{For } \frac{dy}{dx} = p(x)y + Q(x) \quad (8)$$

Put $y = c(x)e^{\int p(x)dx}$. By differentiating, it gives $\frac{dy}{dx} = \frac{dc(x)}{dx}e^{\int p(x)dx} + \underbrace{c(x)p(x)e^{\int} p(x)dx}_{p(x)y}$. Substitute it to the original ODE $\frac{dc(x)}{dx} = Q(x)e^{-\int p(x)dx}$. Comparing the terms, it gives

$$c(x) = \int Q(x)e^{-\int p(x)dx}dx + \bar{c}. \quad (9)$$

Example:

$$(x+1)\frac{dy}{dx} - ny = e^x(x+1)^{n+1}$$

This equation is now expressed as

$$\frac{dy}{dx} = p(x)y + Q(x)$$

$$\frac{dy}{dx} = \frac{n}{x+1}y + \underbrace{e^x(x+1)^n}_{Q(x)}$$

For $x \neq -1$

Solving the homogeneous part of the ODE

$\frac{dy}{dx} = \frac{n}{x+1}y$ then $\frac{dy}{y} = \frac{n}{x+1}dx$

$$ln|y| = nln|x+1| + c_1$$
$$y = c(x+1)^n$$

Look for solution $y = c(x)(x+1)^n$, where $c(x)$ is the variation of parameters. Substitute it to the ODE

$$\frac{dc(x)}{dx}(x+1)^n + nc(x)(x+1)^{n-1} = nc(x)(x+1)^{n-1} + e^x(x+1)^n$$

$$\frac{dy}{dx} = \frac{n}{x+1}y + e^x(x+1)^n$$

Comparison gives $\frac{dc(x)}{dx} = e^x$

Integration of this equation gives $c(x) = e^x + \overline{C}$

General solution is hence given by $y = (x+1)^n(e^x + \overline{C})$

The Bernoulli equation is an important equation type which can be solved in a similar way by variation of parameters. Consider the following form of equation

$$\frac{dy}{dx} = p(x)y + Q(x)y^n \tag{10}$$

$$\textbf{Step } 1: \text{ Put } z = y^{1-n} \tag{11}$$

$$\begin{aligned} \textbf{Step } 2: \text{ Then } \frac{dz}{dx} &= (1-n)y^{-n}\frac{dx}{dy} \\ \frac{dz}{dx} &= (1-n)P(x)z + (1-n)Q(x) \end{aligned} \tag{12}$$

The non-linear ODE now becomes linear ODE. It can be solved by formula

Step 3: $n = -1$, $z = y^2$. Inverting z to get y

$$\frac{dy}{dx} = \frac{y}{2x} + \frac{x^2}{2y} \tag{13}$$

$$\frac{dz}{dx} = \frac{1}{x}z + x^2 \tag{14}$$

$$z = e^{\int \frac{1}{x}dx}\left(\int x^2 e^{-\int \frac{1}{x}dx}dx + c\right) = cx + \frac{1}{2}x^3 \tag{15}$$

Back substitution of $z = y^2$

$$y^2 = cx + \frac{1}{2}x^3 \tag{16}$$

1.5. Homogeneous equations

For equation of the following type, where all the coefficients are constant, it can be evaluated according to different conditions.

$$\frac{dy}{dx} = \frac{a_1x + b_1y + c_1}{a_2x + b_2y + c_2} \tag{17}$$

Case 1: $c_1 = c_2 = 0$

$$\frac{dy}{dx} = \frac{a_1x + b_1y}{a_2x + b_2y} = \frac{a_1 + b_1\frac{y}{x}}{a_2 + b_2\frac{y}{x}} = g\left(\frac{y}{x}\right) \tag{18}$$

Step 1: Set $u = \frac{y}{x}$, then $\frac{dy}{dx} = x\frac{du}{dx} + u$

Step 2: $\frac{du}{dx} = \frac{g(u)-u}{x}$. The resulting non-linear ODE is hence separable and can be solved implicitly.

Step 3: Inverting u to get y.

Case 2: $\begin{vmatrix} a_1 & a_2 \\ b_1 & b_2 \end{vmatrix} = 0$

$a_1b_2 - a_2b_1 = 0$ then $\frac{a_1}{a_2} = \frac{b_1}{b_2} = k$

$$\frac{dy}{dx} = \frac{a_1x + b_1y + c_1}{a_2x + b_2y + c_2} = \frac{k(a_2x + b_2y) + c_1}{a_2x + b_2y + c_2} = f(a_2x + b_2y) \tag{19}$$

By change of variables as $u = a_2x + b_2y$

$\frac{du}{dx} = a_2 + b_2\frac{dy}{dx} = a_2 + b_2f(u)$, then

$$\frac{du}{a_2 + b_2f(u)} = dx \tag{20}$$

The resulting non-linear ODE is now separable and can be solved.

Case 3: $\begin{vmatrix} a_1 & a_2 \\ b_1 & b_2 \end{vmatrix} \neq 0$ $c_1 \neq 0$ and $c_2 \neq 0$

Set $\begin{cases} a_1x + b_1y + c_1 = 0 \\ a_2x + b_2y + c_2 = 0 \end{cases}$. Intersecting point of these two lines on xy - plane and (α, β) ≠ 0

$$xy - \text{plane and } (\alpha, \beta) \neq (0, 0) \tag{21}$$

Apply change of variables

$$\begin{cases} X = x - \alpha \\ Y = y - \beta \end{cases} \begin{cases} x = X + \alpha \\ y = Y + \beta \end{cases} \tag{22}$$

$$\begin{aligned} a_1x + b_1y + c_1 &= a_1(X + \alpha) + b_1(Y + \beta) + c_1 = a_1X + b_1Y + (a_1\alpha + b_1\beta + c_1) \\ a_2x + b_2y + c_2 &= a_2(X + \alpha) + b_2(Y + \beta) + c_2 = a_2X + b_2Y + (a_2\alpha + b_2\beta + c_2) \end{aligned} \tag{23}$$

The original ODE will now become $\frac{dY}{dX} = \frac{a_1X+b_1Y}{a_2X+b_2Y}$ which is homogeneous and separable!

Example: $\frac{dy}{dx} = \frac{x+y-1}{x-y+3}$

Solve for $\begin{cases} x + y - 1 = 0 \\ x - y + 3 = 0 \end{cases}$ we have $\alpha = -1, \beta = 2$

Change of variables $X = x + 1$, $Y = y - 2$

Then, $\frac{dY}{dX} = \frac{dy}{dx} = \frac{x+y-1}{x-y+3} = \frac{X+Y}{X-Y} = \frac{1+\frac{Y}{X}}{1-\frac{Y}{X}}$

Use a change of variable $u = \frac{Y}{X}$ $X\frac{du}{dX} = \frac{1+u^2}{1-u}$ $\frac{(1-u)du}{1+u^2} = \frac{dX}{X}$

$$\Rightarrow tan^{-1}u - \frac{1}{2}ln(1 + u^2) = ln|X| + c$$

$$\Rightarrow tan^{-1}u = ln[\sqrt{1 + u^2}X] + c = ln[\sqrt{(X^2 + Y^2)}] + c$$

$$\Rightarrow tan^{-1}\left(\frac{y-2}{x+1}\right) = ln\sqrt{(x+1)^2 + (y-2)^2} + c$$

There are various tricks to solve the differential equations, like integration factors and other techniques. A very good coverage has been given by Polyanin and Nazaikinskii [29] and will not be repeated here. The purpose of this section is just for illustration that various tricks have been developed for the solution of simple differential equations in homogeneous medium, that is, the coefficients are constants inside a continuous solution domain. The readers are also suggested to read the works of Greenberg [14], Soare et al. [34], Nagle et al. [28], Polyanin et al. [30], Bronson and Costa [4], Holzner [18], and many other published books. There are many elegant tricks that have been developed for the solution of different forms of differential equations, but only very few techniques are actually used for the solution of real life problems.

1.6. Partial differential equations

In many engineering or science problems, such as heat transfer, elasticity, quantum mechanics, water flow and others, the problems are governed by partial differential equations. By nature, this type of problem is much more complicated than the previous ordinary differential equations. There are several major methods for the solution of PDE, including separation of variables, method of characteristic, integral transform, superposition principle, change of variables, Lie group method, semianalytical methods as well as various numerical methods. Although the existence and uniqueness of solutions for ordinary differential equation is well established with the Picard-Lindelöf theorem, but that is not the case for many partial differential equations. In fact, analytical solutions are not available for many partial differential equations, which is a well-known fact, particularly when the solution domain is nonregular or homogeneous, or the material properties change with the solution steps.

1.6.1. Classification of second-order PDE

Refer to the following general second-order partial differential equation:

$$A\frac{\partial^2 u}{\partial x^2} + B\frac{\partial^2 u}{\partial x \partial y} + C\frac{\partial^2 u}{\partial y^2} + D\frac{\partial u}{\partial x} + E\frac{\partial u}{\partial y} + Fu + G = 0 \tag{24}$$

To begin with, let us consider a review of conic curves (ellipse, parabola and hyperbola)

$$Ax^2 + Bxy + Cy^2 + Dx + Ey + F = 0 \tag{25}$$

The conic curve can be classified with the following criterion.

$$B^2 - 4AC = \begin{cases} > 0 \text{ hyperbola} \\ = 0 \text{ parabola} \\ < 0 \text{ ellipse} \end{cases} \tag{26}$$

Following the conic curves, the general partial differential is also classified according to similar criterion as

$$\text{Classification}\begin{cases} B^2 - 4AC > 0 : \text{elliptic} \\ B^2 - 4AC = 0 : \text{parabolic} \\ B^2 - 4AC < 0 : \text{hyperbolic} \end{cases} \tag{27}$$

This classification was proposed by Du Bois-Reymond [41] in 1839. In this section, only some of the more common techniques are discussed, and the readers are suggested to read the works of Hillen et al. [16], Salsa [33], Polyanin and Zaitsev [31], Bertanz [2], Haberman [15] and many other published texts.

1.7. Parabolic type: heat conduction/soil consolidation/diffuse equation

The following equation form is commonly found in many engineering applications.

$$\alpha^2 \frac{\partial^2 u}{\partial x^2} = \frac{\partial u}{\partial t}, 0 < x < L, t > o \tag{28}$$

Initial condition: $u(x,0) = f(x), 0 \leq x \leq L$

Boundary condition: $u(0,t) = 0, u(L,t) = 0, t > 0$

α^2 is a constant known as the thermal diffusivity or coefficient of consolidation. For soil consolidation problem, the governing conditions are given by

Initial excess pore pressure

$$\begin{aligned} u_e(z,0) &= u_i(z), 0 \leq z \leq 2d \\ u_e(0,t) &= 0, u_e(2d,t) = 0, t > 0 \end{aligned} \tag{29}$$

Drained boundary

$$\begin{aligned} &\alpha^2 u_{xx} = u_t, 0 < x < L, t > 0 \\ &u(0,t) = 0, u(L,t) = 0, t > 0 \\ &u(x,0) = f(x), 0 \leq x \leq L \end{aligned} \tag{30}$$

Assuming variable $u(x, t)$ can be separated, using separation of variables

$$u(x,t) = X(t)T(t) \tag{31}$$

$$\begin{aligned} &\alpha^2 X'' T = XT' \\ &\frac{X''}{X} = \frac{1}{\alpha^2}\frac{T'}{T} \\ &\frac{X''}{X} = \frac{1}{\alpha^2}\frac{T'}{T} = -\lambda \rightarrow \begin{cases} X'' + \lambda X = 0 \\ T' + \alpha^2 \lambda T = 0 \end{cases} \end{aligned} \tag{32}$$

A PDE now becomes two ODE which can be solved readily. Based on the boundary condition $u(0,t) = 0, u(L,t) = 0, t > 0$

$$\begin{aligned} u(0,t) = X(0), T(t) &= 0 \\ X(0) = 0, X(L) &= 0 \\ X'' + \lambda X = 0, X(0) = 0, X(L) &= 0 \end{aligned} \tag{33}$$

This is an eigenvalue problem which has solution only for certain λ. The eigenvalues are given by

$$\lambda_n = \frac{n^2 \pi^2}{L^2}, n = 1, 2, 3, \ldots \tag{34}$$

Hence the eigenfunctions are expressed as

$$X_n(x) = sin\left(\frac{n\pi x}{L}\right), n = 1,2,3... \tag{35}$$

For the time-dependent function T,

$$T' + \alpha^2 \lambda T = 0 \tag{36}$$

$$\frac{dT}{T} = -\alpha^2 \lambda dt$$
$$ln|T| = \frac{-\alpha^2 n^2 \pi^2 t}{L^2} + C \tag{37}$$

hence $T_n = k_n e^{-(n\pi\alpha/L)^2 t}$, k_n constant. The fundamental solutions are then expressed as

$$u(x,t) = e^{-(n\pi\alpha/L)^2 t} sin\left(\frac{n\pi x}{L}\right), n = 1,2,3... \tag{38}$$

The Fourier series expansion in x is given by

$$u(0,t) = f(x), 0 \leq x \leq L \tag{39}$$

$$u(x,t) = \sum_{n=1}^{\infty} c_n u_n(x,t) = \sum_{n=1}^{\infty} c_n e^{-(n\pi\alpha/L)^2 t} sin\left(\frac{n\pi x}{L}\right) \tag{40}$$

Initial condition is given as

$$u(x,0) = f(x) = \sum_{n=1}^{\infty} c_n sin\left(\frac{n\pi x}{L}\right) \tag{41}$$

$$\int_0^L f(x) sin\left(\frac{m\pi x}{L}\right) dx = \sum_{n=1}^{\infty} c_n \int_0^L sin\left(\frac{m\pi x}{L}\right) sin\left(\frac{n\pi x}{L}\right) dx$$
$$\int_0^L f(x) sin\left(\frac{n\pi x}{L}\right) dx = c_n \int_0^L sin^2\left(\frac{m\pi x}{L}\right) dx = c_n \frac{L}{2}$$

Solution of the soil consolidation equation is hence given by

$$u(x,t) = \sum_{n=1}^{\infty} c_n e^{-(n\pi\alpha/L)^2 t} sin\left(\frac{n\pi x}{L}\right) \tag{42}$$

$$c_n = \frac{2}{L} \int_0^L f(x) sin\left(\frac{n\pi x}{L}\right) dx \quad \text{(EulerFourier formulas)} \tag{43}$$

1.8. One-dimensional wave equation

One-dimensional (1D) wave equation appears in many physical and engineering problems. For example, a vibrating string or pile driving process is given by this type of differential equation. This problem is also commonly solved by the method of separation of variables

$$\begin{aligned} &a^2 u_{xx} = u_{tt}, 0 < x < L, t > 0 \\ &u(0,t) = 0, u(L,t) = 0, t \geq 0 \\ &u(x,0) = f(x), u(x,0) = 0, 0 \leq x \leq L \end{aligned} \tag{44}$$

Consider $u(x, t)$ is given by $X(x)T(t)$. The wave equation will give

$$\frac{X''}{X} = \frac{1}{a^2}\frac{T'}{T} = -\lambda \rightarrow \begin{cases} X'' + \lambda x = 0 \\ T' + a^2 \lambda t = 0 \end{cases} \tag{45}$$

The partial differential equation will then be given by two equivalent ODEs.

$$\begin{aligned} &u_t(x,0) = X(x)T'(0) = 0, 0 \leq x \leq L \rightarrow T'(0) = 0 \\ &u(0,t) = X(0)T(t) = 0, u(L,t) = X(L)T(t)\ 0 \leq x \leq L \rightarrow T'(0) = 0 \end{aligned} \tag{46}$$

$$X'' + \lambda X = 0, X(0) = X(L) = 0 \tag{47}$$

$$X_n(x) = sin\left(\frac{n\pi x}{L}\right), n = 1, 2, 3, \ldots \tag{48}$$

$$\lambda_n = \frac{n^2 \pi^2}{L^2}, n = 1, 2, 3, \ldots \tag{49}$$

For the time-dependent function T,

$$T' + a^2 \lambda T = 0 \tag{50}$$

$$\begin{gathered} T'(0) = 0\ \lambda_n = n\pi/L \\ \text{Then } T(t) = k_1 cos(n\pi at/L) - k_2 sin(n\pi at/L) \end{gathered} \tag{51}$$

Since $T'(0) = 0\ k_2 = 0$

Therefore, $T(t) = k_1 cos(n\pi at/L)$

Fundamental solution is given by

$$u_n(x,t) = sin\left(\frac{n\pi x}{L}\right) cos\left(\frac{n\pi at}{L}\right), n = 1, 2, 3\ldots \tag{52}$$

The general solution is then given by

$$u(x,t) = \sum_{n=1}^{\infty} c_n u_n(x,t) = \sum_{n=1}^{\infty} c_n sin\left(\frac{n\pi x}{L}\right) cos\left(\frac{n\pi at}{L}\right) \tag{53}$$

Applying the boundary condition

$$\begin{aligned} &u(x,0) = f(x), 0 \leq x \leq L \\ &u(x,0) = f(x) = \sum_{n=1}^{\infty} c_n sin\left(\frac{n\pi x}{L}\right) \rightarrow c_n = \frac{2}{L}\int_0^L f(x) sin\left(\frac{n\pi x}{L}\right) dx \end{aligned} \tag{54}$$

The final solution is then given by

$$u(x,t) = \sum_{n=1}^{\infty} c_n sin\left(\frac{n\pi x}{L}\right) cos\left(\frac{n\pi at}{L}\right) \tag{55}$$

$$c_n = \frac{2}{L}\int_0^L f(x) sin\left(\frac{n\pi x}{L}\right) dx \tag{56}$$

1.9. Laplace equation

Laplace equation forms an important governing condition for many types of problems. Some of the more common forms are given by

three-dimensional Laplace equation $u_{xx} + u_{yy} + u_{zz} = 0$

two-dimensional heat conduction $\alpha^2(u_{xx} + u_{yy}) = u_t$

two-dimensional seepage problem $(k_x u_{xx} + k_y u_{yy}) = 0$

There are two major types of boundary conditions to this problem:

Dirichlet problem: boundary conditions prescribed as u

Neumann problem: normal derivative u_x or u_y are usually prescribed on the boundary for many mathematical problems. This case can be solved by the use of complex analysis or series method for which many analytical solutions are available in the literature. In many anisotropic seepage problems, however, the normal of a derived quantity at any arbitrary direction (seepage flow normal to an impermeable surface) is 0 instead of u_x or u_y being zero. For such cases, it is very difficult to obtain the analytical solution if the solution domain is nonhomogeneous, and the use of numerical method such as the finite element method appears to be indispensable.

Consider the given Laplace equation, using separation of variables for the analysis.

$$\begin{aligned} &u_{xx} + u_{yy} = 0, 0 < x < a, 0 < y < b \\ &u(x,0) = 0, u(x,b) = 0, 0 < x < a \\ &u(0,y) = 0, u(a,y) = f(y), 0 < y \leq b \end{aligned} \tag{57}$$

Using separation of variables, $u(x,t) = X(x)Y(y)$

$$\begin{aligned} &X''Y + XY'' = 0 \\ &\frac{X''}{X} = -\frac{Y''}{Y} = \lambda \rightarrow \begin{matrix} X'' - \lambda X = 0 \\ Y' + \lambda Y = 0 \end{matrix} \end{aligned} \tag{58}$$

$$u_{xx} + u_{yy} = 0, 0 < x < a, 0 < y < b \tag{59}$$

$$\begin{aligned} &u(x,0) = 0, u(x,b) = 0, 0 < x < a \\ &u(0,y) = 0, u(a,y) = f(y), 0 < y \leq b \end{aligned} \tag{60}$$

$$\begin{aligned} u(0,y) &= X(0)Y(y) = 0, 0 < y < b \rightarrow X(0) = 0, \\ u(x,0) &= X(x)Y(0) = 0, 0 < x < a \rightarrow Y(0) = 0, \\ u(x,b) &= X(x)Y(b) = 0, 0 < x < a \rightarrow Y(b) = 0, \end{aligned} \tag{61}$$

$$\begin{aligned} X^{''} - \lambda X = 0, X(0) = 0 \\ Y^{''} + \lambda Y = 0, Y(0) = 0, Y(b) = 0 \end{aligned} \tag{62}$$

$$\lambda_n = \frac{n^2\pi^2}{b^2}, Y_n(y) = sin\left(\frac{n\pi y}{b}\right), n = 1, 2, 3, \ldots \tag{63}$$

$X^{''} - \lambda X = 0$, hence $X(x) = k_1 cosh(n\pi x/b) - k_2 sin(n\pi x/b)$

Since $X(0) = 0$, $k_1 = 0$

$$X(x) = k_2 sinh\left(\frac{n\pi x}{b}\right) \tag{64}$$

$$u_n(x,y) = sinh\left(\frac{n\pi x}{b}\right) sin\left(\frac{n\pi y}{b}\right) n = 1, 2, 3\ldots \tag{65}$$

$$\begin{aligned} & u(a,y) = f(y), 0 \leq y \leq b \\ & u(x,y) = \sum_{n=1}^{\infty} c_n u_n(x,y) = \sum_{n=1}^{\infty} c_n sin\left(\frac{n\pi x}{b}\right) cos\left(\frac{n\pi y}{b}\right) \end{aligned} \tag{66}$$

Based on the Fourier expansion as given by

$$\begin{aligned} & \int_0^b f(y) sin\left(\frac{m\pi y}{b}\right) dy = \sum\nolimits_{n=1}^{\infty} c_n sinh\left(\frac{n\pi a}{b}\right) \int_0^b sin\left(\frac{m\pi y}{b}\right) sin\left(\frac{n\pi y}{b}\right) dy \\ & \int_0^b f(x) sin\left(\frac{n\pi x}{b}\right) dx = sinh\frac{m\pi a}{b} c_n \int_0^b sin^2\left(\frac{n\pi x}{b}\right) dx = sinh\frac{m\pi a}{b} c_n \frac{b}{2} \\ & u(a,y) = f(y) = \sum\nolimits_{n=1}^{\infty} c_n sinh\left(\frac{n\pi a}{b}\right) sin\left(\frac{n\pi y}{b}\right) \end{aligned} \tag{67}$$

$$c_n sinh\left(\frac{n\pi a}{b}\right) = \frac{2}{b}\int_0^b f(y) sin\left(\frac{n\pi y}{b}\right) dy$$

$$c_n = \frac{2}{b} sinh\left(\frac{n\pi a}{b}\right)^{-1} \int_0^b f(y) sin\left(\frac{n\pi y}{b}\right) dy$$

1.10. Introduction to numerical methods

In general, analytical solutions are not available for most of the practical differential equations, as regular solution domain and homogeneous conditions may not be present for practical problems. Moreover, the solution domain may be indeterminate (free surface seepage flow), the displacement is large so that the solution may deform under motion, or in an extreme case part of the material may tear off from the main body with continuous formation and removal of contacts. Many engineering problems fall into such category by nature, and the use of numerical methods will be necessary. Currently, there are several major numerical methods commonly used by the engineers: finite difference method, finite element method, boundary element method and distinct element. There are also other less common numerical methods available for practical problems, and many researchers also try to combine two or even more fundamental numerical methods so as to achieve greater efficiency in the analysis. In general, the solution domain is discretized into series of subdomains with many degrees of freedom. The number of variables or degrees of freedom may even exceed millions for large-scale problems, and sometimes very special material properties are encountered so that the system is highly sensitive to the method of discretization and the method of solution. Similar to the ODE and PDE, it is impossible to discuss the details of all the numerical methods and the author choose to discuss the finite element method due to the wide acceptance of the method and this method is more suitable for general complicated methods.

Except for some simple problems with regular geometry and loading, it is very difficult to solve most of the boundary value problems with the yield of analytical solutions. Towards this, the use of numerical method seems indispensable, and the finite element is one of the most popular methods used by the engineers [32, 38]. There are two fundamental approaches to FEM, which are the weighted residual method (WRM) and variational principle, but there are also other less popular principles which may be more effective under certain special cases. In finite element analysis of an elastic problem, solution is obtained from the weak form of the equivalent integration for the differential equations by WRM as an approximation. Alternatively, different approximate approaches (e.g. collocation method, least square method and Galerkin method) for solving differential equations can be obtained by choosing different weights based on the WRM and the Galerkin method appears to be the most popular approach in general.

Specifically, in elasticity for instance, the principle of virtual work (including both principle of virtual displacement and virtual stress) is considered to be the weak form of the equivalent integration for the governing equilibrium equations. Furthermore, the aforementioned weak form of equivalent integration on the basis of the Galerkin method can also be evolved to a variation of a functional if the differential equations have some specific properties such as linearity and selfadjointness. Principles of minimum potential energy and complementary energy are two variational approaches equivalent to the fundamental equations of elasticity.

Since displacement is usually the basic unknown quantity in FEM, only the principle of virtual displacement and minimum potential energy will be introduced in the following section. In this case, the FEM introduced herein is also called displacement finite element method (DFEM). There are other ways to form the basis of FEM with advantages in some cases, but these approaches are less general and will not be discussed here.

1.11. Principle of virtual displacement

The principle of virtual displacement is the weak form of the equivalent integration for equilibrium equations and force boundary conditions. Given the equilibrium equations and force boundary conditions in index notation,

$$\sigma_{ij,j} + f_i = 0, \text{(in domain } V) \tag{68}$$

$$\sigma_{ij} n_j - T_i = 0, \text{(on domain boundary } S_\sigma) \tag{69}$$

In WRM, without loss of generality, the variation of true displacement δu_i and its boundary value (i.e. $-\delta u_i$) can be selected as the weight functions in the equivalent integration

$$\int_V \delta u_i(\sigma_{ij,j} + f_i)dV - \int_{S_\sigma} \delta u_i(\sigma_{ij}n_j - T_i)dS = 0 \tag{70}$$

The weak form of Eq. (70) is given as

$$\int_V (-\delta\varepsilon_{ij}\sigma_{ij} + \delta u_i f_i)dV + \int_{S_\sigma} \delta u_i T_i dS = 0 \tag{71}$$

It can be seen clearly from Eq. (71) that the first item in the volume integral indicates the work done by the stresses under the virtual strain (i.e. internal virtual work), while the remaining items indicate the work done by the body force and surface force under the virtual displacement (i.e. external virtual work). In other words, the summation of the internal and external virtual works is equal to 0, which is called the principle of virtual displacement. Under this case, we can conclude that a force system will satisfy the equilibrium equations if the summation of the work done by it under any virtual displacement and strain is equal to 0.

1.12. Principle of minimum potential energy (PMPE)

Based on Eq. (71), we can deduce that

$$\int_V (\delta\varepsilon_{ij}D_{ijkl}\varepsilon_{kl} - \delta u_i f_i)dV + \int_{S_\sigma} \delta u_i T_i dS = 0 \tag{72}$$

Due to the symmetry of the constitutive matrix D_{ijkl}, we can further obtain

$$(\delta\varepsilon_{ij})D_{ijkl}\varepsilon_{kl} = \delta\left(\frac{1}{2}D_{ijkl}\varepsilon_{ij}\varepsilon_{kl}\right) = \delta U(\varepsilon_{mn}) \tag{73}$$

where $U(\varepsilon_{mn})$ is the unit volume strain energy. Given the assumptions in linear elasticity

$$-\delta\phi(u_i) = f_i\delta u_i, \ -\delta\psi(u_i) = T_i\delta u_i \tag{74}$$

Eq. (72) is further simplified to

$$\delta\Pi_P = 0 \tag{75}$$

Π_P is the total potential energy of the system, which is equal to the summation of the potential energy of deformation and external force and can be expressed as

$$\Pi_P = \Pi_P(u_i) = \int_V \left(\frac{1}{2}D_{ijkl}\varepsilon_{ij}\varepsilon_{kl} - f_i u_i\right)dV - \int_{S_\sigma} T_i u_i dS \tag{76}$$

Eq. (75) shows that, among all the potential displacements, the total potential energy of system will take stationary value at the real displacement, and it can be further verified that this stationary value is exactly the minimum value which is the principle of minimum potential energy.

1.13. General expressions and implementation procedure of FEM

The solution of a general continuum problem by FEM always follows an orderly step-by-step process which is easy to be programmed and used by the engineers. For illustration, a three-node triangular element for plane problems is taken as an example to illustrate the general expressions and implementation procedures of FEM.

1.13.1. Discretization of domain

The first step in the finite element method is to divide the structure or solution region into subdivisions or elements. Hence, the structure is to be modelled with suitable finite elements. In general, the number, type, size, and arrangement of the elements are critical towards good performance of the numerical analysis. A typical discretization with three-node triangular element is shown schematically in **Figure**.

Mesh generation can be a difficult process for a general irregular domain. If only triangular element is to be generated, this is a relatively simple work, and many commercial programs can perform well in this respect. There are also some public domain codes (EasyMesh or Triangle written in C) which are sufficient for normal purposes. For quadrilateral or higher elements, mesh generation is not that simple, and it is preferable to rely on the use of commercial programs for such purposes.

1.13.2. Interpolation or displacement model

As can be seen from **Figure (b)**, the nodal number of a typical three-node triangular element is coded in anticlockwise order (i.e. in the order of *i, j* and *m*), and each node has two degrees of freedom (DOFs) or two displacement components which is stored in a column vector in index notation as follows:

$$a_i = \begin{bmatrix} u_i \\ v_i \end{bmatrix} (i, j, m) \tag{77}$$

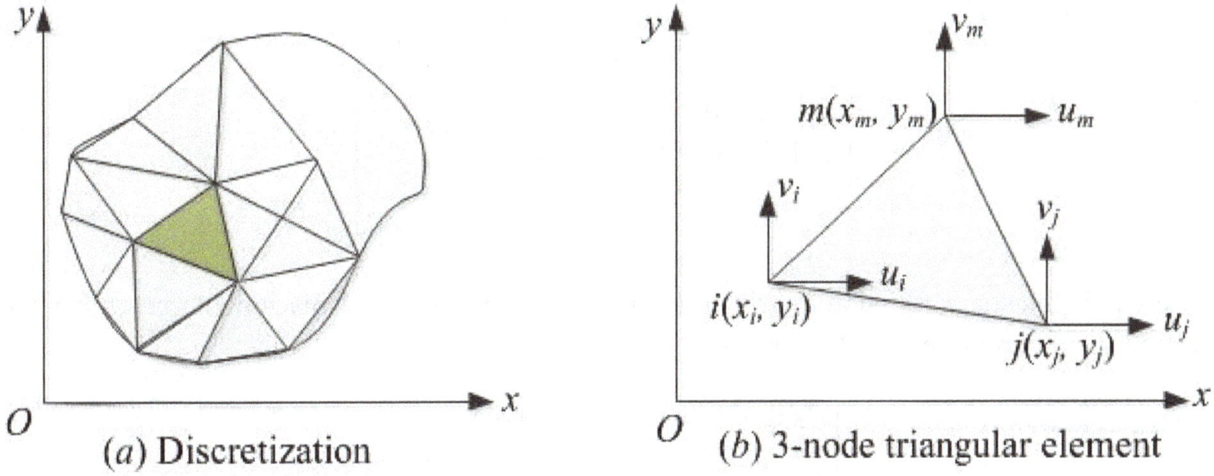

Figure. Discretization of a two-dimensional domain with three-node triangular element.

Totally, each element has six nodal displacements, i.e. six DOFs. Putting all the displacements in a column vector, we can obtain the element nodal displacement column matrix as

$$a^e = \begin{bmatrix} a_i \\ a_j \\ a_m \end{bmatrix} = \begin{bmatrix} u_i & v_i & u_j & v_j & u_m & v_m \end{bmatrix}^T \tag{78}$$

In FEM, a nodal displacement is chosen as the basic unknowns, so interpolation at any arbitrary point is based on the three nodal displacements of each element, which is called a displacement mode. For a three-node triangular element, linear polynomial is utilized, and the element displacement in both x -direction and y-direction are

$$u = \beta_1 + \beta_2 x + \beta_3 y \tag{79}$$

$$v = \beta_4 + \beta_5 x + \beta_6 y \tag{80}$$

Obviously, displacements of all the three nodes should satisfy Eqs. (79) and (80). By substituting the six nodal displacement components into these equations, it is easy to obtain another form of displacement mode as

$$u = N_i u_i + N_j u_j + N_m u_m \tag{81}$$

$$v = N_i v_i + N_j v_j + N_m v_m \tag{82}$$

where

$$N_i = \frac{1}{2A}(a_i + b_i x + c_i y)(i, j, m). \tag{83}$$

In Eq. (81), N_i, N_j and N_m denote the interpolation function or shape function for the three nodes, respectively. A is the area of the element, and $a_i, b_i, c_i \cdots, c_m$ are constants related to the coordinates of the three nodes. Similarly, Eqs. (81) and (82) can also be expressed in the form of matrix as

$$u = \begin{bmatrix} u \\ v \end{bmatrix} = \begin{bmatrix} N_i & 0 & N_j & 0 & N_m & 0 \\ 0 & N_i & 0 & N_j & 0 & N_m \end{bmatrix} \begin{bmatrix} u_i \\ v_i \\ u_j \\ v_j \\ u_m \\ v_m \end{bmatrix} = Na^e \tag{84}$$

where N is the shape function matrix and a^e is the element nodal displacement vector. For the geometric equations, element strains are

$$\begin{aligned} \varepsilon = \begin{bmatrix} \varepsilon_x \\ \varepsilon_y \\ \gamma_{xy} \end{bmatrix} &= Lu = LNa^e = L\begin{bmatrix} N_i & N_j & N_m \end{bmatrix} a^e \\ &= \begin{bmatrix} B_i & B_j & B_m \end{bmatrix} a^e = Ba^e \end{aligned} \tag{85}$$

where L is the differential operator and B is the element strain displacement matrix which can be given as

$$B_i = LN_i = \begin{bmatrix} \frac{\partial}{\partial x} & 0 \\ 0 & \frac{\partial}{\partial y} \\ \frac{\partial}{\partial y} & \frac{\partial}{\partial x} \end{bmatrix} \begin{bmatrix} N_i & 0 \\ 0 & N_i \end{bmatrix} = \begin{bmatrix} \frac{\partial N_i}{\partial x} & 0 \\ 0 & \frac{\partial N_i}{\partial y} \\ \frac{\partial N_i}{\partial y} & \frac{\partial N_i}{\partial x} \end{bmatrix} (i, j, m) \tag{86}$$

Substitute Eq. (85) by the stress-strain relation,

$$\sigma = \begin{bmatrix} \sigma_x \\ \sigma_y \\ \tau_{xy} \end{bmatrix} = D\varepsilon = DBa^e = Sa^e \tag{87}$$

where

$$S = DB = D[\,B_i \quad B_j \quad B_m\,] = [\,S_i \quad S_j \quad S_m\,] \tag{88}$$

S is called the element stress matrix. It should be noted that both the strain and stress matrices are constant for each element, because in a three-node triangular element, the displacement mode is a first-order function, and differentiating this function will give a constant function.

1.13.3. Stiffness equilibrium equation (SEE) of FEM derived from PMPE

For elastic plane problems, the total potential energy Π_P in Eq. (76) can be expressed in matrix formulation as follows:

$$\prod_P = \int_\Omega \frac{1}{2}\varepsilon^T D\varepsilon t dxdy - \int_\Omega u^T f t dxdy - \int_{S_\sigma} u^T T t dS \tag{89}$$

where t, f, and T denote the thickness, body force and surface force, respectively. For an FEM problem, the total potential energy is the summation of that from all the elements. Therefore, substituting Eqs. (84) and (85) into Eq. (89) gives

$$\Pi_P = \sum_e \Pi_P^e = \sum_e \left(a^{eT} \int_{\Omega_e} \frac{1}{2} B^T DBtdxdya^e \right) - \sum_e (a^{eT} \int_{\Omega_e} N^T f t dxdy) - \sum_e (a^{eT} \int_{S_\sigma{}^e} N^T T t dxdy) \tag{90}$$

Eq. (90) can be viewed as

$$\begin{aligned} K^e &= \int_{\Omega_e} B^T DBtdxdy, P_f^e = \int_{\Omega_e} N^T f t dxdy \\ P_S^e &= \int_{S_\sigma{}^e} N^T T t dxdy, P^e = P_f^e + P_S^e \end{aligned} \tag{91}$$

where K^e and P^e are named as the element stiffness matrix and equivalent element nodal load matrix, respectively. Substitute Eq. (91) to Eq. (90), the total potential energy of the structure can be simplified as

$$\Pi_P = a^T \frac{1}{2} \sum_e (K^e) a - a^T \sum_e (P^e) \tag{92}$$

Given

$$K = \sum_e K^e, P = \sum_e P^e \tag{93}$$

Eq. (92) is further simplified as

$$\Pi_P = \frac{1}{2} a^T K a - a^T P a \tag{93a}$$

where $K$and P are global stiffness matrix and global nodal load matrix, respectively.

For PMPE, the variation of Π_P is equal to 0 and the unknown variable is a, thus Eq. (75) gives

$$\frac{\partial \Pi_P}{\partial a} = 0 \tag{94}$$

which finally comes to the SEE of FEM as

$$Ka = P \tag{95}$$

From Eq. (93), we know that the global stiffness matrix and the global load matrix are the assemblage of the element stiffness matrices and equivalent element nodal load matrices, respectively. Specifically, in order to solve Eq. (93), element stiffness matrix, element equivalent nodal load vector, global stiffness matrices and global nodal load vector are all determined together with some given displacement boundary conditions. Without the provision of adequate boundary condition, the system is singular as rigid body motion will produce no stress in the system and such mode will be present in the SEE.

1.13.4. Derivation of element stiffness matrices (ESM)

For a three-node triangular element, the element strain matrix $\boldsymbol{B}$ is constant, thus Eq. (91) gives

$$K^e = B^T DBtA = \begin{bmatrix} K_{ii} & K_{ij} & K_{im} \\ K_{ji} & K_{jj} & K_{jm} \\ K_{mi} & K_{mj} & K_{mm} \end{bmatrix} \tag{96}$$

of which the submatrix

$$K_{ij} = \begin{bmatrix} k_{ij}^{xx} & k_{ij}^{xy} \\ k_{ij}^{yx} & k_{ij}^{yy} \end{bmatrix} \tag{97}$$

K_{ij} indicates the ith nodal force along the x- and y-directions in the Cartesian coordinate system when the displacement of the jth node is unit along the x- and y-directions, which can be easily obtained. Moreover, the element stiffness matrix is symmetric, and the computational memory required in an FEM program can be reduced by using this property.

It should be noted that for a higher order triangular element (e.g. six-node triangular element) or quadrilateral element for which higher order terms are involved, the strain matrix $\boldsymbol{B}$ is not constant any more so that the element stiffness matrix needs to be evaluated by numerical integration (direct integration is seldom adopted). Towards this, numerical integration methods such as the Gaussian integration or the Newton-Cotes integration can be utilized.

1.13.5. Assembling of ESMs and ENLMs

For an FEM process, we need to solve Eq. (95) which is the global equilibrium equation. Most of the elements in the matrix $K$are 0 simply because each node is only shared by a few surrounding elements. In view of that, a rectangular matrix can represent the global stiffness matrix (which is a square matrix), and the half bandwidth D can be defined as

$$D = (1 + NDIF) \times NDOF \tag{98}$$

where $NDIF$ denotes the largest absolute difference between the element node numbers among all the elements in the finite element mesh.

In conclusion, the properties of the global stiffness matrix can be summarized as: symmetric, banded distribution, singularity and sparsity. Among all the properties, singularity will vanish by introducing appropriate boundary conditions to Eq. (95) to eliminate the rigid body motion. Also, other properties like banded distribution should be fully taken into consideration to reduce the computational memory and enhance the computation efficiency.

1.13.6. Isoparametric element and numerical integration

Most of the engineering structure is not regular in shape, and some of them even have very complicated boundary shapes. Although the use of triangular element can always fit a complicated boundary, the accuracy of this element is low in general. To cope with the irregular boundary shape with a higher accuracy in analysis, one of the most common approaches is the use of higher-order element, and the isoparametric formulation is the most commonly used at present. Consider an arbitrary four-node quadrilateral element as an example which is schematically shown in **Figure**. If we can find the transformation from **Figure (a) to (b)**, then it will become easier to carry numerical integration with complicated shapes for an arbitrary element. In **Figure (a)**, we define the Cartesian coordinate system, while in **Figure (b)**, we define the local coordinate system (or natural coordinate system) within a specific domain (i.e. $\xi, \eta \in (-1, 1)$). The relation between these two kinds of coordinate system can be described as

$$\begin{Bmatrix} x \\ y \end{Bmatrix} = f \begin{Bmatrix} \xi \\ \eta \end{Bmatrix} \tag{99}$$

which can be further modified by the interpolation function at nodes in the local coordinate system as follows:

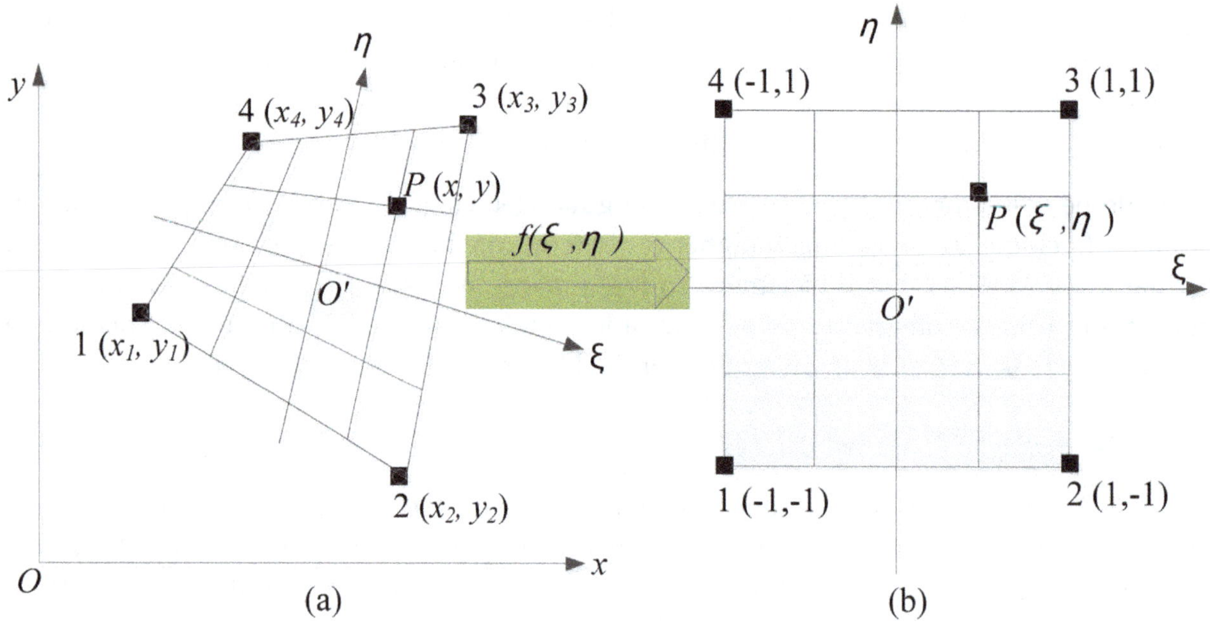

Figure. Isoparametric transition.

$$x = \Sigma_{i=1}^{m} N_i' x_i, \quad y = \sum_{i=1}^{m} N'_i y_i \tag{100}$$

where (x_i, y_i) are coordinates in the Cartesian coordinate system corresponding to the ith node in local coordinate system, N_i' is interpolation function of the ith node in local coordinate system and m is the number of nodes chosen to transform the coordinates. Therefore, the regular element in the natural coordinate system can be transformed to the irregular element in the Cartesian coordinate system. The former element is called the parent element, while the latter is called the subelement. Specifically, Eq. (101) can be further expanded as

$$\begin{Bmatrix} x \\ y \end{Bmatrix} = \begin{bmatrix} N_1 & 0 & N_2 & 0 & N_3 & 0 & N_4 & 0 \\ 0 & N_1 & 0 & N_2 & 0 & N_3 & 0 & N_4 \end{bmatrix} \begin{Bmatrix} x_1 \\ y_1 \\ x_2 \\ y_2 \\ x_3 \\ y_3 \\ x_4 \\ y_4 \end{Bmatrix} \tag{101}$$

Using the same interpolation functions, the element displacement model can be written as

$$\begin{Bmatrix} u \\ v \end{Bmatrix} = \begin{bmatrix} N_1 & 0 & N_2 & 0 & N_3 & 0 & N_4 & 0 \\ 0 & N_1 & 0 & N_2 & 0 & N_3 & 0 & N_4 \end{bmatrix} \begin{Bmatrix} u_1 \\ v_1 \\ u_2 \\ v_2 \\ u_3 \\ v_3 \\ u_4 \\ v_4 \end{Bmatrix} \tag{102}$$

where J denotes the Jacobi matrix while the interpolation functions are given by

$$\begin{aligned} N_1 &= \frac{1}{4}(1+\xi)(1+\eta), N_2 = \frac{1}{4}(1-\xi)(1+\eta) \\ N_3 &= \frac{1}{4}(1+\xi)(1+\eta), N_2 = \frac{1}{4}(1-\xi)(1+\eta) \end{aligned} \tag{103}$$

As mentioned before, during the derivation of the element stiffness matrix and the equivalent load vector, the derivative of the shape function and the integration in element surface or volume in the Cartesian coordinate system are required. Since the shape functions adopted herein are expressed in natural coordinates, therefore, derivative and integration transformation relationships are essential when isoparametric element is used.

1.13.7. *Derivative and integral transformation*

According to the law of partial differential,

$$\begin{aligned} \frac{\partial N_i}{\partial \xi} &= \frac{\partial N_i}{\partial x}\frac{\partial x}{\partial \xi} + \frac{\partial N_i}{\partial y}\frac{\partial y}{\partial \xi}, \\ \frac{\partial N_i}{\partial \eta} &= \frac{\partial N_i}{\partial x}\frac{\partial x}{\partial \eta} + \frac{\partial N_i}{\partial y}\frac{\partial y}{\partial \eta}, \end{aligned} \tag{104}$$

or in matrix form

$$\begin{Bmatrix} \dfrac{\partial N_i}{\partial \xi} \\ \dfrac{\partial N_i}{\partial \eta} \end{Bmatrix} = \begin{bmatrix} \dfrac{\partial x}{\partial \xi} & \dfrac{\partial y}{\partial \xi} \\ \dfrac{\partial x}{\partial \eta} & \dfrac{\partial y}{\partial \eta} \end{bmatrix} \begin{Bmatrix} \dfrac{\partial N_i}{\partial x} \\ \dfrac{\partial N_i}{\partial y} \end{Bmatrix} = J \begin{Bmatrix} \dfrac{\partial N_i}{\partial x} \\ \dfrac{\partial N_i}{\partial y} \end{Bmatrix} \tag{105}$$

Inverse of Eq. (105) gives

$$\begin{Bmatrix} \dfrac{\partial N_i}{\partial x} \\ \dfrac{\partial N_i}{\partial y} \end{Bmatrix} = J^{-1} \begin{Bmatrix} \dfrac{\partial N_i}{\partial \xi} \\ \dfrac{\partial N_i}{\partial \eta} \end{Bmatrix} \tag{106}$$

where

$$\begin{aligned} J &= \begin{bmatrix} \dfrac{\partial x}{\partial \xi} & \dfrac{\partial y}{\partial \xi} \\ \dfrac{\partial x}{\partial \eta} & \dfrac{\partial y}{\partial \eta} \end{bmatrix} = \begin{bmatrix} \displaystyle\sum_{i=1}^{4} \frac{\partial N_i}{\partial \xi} x_i & \displaystyle\sum_{i=1}^{4} \frac{\partial N_i}{\partial \xi} y_i \\ \displaystyle\sum_{i=1}^{4} \frac{\partial N_i}{\partial \eta} x_i & \displaystyle\sum_{i=1}^{4} \frac{\partial N_i}{\partial \eta} y_i \end{bmatrix} \\ &= \begin{bmatrix} \dfrac{\partial N_1}{\partial \xi} & \dfrac{\partial N_2}{\partial \xi} & \dfrac{\partial N_3}{\partial \xi} & \dfrac{\partial N_4}{\partial \xi} \\ \dfrac{\partial N_1}{\partial \eta} & \dfrac{\partial N_2}{\partial \eta} & \dfrac{\partial N_3}{\partial \eta} & \dfrac{\partial N_4}{\partial \eta} \end{bmatrix} \begin{bmatrix} x_1 & y_1 \\ x_2 & y_2 \\ x_3 & y_3 \\ x_4 & y_4 \end{bmatrix} \end{aligned} \tag{107}$$

For an infinitely small element, the area under the Cartesian coordinate system and the natural coordinate system are related by

$$ds = dxdy = |J|d\xi d\eta, \tag{108}$$

where $|J|$ is the determinant of the Jacobian matrix $\boldsymbol{J}$. Therefore, element stiffness matrix and equivalent nodal load matrix in Eq. (91) can be transformed to

$$\begin{aligned} K^e &= \int_{\Omega_e} B^T DBt|J|d\xi d\eta, P_f^e = \int_{\Omega_e} N^T f|J|d\xi d\eta \\ P_S^e &= \int_{S_\sigma{}^e} N^T T|J|d\xi d\eta \end{aligned} \tag{109}$$

For solving the integral equation, usually the Gaussian integration method is employed. In practice, both two and three integration points along each direction of integration are commonly used. Since the discretized system is usually overstiff, it is commonly observed that the use of two integration points along each direction of integration will slightly reduce the stiffness of the matrix and give better results as compared with the use of three integration points. The use of exact integration is possible for some elements, but such approaches are usually tedious and are seldom adopted. The advantage in using the exact integration is that the integration is not affected by the shape of the element while the transformation as shown in Eq. (109) may be affected if the poor shape of the element is poor. The author has developed many finite element programs for teaching and research purposes which can be obtained at ceymchen@polyu.edu.hk. The programs available include plane stress/strain problem, thin/thick plate bending problem, consolidation in 1D and 2D (Biot), seepage problem, slope stability problem, pile foundation problems and others.

1.14. Distinct element method

In practical applications, a limit equilibrium method based on the method of slices or method of columns and strength reduction method based on the finite element method or finite difference method are used for many types of stability problems. These two major analysis methods take the advantage that the *in situ* stress field which is usually not known with good accuracy is not required in the analysis. The uncertainties associated with the stress-strain relation can also be avoided by a simple concept of factor of safety or the determination of the ultimate limit state. In general, this approach is sufficient for engineering analysis and design. If the condition of the system after failure has initiated is required to be assessed, these two methods will not be applicable. Even if the *in situ* stress field and the stress-strain relation can be defined, the post-failure collapse is difficult to be assessed using the conventional continuum-based numerical method, as sliding, rotation and collapse of the slope involve very large displacement or even separation without the requirement of continuity.

The most commonly used numerical methods for continuous systems are the FDM, the FEM and the boundary element method (BEM). The basic assumption adopted in these numerical methods is that the materials concerned are continuous throughout the physical processes. This assumption of continuity requires that, at all points in a problem domain, the material cannot be torn open or broken into pieces. All material points originally in the neighbourhood of a certain point in the problem domain remain in the same neighbourhood throughout the whole physical process. Some special algorithms have been developed to deal with material fractures in continuum mechanics-based methods, such as the special joint elements by Goodman [13] and the displacement discontinuity technique in BEM by Crouch and Starfield [5]. However, these methods can only be applied with limitations [21]:

1. large-scale slip and opening of fracture elements are prevented in order to maintain the macroscopic material continuity;

2. the amount of fracture elements must be kept to relatively small so that the global stiffness matrix can be maintained well-posed, without causing severe numerical instabilities; and

3. complete detachment and rotation of elements or groups of elements as a consequence of deformation are either not allowed or treated with special algorithms.

Before a slope starts to collapse, the factor of safety serves as an important index in both the LEM and SRM to assess the stability of the slope. The movement and growth after failure have launched which is also important in many cases that cannot be simulated on the continuum model, and this should be analyzed by the distinct element method (DEM).

In continuum description of soil material, the well-established macro-constitutive equations whose parameters can be measured experimentally are used. On the other hand, a discrete element approach will consider that the material is composed of distinct grains or particles that interact with each other. The commonly used distinct element method is an explicit method based on the finite difference principles which is originated in the early 1970s by a landmark work on the progressive movements of rock masses as 2D rigid block assemblages [6]. Later, the works by Cundall are developed to the early versions of the UDEC and 3DEC codes [9, 10, 12]. The method has also been developed for simulating the mechanical behaviour of granular materials [8], with a typical early code BALL [7], which later evolved into the codes of the PFC group for 2D and 3D problems of particle systems (Itasca, 1995). Through continuous developments and extensive applications over the last three decades, there has accumulated a great body of knowledge and a rich field of literature about the distinct element method. The main trend in the development and application of the method in rock engineering is represented by the history and results of the code groups UDEC/3DEC. Currently, there are many open source (Oval, LIGGGHTS, ESyS, Yade, ppohDEM, Lammps) as well as commercial DEM programs, but in general, this method is still limited to basic research instead of practical application as there are many limitations which include: (1) difficult to define and determine the microparameters; (2) there are still many drawbacks in the use of matching with the macro response to determine the microparameters; (3) not easy to set up a computer model; (4) not easy to include structural element or water pressure; (5) extremely time consuming to perform an analysis; and (6) postprocessing is not easy or trivial. It should also be noted that DEM can be formulated by an energy-based implicit integration scheme which is the discontinuous deformation analysis (DDA) method. This method is similar in many respect to the force-based explicit integration scheme as mentioned previously.

In DEM, the packing of granular material can be defined from statistical distributions of grain size and porosity, and the particles are assigned normal and shear stiffness and friction coefficients in the contact relation. Two types of bonds can be represented either individually or simultaneously; these bonds are referred to the contact and parallel bonds, respectively (Itasca, 1995). Although the individual particles are solid, these particles are only partially connected at the contact points which will change at different time step. Under low normal stresses, the strength of the tangential bonds of most granular materials will be weak and the material may flow like a fluid under very small shear stresses. Therefore, the behaviour of granular material in motion can be studied as a fluid-mechanical phenomenon of particle flow where individual particles may be treated as "molecules" of the flowing granular material. In many particle models for geological materials in practice, the number of particles contained in a typical domain of interest will be very large, similar to the large numbers of molecules.

One of the primary objectives of the particle model is the establishment of the relations between microscopic and macroscopic variables/parameters of the particle systems, mainly through micromechanical constitutive relations at the contacts. Compared with a continuum, particles have an additional degree of freedom of rotation which enables them to transmit couple stresses, besides forces through their translational degrees of freedom. At certain moment, the positions and velocities of the particles can be obtained by translational and rotational movement equations and any special physical phenomenon can be traced back from every single particle interactions. Therefore, it is possible for DEM to analyze large deformation problems and a flow process which will occur after slope failure has initiated. The main limitation of DEM is that there is great difficulty in relating the microscopic and macroscopic variables/parameters; hence, DEM is mainly tailored towards qualitative instead of quantitative analysis.

DEM runs according to a time-difference scheme in which calculation includes the repeated application of the law of motion to each particle, a force-displacement law to each contact, and a contact updating scheme. Generally, there are two types of contact in the program which are the ball-wall contact and the ball-ball contact. In each cycle, the set of contacts is updated from the known particles and known wall positions. Force-displacement law is firstly applied on each contact, and new contact force is then calculated according to the relative motion and constitutive relation. Law of motion is then applied to each particle to update the velocity, the direction of travel based on the resultant force, and the moment and contact acting on the particles. Although every particle is assumed as a rigid material, the behaviour of the contacts is characterized using a soft contact approach in which finite normal stiffness is taken to represent the stiffness which exists at the contact. The soft contact approach allows small overlap between the particles which can be easily observed. Stress on particles is then determined from this overlapping through the particle interface.

1.15. General formulation of DEM

The PFC runs according to a time-difference scheme in which calculation includes the repeated application of the law of motion to each particle, a force-displacement law to each contact, and a contact updating a wall position. Generally, there are two types of contact exist in the program which are ball-to-wall contact and ball-to-ball contact. In each cycle, the set of contacts is updated from the known particle and the known wall position. The force-displacement law is first applied on each contact. New contact force is calculated and replaces the old contact force. The force calculations are based on preset parameters such as normal stiffness, density, and friction. Next, a law of motion is applied to each particle to update its velocity, direction of travel based on the resultant force, moment and contact acting on particle. The force-displacement law is then applied to continue the circulation.

1.16. The force-displacement law

The force-displacement law is described for both the ball-ball and ball-wall contacts. The contact arises from contact occurring at a point. For the ball-ball contact, the normal vector is directed along the line between the ball centres. For the ball-wall contact, the normal vector is directed along the line defining the shortest distance between the ball centre and the wall. The contact force vector F_i is composed of normal and shear component in a single plane surface

$$F_i = F_{ij}^n(t) + F_{ij}^s(t + \Delta t) \tag{110}$$

The force acting on particle i in contact with particle j at time t is given by

$$F_{ij}^n(t) = k_n\left(r_i + r_j - l_{ij}(t)\right) \tag{111}$$

where r_j and r_i stand for particle i and particle j radii, $l_{ij}(t)$ is the vector joining both centres of the particles and k_n represents the normal stiffness at the contact. The shear force acting on particle i during a contact with particle j is determined by

$$F_{ij}^s(t + \Delta t) = \pm\min(F_{ij}^s(t) + k_s\Delta s_{ij}, f|F_{ij}^n(t + \Delta t)|) \tag{112}$$

where f is the particle friction coefficient, k_s represents the tangent shear stiffness at the contact. The new shear contact force is found by summing the old shear force (min $F_{ij}(t)$) with the shear elastic force. Δs_{ij} stands for the shear contact displacement-increment occurring over a time step Δt.

$$\Delta s_{ij} = v_{ij}^s \Delta t \tag{113}$$

where V_{ij}^s is the shear component of the relative velocity at contact between particles i and j over the time step Δt.

1.17. Law of motion

The motion of the particle is determined by the resultant force and moment acting on it. The motion induced by resultant force is called translational motion. The motion induced by resulting moment is rotational motion. The equations of motion are written in vector form as follows:

- (Translational motion)

$$\sum_j F_{ij} + m_i g + F_i^d = m_i x_i^n \tag{114}$$

- (Rotational motion)

$$\sum_j r_i F_{ij} + M_i^d = I_r \theta_i^n \tag{115}$$

where x''_i and θ''_i stand for the translational acceleration and rotational acceleration of particles i. I_r stands for moment of inertia. F_i^d and M_i^d stand for the damping force and damping moment. Unlike finite element formulation, there are now three degree of freedom for 2D problem and six degree of freedom for 3D problems. In Cundall and Strack's explicit integration distinct element approach, solution of the global system of equation is avoided by considering the dynamic equilibrium of the individual particles rather than solving the entire system simultaneously. That means, Newton's law of motion is applied directly. This approach also avoids the generation and storage of the large global stiffness matrix that will appear in finite element analysis. On the other hand, the implicit DDA approach will generate a global stiffness matrix which is even larger than that in finite element analysis, as the rotation is involved directly in the stiffness matrix.

In a typical DEM simulation, if there is no yield by contact separation or frictional sliding, the particles will vibrate constantly and the equilibrium is difficult to be achieved. To avoid this phenomenon which is physically incorrect, numerical or artificial damping is usually adopted in many DEM codes, and the two most common approaches to damping are the mass damping and non-viscous damping. For mass damping, the amount of damping that each particle "feels" is proportional to its mass, and the proportionality constant depends on the eigenvalues of the stiffness matrix. This damping is usually applied equally to all the nodes. As this form of damping introduces body forces, which may not be appropriate in flowing regions, it may influence the mode of failure. Alternatively, Cundall [11] proposed an alternative method where the damping force at each node is proportional to the magnitude of the out-of-balance-force, with a sign to ensure that the vibrational modes are damped rather than the steady motion. This form of damping has the advantage that only accelerating motion is damped and no erroneous damping forces will arise from steady-state motion. The damping constant is also non-dimensional and the damping is frequency independent. As suggested by Itasca [20], an advantage of this approach is that it is similar to the hysteretic damping, as the energy loss per cycle is independent of the rate at which the cycle is executed. While damping is one way to overcome the non-physical nature of the contact constitutive models in DEM simulations, it is quite difficult to select an appropriate and physically meaningful value for the damping. For many DEM simulations, particles are moving around each other and the dominant form of energy dissipation is for frictional sliding and contact breakages. The choice of damping may affect the results of computations. Currently, most of the DEM codes allow the use of automatic damping or manually prescribed the damping if necessary.

To capture the inherent non-linearity behaviour of the problem (with generation and removal of contacts, non-linear contact response and stress-strain behaviour and others), the displacement and contact forces in a given time step must be small enough so that in a single time step, the disturbances cannot propagate from a particle further than its nearest neighbours. For most of the DEM programs, this can be achieved automatically and the default setting is usually good enough for normal cases. It is, however, sometimes necessary to manually adjust the time step in some special cases when the input parameters are unreasonably high or low. Most of the DEM codes use the central difference time integration algorithm which is a second-order scheme in time step.

1.18. Measuring logic

If the local results in DEM are analyzed, it is found that there will be large fluctuations with respect to both locations and time. Such results are not surprising, as the results are highly sensitive to the interaction between particles and hence the time step under which the results are monitored. It can be viewed that such local results can be meaningless unless the results are monitored over a long time span or region. A number of quantities in a DEM model are defined with respect to a specified measurement circle. These quantities include coordinate number, porosity, sliding fraction, stress and strain rate. The coordination number and stress are defined as the average number of contacts per particle. Only particles with centroids that are contained within the measurement circle are considered in computation. In order to account for the additional area of particles that is being neglected, a corrector factor based on the porosity is applied to the computed value of stress.

Since measurement circle is used, stress in particle is described as the two in-plane force acting on each particle per volume of particle. Average stress is defined as the total stress in particle divided by the volume of measurement circle. Thus, shape of particle is regardless of the average stress measurement because the reported stress is easily scaled by volume unity. The reported stress is interpreted as the stress per volume of measurement circle.

1.19. Discussion and conclusion

There are also various publications on the numerical solutions of differential equations, and the readers are suggested to the works of Lee and Schiesser [24], Jovanoic and Suli [22], Veiga et al. [37], Sewell [35], Morton and Mayers [27], Logg et al. [25], Holmes [17], Lui [26], Lapids and Pinder [23] and Iserles [19]. It is impossible for the author to cover every available analytical or numerical method; hence, the author has chosen some methods that are actually used for teaching and research. The readers are strongly encouraged to consult the numerous resources available in various books and publications. There are still new developments available for the solutions of specific differential equations in large-scale problems, and this is also the current trend in the development of differential equation solution.

Due to the importance of the solution of differential equations, there are other important numerical methods that are used by different researchers but are not discussed here, which include the finite difference and boundary element methods (computer codes for learning can also be obtained from the author). Differential equations rely on the Taylor's series, and the derivatives in the differential equation can be replaced with finite difference approximations on a discretized domain. This will result in a system of algebraic equations that can be solved implicitly or explicitly. There are various ways to form the derivatives, and the most common methods are the forward difference, backward difference and the central difference schemes. While the finite difference methods may be more suitable for different types of differential equations, this method is less convenient to deal with irregular boundary conditions as compared with the finite element method. For highly irregular domain where it is not easy to form a nice discretization, the finite element method will also be much easier and natural to deal with for such condition. In this respect, it is not surprising that many engineering programs are written by the use of the finite element method than the finite difference method.

The boundary element method (BEM) is another numerical method for solving linear partial differential equations which can be formulated as integral equations. The boundary element method uses the given boundary conditions to fit boundary values into the integral equation. In the post-processing stage, the integral equation will be used to calculate the solution directly at any given point inside the solution domain numerically. BEM is applied to problems for which Green's functions can be calculated, thus this method is initially designed for problems in linear homogeneous media. The dimension of the problem will then be reduced by one. For example, two-dimensional problem will be effectively reduced to one-dimensional problem along the boundary, and this will greatly improve the efficiency of computation. The requirement from the boundary element method imposes considerable restrictions on the range and generality of problems to which the boundary element method can usefully be applied. There are some new developments to the boundary element method so that it can be used for non-linear problem or problems with several major materials (problems with random distribution of material properties are still not applicable). The fundamental solutions are often difficult to integrate. One important property of boundary element analysis is the solution of a fully populated matrix as compared with that in the finite element/difference method. For complicated problems, the boundary element will lose its advantage as compared with other numerical methods. Due to the various limitations, there are only limited boundary element programs available to the researchers. Interested readers can consult the works of Banerjee [1], Brebbia et al. [3] and Trevelyan [36]. It appears that there are less interest in the use and development of the boundary element method in the recent years, due to the various limitations of this method in general non-linear non-homogeneous problem.

In history, various techniques have been developed for ordinary differential equations and partial differential equations under different boundary conditions. While these tricks appear to be elegant, they are not readily adopted for normal engineering use due to various limitations. Being an engineer, the author seldom adopted the methods as outlined in this chapter in actual applications (but do adopt for teaching), except the numerical methods as outlined in this chapter. At present, there are many proprietary or open source finite elements or distinct element codes being used for many complicated real problems. The computer codes (usually in Fortran or C) are usually difficult to be read (if available), and the computer codes for all the partial differential forms (including some extended formats) that have been discussed in this chapter can be readily available from the author for learning purposes. There are also very powerful and general finite element tools or differential equations solver such as FreeFem++, Comsol, Matlab, Mathematica, Maple and Maxima which are used by many scientists and engineers [39, 40]. The use of parallel computing is also strongly influenced by the needs to solve complicated partial differential equations over large solution domain.

Author details

Cheng Yung Ming

Address all correspondence to: ceymchen@polyu.edu.hk

Department of Civil and Environmental Engineering, Hong Kong Polytechnic University, Hong Kong SAR, China

References

[1] Banerjee P.K. (editor). Developments in boundary element methods, Vols. 1,2,3, Applied Science Publishers, London, 1979.

[2] Bertanz R. Fourier series and numerical methods for partial differential equations, John Wiley, USA, 2010.

[3] Brebbia C.A., Telles J.C.F. and Wrobel L.C. Boundary element techniques, Springer-Verlag, Berlin, 1984.

[4] Bronson R. and Costa G.B. Differential equations, McGraw-Hill, USA, 2006.

[5] Crouch S.L. and Starfield A.M. (editors). Boundary element methods in solid mechanics, George Allen & Unwin, London, 1983.

[6] Cundall P. A. A computer model for simulating progressive, large-scale movements in blocky rock systems. In: Proceedings of the International Symposium on Rock Mechanics. Nancy, France, 1971: 129–136.

[7] Cundall P.A. Ball – A computer program to model granular medium using the distinct element method, Technical note TN-LN-13, Advanced Technology Group, Dames and Moore, London, 1978:129–163.

[8] Cundall P.A. and Strack O.D.L. A discrete model for granular assemblies, Geotechnique, 1979, 29(1):47–65.

[9] Cundall P.A. UDEC-A generalized distinct element program for modelling jointed rock, Final Tech. Rep. Eur. Res. Office (US Army Contract DAJA37-79-C-0548), Springer, Netherland, 1980.

[10] Cundall P.A. and Hart R.D. Development of generalized 2-D and 3-D distinct element programs for modeling jointed rock, Misc. Paper SL-85-1, US Army Corps of Engineers, USA, 1985.

[11] Cundall, P.A. Distinct element models of rock and soil structure, in Analytical and Computational Methods in Engineering Rock Mechanics, Brown, E.T. ed., George Allen and Unwin, London, 1987:129–163.

[12] Cundall P.A. and Hart D.H. Numerical modelling of discontinua, Engineering with Computers, 1992, 9:101–11.

[13] Goodman, R. E. Methods of geological engineering in discontinuous rocks. West Publishing Company, San Francisco, CA, USA, 1976.

[14] Greenberg M.D. Solutions manual to accompany ordinary differential equations, John Wiley, USA, 2012.

[15] Haberman R. Applied partial differential equations, Pearson Education, USA, 2005.

[16] Hillen T., Leonard I.E. and Roessel H.V. Partial differential equations, John Wiley, USA, 2012.

[17] Holmes M.H. Introduction to numerical methods in differential equations, Springer, USA, 2007.

[18] Holzner S. Differential equations for dummies, John Wiley, USA, 2008.

[19] Iserles A. A first course in the numerical analysis of differential equations, Cambridge University Press, UK, 2009.

[20] Itasca Consulting Group Inc, a company based in US. PFC3D 3.1, User Guide, 2004.

[21] Jing L.R., Stephansson O. and Erling N. Study of rock joints under cyclic loading conditions, Rock Mechanics and Rock Engineering, 1993, 26, 3:215–232.

[22] Jovanoic B.S. and Suli E. Analysis of finite difference schemes, Springer, USA, 2014.

[23] Lapids L. and Pinder G.F. Numerical solution of partial differential equations in science and engineering, John Wiley, Italy, 1999.

[24] Lee H.J. and Schiesser W.E. Ordinary and partial differential equation routines in C, C++, Fortran, Java, Maple, and MATLAB, Chapman and Hall, USA, 2004.

[25] Logg A., Mardal K.A. and Wells G.N. Automated solution of differential equations by the finite element method, Springer, France, 2012.

[26] Lui S.H. Numerical analysis of partial differential equations, John Wiley, USA, 2011.

[27] Morton K.W. and Mayers D.F. Numerical solution of partial differential equations, Cambridge University Press, UK, 2005.

[28] Nagle R.K., Saff E.B. and Snider A.D. Fundamentals of differential equations and boundary value problems, Addison Wesley, USA, 2012.

[29] Polyanin A.D. and Nazaikinskii V.E. Handbook of linear partial differential equations for engineers and scientists, 2nd edition, CRC Press, USA, 2016.

[30] Polyallin A.D., Zaitsev V.F. and Moussiaux A. Handbook of first order partial differential equations, Taylor and Francis, UK, 2002.

[31] Polyanin A.D. and Zaitsev V.F. Handbook of nonlinear Partial Differential Equations, 2nd edition, Chapman & Hall, USA, 2004.

[32] Rao S.S. The finite element method in engineering, Butterworth Heinemann, London, 2011.

[33] Salsa S. Partial differential equations in action - from modelling to theory, Springer, Italy, 2008.

[34] Soare M.V., Teodorescu P.P. and Toma I. Ordinary differential equations with applications to mechanics, Springer, Netherland, 2000.

[35] Sewell G. The numerical solution of ordinary and partial differential equations, John Wiley, USA, 2005.

[36] Trevelyan J. Boundary elements for engineers: theory and applications, Computational Mechanics Publications, Boston, USA, 1994.

[37] Veiga L.B.D., Lipnikov K. and Manzini G. The mimetic finite difference method for elliptic problems, Springer, Switzerland, 2014.

[38] Zienkiewicz O.C., Taylor R.L. and Zhu J.Z. The finite element method: Its basis and fundamentals, 6th edition, Butterworth Heinemann, London, 2011.

[39] Comsol, https://www.comsol.com/

[40] FreeFem++, http://www.freefem.org/

[41] H Weber, Paul du Bois-Reymond, Mathematische Annalen 35 (1890), 457–462

9

Integral-Equation Formulations of Plasmonic Problems in the Visible Spectrum and Beyond

Abdulkerim Çekinmez,
Barişcan Karaosmanoğlu and Özgür Ergül

Abstract

Computational modeling of nano-plasmonic structures is essential to understand their electrodynamic responses before experimental efforts in measurement setups. Similar to the other ranges of the electromagnetic spectrum, there are alternative methods for the numerical analysis of nano-plasmonic problems, while the optics literature is dominated by differential equations that require discretizations of the host media with artificial truncations. These approaches often need serious assumptions, such as periodicity, infinity, or self-similarity, in order to reduce the computational load. On the other hand, surface integral equations based on integro-differential operators can bring important advantages for accurate and efficient modeling of nano-plasmonic problems with arbitrary geometries. Electrical properties of materials, which may be obtained either experimentally or via physical modeling, can easily be inserted into integral-equation formulations, leading to accurate predictions of electromagnetic responses of complex structures. This chapter presents the implementation of such accurate, efficient, and reliable solvers based on appropriate combinations of surface integral equations, discretizations, numerical integrations, fast algorithms, and iterative techniques. As a case study, nanowire transmission lines are investigated in wide-frequency ranges, demonstrating the capabilities of the developed implementations.

Keywords: surface integral equations, multilevel fast multipole algorithm, surface plasmons, computational electromagnetics

1. Introduction

As in all areas of electrodynamics, numerical study of plasmonic problems is essential to understand interactions between electromagnetic waves and matter at the higher range of the spectrum. Applications include nanowires for negative refraction, imaging, and super-resolution

[1, 2], and nanoantennas for energy harvesting, single-molecule sensing, and optical links [3–9], to name a few. At optical frequencies, some metals are known to possess strong plasmonic properties [10] that are crucial for a majority of such applications, while their accurate analysis requires more than perfectly conducting models that are common in radio and microwave regimes. In the infrared region, it may not be obvious when perfect conductivity or impedance approximation methods can safely be used. Hence, it is desirable to extend the plasmonic-modeling capabilities across wide ranges of frequencies until they converge to the other forms. While, in the literature, experimental studies are often supported by differential solvers, their applicability to complex problems is usually limited to small-scale and/or simplified models due to well-known drawbacks, such as need for space (host-medium) discretizations that are accompanied with artificial truncations. Major tools of computational electromagnetics, that is, surface integral equations [11, 12] employing integro-differential operators, are recently applied to plasmonic problems with promising results for realistic simulations of complex structures [13–23]. In fact, surface integral equations need only the discretization of boundaries between different media, which usually correspond to the surface of the plasmonic object. In addition to homogeneous bodies, they are also applicable to piecewise homogeneous cases, making it possible to analyze structures with coexisting multiple materials [24].

Using surface integral equations, it is possible to solve plasmonic problems involving finite models with arbitrary geometries, without periodicity, self-similarity, and infinity assumptions. When the object is large in terms of wavelength, fast and efficient methods, such as the multilevel fast multipole algorithm (MLFMA) [25], are available to accelerate solutions [26–28]. For plasmonic modeling, effective permittivity values with negative real parts are required, while they are already available via theoretical and experimental studies [10]. In the phasor domain with time-harmonic sources, which is considered in this chapter, permittivity is a simulation parameter with a fixed value at a given frequency. Then, frequency sweeps can be performed by using the discrete values of the permittivity with respect to frequency. As theoretical models, Drude (D) or Lorentz-Drude (LD) models are commonly used. While these models (especially the Lorentz-Drude model) provide reliable permittivity values in wide-frequency ranges, they deviate from experimental data at higher frequencies of the optical spectrum. From the perspective of surface integral equations, it does not matter where the permittivity values are obtained from. Besides, there is a great flexibility in geometric modeling, allowing sharp edges and corners, tips, and subwavelength details [29]. On top of these, the background of surface integral equations provides self-consistency and accuracy-check mechanisms, such as based on the equivalence theorem, enabling accuracy analysis without resorting to alternative solvers [30].

From numerical point of view, surface integral equations bring their own challenges when they are applied to plasmonic problems. In free space, plasmonic objects are naturally high-contrast problems [15], leading to difficulties in maintaining the accuracy and/or efficiency. Considering the equivalence theorem, ideal mesh size for surface formulations can be selected based on wavenumber of the host medium, where the impressed sources are located [26]. Therefore, the source of the inaccuracy is not directly the discretization size, but a combination of geometric deviation (for smooth objects), numerical integration, and imbalanced contributions from inner/outer media. Efficiency of iterative solutions may also deteriorate due to imbalanced matrix blocks that lead to ill-conditioned matrix equations [31]. On the other hand, numerical challenges are not only due to the high contrasts of plasmonic objects. The effective permittivity of a plasmonic medium is typically negative, which becomes increasingly large at lower frequencies. In numerical solutions, integro-differential operators become localized with exponentially decaying Green's function. This localization is responsible for the evolution of plasmonic formulations into perfectly conducting types, while this process may not be achieved smoothly in discrete forms. Some traditional formulations break down due to dominant inner contributions, which are difficult to compute accurately [32], if not impossible. Classical singularity extractions may fail to provide smooth integrands, leading to increasingly inaccurate near-zone interactions. While all formulations may be improved by manipulating integrations into more suitable forms, our focus is to develop new formulations that reduce into perfectly conducting formulations in the limit. All results presented in this chapter are obtained by such a stabilized integral-equation formulation, namely a modified combined tangential formulation (MCTF), which provides accurate results using the conventional Rao-Wilton-Glisson (RWG) discretizations [33].

The chapter is organized as follows. In Section, we present surface integral equations, with the emphasis on MCTF. Discretization is presented in Section, including implementation details that may be followed by the readers to develop their own solvers. MLFMA is further discussed in Section, demonstrating how to accelerate numerical solutions. Finally, we present an extensive case study, involving nanowire transmission lines in a wide range of frequency to illustrate the significant differences between the analytical models and measurement data for the permittivity values. In the following, time-harmonic electrodynamic problems are considered with $\exp(-i\omega t)$ time dependency, where $i^2 = -1$ and $\omega = 2\pi f$ is the angular frequency.

2. Surface integral equations

For deriving surface formulations, we consider a plasmonic object with permittivity/permeability (ε_p/μ_p) located in unbounded free space with permittivity/permeability (ε_o/μ_o). Alternative surface integral equations can be obtained by considering the boundary conditions on the surface of the object. In a general form, we have

$$\begin{bmatrix} \mathcal{Z}_{11} & \mathcal{Z}_{12} \\ \mathcal{Z}_{21} & \mathcal{Z}_{22} \end{bmatrix} \cdot \begin{bmatrix} \boldsymbol{J} \\ \boldsymbol{M} \end{bmatrix}(\boldsymbol{r}) = \begin{bmatrix} a\hat{\boldsymbol{n}} \times \hat{\boldsymbol{n}} \times \boldsymbol{E}^{\text{inc}} - e\hat{\boldsymbol{n}} \times \boldsymbol{H}^{\text{inc}} \\ c\hat{\boldsymbol{n}} \times \hat{\boldsymbol{n}} \times \boldsymbol{H}^{\text{inc}} + g\hat{\boldsymbol{n}} \times \boldsymbol{E}^{\text{inc}} \end{bmatrix}(\boldsymbol{r}), \tag{1}$$

where $\boldsymbol{J} = \hat{\boldsymbol{n}} \times \boldsymbol{H}$ and $\boldsymbol{M} = -\hat{\boldsymbol{n}} \times \boldsymbol{E}$ are the equivalent currents written in terms of the tangential electric field intensity $\boldsymbol{E}$ and the magnetic field intensity $\boldsymbol{H}$ on the closed surface $(\boldsymbol{r} \in S)$. In the above, $\hat{\boldsymbol{n}}$ is the unit vector outward the object, and $\boldsymbol{E}^{\text{inc}}$ and $\boldsymbol{H}^{\text{inc}}$ are the incident electric and magnetic fields, respectively, created by impressed sources located in the host medium. At an observation point on a locally planar surface (solid angle = 2π), the combined operators can be written as

$$\mathcal{Z}_{11} = -\hat{\boldsymbol{n}}\times\hat{\boldsymbol{n}}\times(a\eta_o\mathcal{T}_o + b\eta_p\mathcal{T}_p) + \hat{\boldsymbol{n}}\times(e\mathcal{K}_{\mathrm{PV},o} - f\mathcal{K}_{\mathrm{PV},p}) - (e+f)\mathcal{I}/2 \tag{2}$$

$$\mathcal{Z}_{12} = \hat{\boldsymbol{n}}\times\hat{\boldsymbol{n}}\times(a\mathcal{K}_{\mathrm{PV},o} + b\mathcal{K}_{\mathrm{PV},p}) - (a-b)\hat{\boldsymbol{n}}\times\mathcal{I}/2 + \hat{\boldsymbol{n}}\times(e\eta_o^{-1}\mathcal{T}_o - f\eta_p^{-1}\mathcal{T}_p) \tag{3}$$

$$\mathcal{Z}_{21} = -\hat{\boldsymbol{n}}\times\hat{\boldsymbol{n}}\times(c\mathcal{K}_{\mathrm{PV},o} + d\mathcal{K}_{\mathrm{PV},p}) + (c-d)\hat{\boldsymbol{n}}\times\mathcal{I}/2 - \hat{\boldsymbol{n}}\times(g\eta_o\mathcal{T}_o - h\eta_p\mathcal{T}_p) \tag{4}$$

$$\mathcal{Z}_{22} = -\hat{\boldsymbol{n}}\times\hat{\boldsymbol{n}}\times(c\eta_o^{-1}\mathcal{T}_o + d\eta_p^{-1}\mathcal{T}_p) + \hat{\boldsymbol{n}}\times(g\mathcal{K}_{\mathrm{PV},o} - h\mathcal{K}_{\mathrm{PV},p}) - (g+h)\mathcal{I}/2, \tag{5}$$

where $\{a, b, c, d, e, f, g, h\}$ are generalized coefficients. In the above, $\eta_o = \sqrt{\mu_o}/\sqrt{\varepsilon_o}$ is the intrinsic impedance of the host medium, whereas $\eta_p = \sqrt{\mu_p}/\sqrt{\varepsilon_p}$ is the complex intrinsic impedance of the plasmonic object. The integro-differential and identity operators are derived as

$$\mathcal{T}_u\{\boldsymbol{X}\}(\boldsymbol{r}) = ik_u \int_S d\boldsymbol{r}'[\boldsymbol{X}(\boldsymbol{r}') + \frac{1}{k_u^2}\nabla' \cdot \boldsymbol{X}(\boldsymbol{r}')\nabla]g_u(\boldsymbol{r}, \boldsymbol{r}') \tag{6}$$

$$\mathcal{K}_{PV,u}\{\boldsymbol{X}\}(\boldsymbol{r}) = \int_{\mathrm{PV},S} d\boldsymbol{r}'\boldsymbol{X}(\boldsymbol{r}')\times\nabla' g_u(\boldsymbol{r}, \boldsymbol{r}') \tag{7}$$

$$\mathcal{I}\{\boldsymbol{X}\}(\boldsymbol{r}) = \boldsymbol{X}(\boldsymbol{r}) \tag{8}$$

for $\boldsymbol{r}\in S$, where PV indicates the principal value of the integral, $\nabla = \hat{x}\partial/\partial x + \hat{y}\partial/\partial y + \hat{z}\partial/\partial z$ is the differential operator, $g_u(\boldsymbol{r}, \boldsymbol{r}') = \exp(ik_u|\boldsymbol{r}-\boldsymbol{r}'|)/(4\pi|\boldsymbol{r}-\boldsymbol{r}'|)$ is the homogeneous-space Green's function, and $k_u = 2\pi/\lambda_u = \omega\sqrt{\mu_u\varepsilon_u}$ is the wavenumber for $u = \{o, p\}$.

The conventional formulations can be obtained by setting the generalized coefficients to suitable values such that the outer and inner problems are coupled while the internal resonances are removed. By using nonzero values for $\{e, f, g, h\}$ while setting $\{a, b, c, d\}$ to zero leads to N-formulations, such as the Müller formulation and the combined normal formulation [12]. These formulations contain the identity operator $\mathcal{I}$, which usually dominates the matrix equations when Galerkin discretization is used. Therefore, matrix equations derived from N-formulations are generally easier to solve iteratively. On the other hand, T-formulations are obtained by selecting $\{a, b, c, d\}$ nonzero, while inserting zero values for $\{e, f, g, h\}$. The Poggio-Miller-Chang-Harrington-Wu-Tsai formulation [34] and the combined tangential formulation [12] are among the well-known T-formulations. As opposed to N-formulations, T-formulations contain either the rotational identity operator $\hat{\boldsymbol{n}}\times\mathcal{I}$ or no identity operator at all (when $a = b$ and $c = d$). Hence, using a Galerkin discretization, T-formulations do not contain a dominant identity operator and they produce matrix equations that are potentially ill-conditioned. Finally, when a mixture of coefficients are used from the sets $\{a, b, c, d\}$ and $\{e, f, g, h\}$, mixed formulations are obtained. For example, the JM combined-field integral equation [35] is a mixed formulation when all coefficients are nonzero. Obviously, mixed formulations always contain a dominant identity operator (due to either $\mathcal{I}$ or $\hat{\boldsymbol{n}}\times\mathcal{I}$).

Discretization is an important stage of numerical solutions. All formulations described above can be discretized in different ways such that the derived matrix equations can be well conditioned, and, at the same time, they may produce accurate results. On the other hand, using a Galerkin scheme employing the same set of basis and testing functions, N-formulations and mixed formulations usually produce better-conditioned matrix equations than T-formulations, as mentioned above. In addition, when low-order discretizations are used, the existence of a dominant identity operator is critical in terms of accuracy. It is well known that a discretized identity operator acts like a discretized integro-differential operator with a Dirac-delta kernel [36]. Therefore, a low-order discretization of the identity operator may produce large errors, leading to inaccurate results if the operator is directly tested such that it dominates the matrix equation. RWG discretizations of N-formulations and mixed formulations have this serious drawback, making them less preferred (despite their faster iterative solutions) in comparison to T-formulations in many applications. The tradeoff between the efficiency and accuracy has been resolved in many studies [37] by improving the accuracy of N-formulations and mixed formulations via alternative discretizations and/or by improving the efficiency of T-formulations via preconditioning.

In the context of plasmonic problems, further challenges appear in surface formulations. First, considering that their permittivity values can be written as $\varepsilon_p = \varepsilon_o(-\varepsilon_R + i\varepsilon_I)$, where both ε_R and ε_I are positive, plasmonic objects are naturally high-contrast structures in free space (except for very high frequencies for which $-\varepsilon_R \to 1$). Then, the matrix equations derived from surface formulations can be unbalanced, leading to efficiency and/or accuracy problems. For planar discretizations of curved surfaces, fine discretizations are needed to capture the geometry of the object. At lower frequencies of the optical range, ε_R can be very large (as large as 1000 and beyond) such that the localization of the operators as $\mathcal{T}_p \to -\mathcal{I}/2$ and $\mathcal{K}_{\mathrm{PV},p}-\mathcal{I}/2 \to -\mathcal{I}/2$ when $\varepsilon_R \to \infty$ leads to numerical problems if the blocks are not weighted properly (that occurs in many conventional formulations). While the well-known perfectly conducting models may be used at lower frequencies, it may not be obvious where the plasmonic model can be omitted for a given structure. Hence, it is desirable to extend the applicability of the surface integral equations in wide-frequency ranges until other kinds of approaches can safely be used. In a recent study, we show that a new tangential formulation, namely MCTF, provides reliable and convergent solutions in wide ranges of frequencies of the optical spectrum [32]. Considering the general form, MCTF is obtained by using $a = b = 1$ and $c = d = \eta_o\eta_p$, while setting $e = f = g = h = 0$. Therefore, we obtain

$$\mathcal{Z}_{11}^{\mathrm{MCTF}} = -\hat{\boldsymbol{n}}\times\hat{\boldsymbol{n}}\times(\eta_o\mathcal{T}_o + \eta_p\mathcal{T}_p) \tag{9}$$

$$\mathcal{Z}_{12}^{\mathrm{MCTF}} = \hat{\boldsymbol{n}}\times\hat{\boldsymbol{n}}\times(\mathcal{K}_{\mathrm{PV},o} + \mathcal{K}_{\mathrm{PV},p}) \tag{10}$$

$$\mathcal{Z}_{21}^{\mathrm{MCTF}} = -\hat{\boldsymbol{n}}\times\hat{\boldsymbol{n}}\times\eta_o\eta_p(\mathcal{K}_{\mathrm{PV},o} + \mathcal{K}_{\mathrm{PV},p}) \tag{11}$$

$$\mathcal{Z}_{22}^{\mathrm{MCTF}} = -\hat{\boldsymbol{n}}\times\hat{\boldsymbol{n}}\times(\eta_p\mathcal{T}_o + \eta_o\mathcal{T}_p). \tag{12}$$

It can be observed that MCTF is completely free of the identity operator, and it can be shown that it smoothly turns into the electric-field integral equation for perfectly conducting objects as the frequency drops and ε_R goes to infinity. In the following, we consider numerical solutions of plasmonic problems formulated with MCTF.

3. Discretization

Similar to the diversity of surface integral equations, discretization can be performed in alternative ways. Using a Galerkin scheme, the basis and testing functions are selected as the same set of N functions locally defined on the surface. As a popular choice for triangular discretizations, which is also considered in this chapter, the RWG functions are defined as [33]

$$\boldsymbol{f}_n(\boldsymbol{r}) = \begin{cases} \dfrac{l_n}{2A_{n1}}(\boldsymbol{r}-\boldsymbol{r}_{n1}), & \boldsymbol{r} \in S_{n1} \\ \dfrac{l_n}{2A_{n2}}(\boldsymbol{r}_{n2}-\boldsymbol{r}), & \boldsymbol{r} \in S_{n2} \\ 0, & \boldsymbol{r} \notin S_n. \end{cases} \tag{13}$$

Each RWG function is located on a pair of triangles sharing an edge. In the above, l_n represents the length of the main edge, A_{n1} and A_{n2} are, respectively, the areas of the first (S_{n1}) and the second (S_{n2}) triangles, and $\boldsymbol{r}_{n1}$ and $\boldsymbol{r}_{n2}$ represent the coordinates of the nodes opposite of the edge. The RWG functions are divergence conforming and their divergence is finite everywhere, that is,

$$\nabla \cdot \boldsymbol{f}_n(\boldsymbol{r}) = \begin{cases} \dfrac{l_n}{A_{n1}}, & \boldsymbol{r} \in S_{n1} \\ -\dfrac{l_n}{A_{n2}}, & \boldsymbol{r} \in S_{n2} \\ 0, & \boldsymbol{r} \notin S_n, \end{cases} \tag{14}$$

while the charge neutrality is satisfied locally as $A_{n1}l_n/A_{n1} - A_{n2}l_n/A_{n2} = 0$.

By selecting the basis and testing functions ($\boldsymbol{b}_n$ and $\boldsymbol{t}_m$ for $\{n, m\} = \{1, 2, \ldots, N\}$) as the same set of the RWG functions, MCTF can be discretized as

$$\begin{bmatrix} \overline{\boldsymbol{Z}}_{11}^{\text{MCTF}} & \overline{\boldsymbol{Z}}_{12}^{\text{MCTF}} \\ \overline{\boldsymbol{Z}}_{21}^{\text{MCTF}} & \overline{\boldsymbol{Z}}_{22}^{\text{MCTF}} \end{bmatrix} \cdot \begin{bmatrix} \boldsymbol{a}_J \\ \boldsymbol{a}_M \end{bmatrix} = \begin{bmatrix} \boldsymbol{w}_1^{\text{MCTF}} \\ \boldsymbol{w}_2^{\text{MCTF}} \end{bmatrix}, \tag{15}$$

where $\boldsymbol{a}_J$ and $\boldsymbol{a}_M$ are vectors containing complex coefficients to expand the current densities. The matrix elements and the elements of the right-hand-side vector are derived as

$$\overline{\boldsymbol{Z}}_{11}^{\text{MCTF}} = \eta_o \overline{\boldsymbol{T}}_o^T + \eta_p \overline{\boldsymbol{T}}_p^T \tag{16}$$

$$\overline{\boldsymbol{Z}}_{12}^{\text{MCTF}} = -\overline{\boldsymbol{K}}_{\text{PV},o}^T - \overline{\boldsymbol{K}}_{\text{PV},p}^T \tag{17}$$

$$\overline{\boldsymbol{Z}}_{21}^{\text{MCTF}} = \eta_o \eta_p (\overline{\boldsymbol{K}}_{\text{PV},o}^T + \overline{\boldsymbol{K}}_{\text{PV},p}^T) \tag{18}$$

$$\overline{\boldsymbol{Z}}_{22}^{\text{MCTF}} = \eta_p \overline{\boldsymbol{T}}_o^T + \eta_o \overline{\boldsymbol{T}}_p^T \tag{19}$$

and

$$w_1^{\text{MCTF}} = -\int_{S_m} d\boldsymbol{r}\boldsymbol{t}_m(\boldsymbol{r}) \cdot \boldsymbol{E}^{\text{inc}}(\boldsymbol{r}) \tag{20}$$

$$w_2^{\text{MCTF}} = -\eta_o \eta_p \int_{S_m} d\boldsymbol{r}\boldsymbol{t}_m(\boldsymbol{r}) \cdot \boldsymbol{H}^{\text{inc}}(\boldsymbol{r}), \tag{21}$$

respectively. Furthermore, the discretized operators can be written as

$$\overline{\boldsymbol{T}}_u^T[m, n] = ik_u \int_{S_m} d\boldsymbol{r}\boldsymbol{t}_m(\boldsymbol{r}) \cdot \int_{S_n} d\boldsymbol{r}' g_u(\boldsymbol{r}, \boldsymbol{r}')\boldsymbol{b}_n(\boldsymbol{r}') + \frac{i}{k_u} \int_{S_m} d\boldsymbol{r}\boldsymbol{t}_m(\boldsymbol{r}) \cdot \int_{S_n} d\boldsymbol{r}' \nabla g_u(\boldsymbol{r}, \boldsymbol{r}') \nabla' \cdot \boldsymbol{b}_n(\boldsymbol{r}') \tag{22}$$

$$\overline{\boldsymbol{K}}_{\text{PV}, u}^T[m, n] = \int_{S_m} d\boldsymbol{r}\boldsymbol{t}_m(\boldsymbol{r}) \cdot \int_{\text{PV}, S_n} d\boldsymbol{r}' \boldsymbol{b}_n(\boldsymbol{r}') \times \nabla' g_u(\boldsymbol{r}, \boldsymbol{r}'), \tag{23}$$

where the integrals are evaluated on the supports of the testing and basis functions (S_m and S_n).

At this stage, we can consider the interaction of two half RWG functions associated with the ath triangle of the mth edge and bth triangle of the nth edge, respectively ($\{a, b\} = \{1, 2\}$). One can obtain

$$\begin{aligned} \overline{\boldsymbol{T}}_u^T[m, n, a, b] = {} & \frac{\gamma_{ma}\gamma_{nb} l_m l_n}{4} ik_u \frac{1}{A_{ma}} \int_{S_{ma}} d\boldsymbol{r}(\boldsymbol{r} - \boldsymbol{r}_{ma}) \cdot \frac{1}{A_{nb}} \int_{S_{nb}} d\boldsymbol{r}'(\boldsymbol{r}' - \boldsymbol{r}_{nb}) g_u(\boldsymbol{r}, \boldsymbol{r}') \\ & - \gamma_{ma}\gamma_{nb} l_m l_n \frac{i}{k_u} \frac{1}{A_{ma}} \int_{S_{ma}} d\boldsymbol{r} \frac{1}{A_{nb}} \int_{S_{nb}} d\boldsymbol{r}' g_u(\boldsymbol{r}, \boldsymbol{r}') \end{aligned} \tag{24}$$

$$\overline{\boldsymbol{K}}_{\text{PV}, u}^T[m, n, a, b] = \frac{\gamma_{ma}\gamma_{nb} l_m l_n}{4} \frac{1}{A_{ma}} \int_{S_{ma}} d\boldsymbol{r}(\boldsymbol{r} - \boldsymbol{r}_{ma}) \cdot (\boldsymbol{r} - \boldsymbol{r}_{nb}) \times \frac{1}{A_{nb}} \int_{\text{PV}, S_{nb}} d\boldsymbol{r}' \nabla' g_u(\boldsymbol{r}, \boldsymbol{r}'), \tag{25}$$

where $\gamma_{nb}, \gamma_{ma} = \pm 1$, depending on the direction of the basis and testing functions on triangles. For the integrations on the testing and basis triangles, alternative methods can be used. Applying Gaussian quadrature is common in the literature, if the singularity of Green's function is extracted from the inner integrals. In any case, the integration methods used on the testing and basis triangles do not have to be the same, that is, different sampling schemes can be used. For the sake of brevity, we consider a single-point testing scheme by using the center point of each triangle $\boldsymbol{r}_{ma}^{\text{cr}}$, leading to

$$\begin{aligned} \overline{\boldsymbol{T}}_u^T[m, n, a, b] & = \frac{\gamma_{ma}\gamma_{nb} l_m l_n}{4} ik_u (\boldsymbol{r}_{ma}^{\text{cr}} - \boldsymbol{r}_{ma}) \cdot \frac{1}{A_{nb}} \int_{S_{nb}} d\boldsymbol{r}'(\boldsymbol{r}' - \boldsymbol{r}_{nb}) g_u(\boldsymbol{r}_{ma}^{\text{cr}}, \boldsymbol{r}') \\ & - \gamma_{ma}\gamma_{nb} l_m l_n \frac{i}{k_u} \frac{1}{A_{nb}} \int_{S_{nb}} d\boldsymbol{r}' g_u(\boldsymbol{r}_{ma}^{\text{cr}}, \boldsymbol{r}') \end{aligned} \tag{26}$$

$$\begin{aligned} \overline{\boldsymbol{T}}_u^T[m, n, a, b] & = \frac{\gamma_{ma}\gamma_{nb} l_m l_n}{4} ik_u (\boldsymbol{\rho}_{ma}^{\text{cr}} - \boldsymbol{\rho}_{ma}) \cdot \frac{1}{A_{nb}} \int_{S_{nb}} d\boldsymbol{r}'(\boldsymbol{\rho}' - \boldsymbol{\rho}_{ma}^{\text{cr}}) g_u(\boldsymbol{r}_{ma}^{\text{cr}}, \boldsymbol{r}') \\ & + \frac{\gamma_{ma}\gamma_{nb} l_m l_n}{4} ik_u (\boldsymbol{\rho}_{ma}^{\text{cr}} - \boldsymbol{\rho}_{ma}) \cdot (\boldsymbol{\rho}_{ma}^{\text{cr}} - \boldsymbol{\rho}_{nb}) \frac{1}{A_{nb}} \int_{S_{nb}} d\boldsymbol{r}' g_u(\boldsymbol{r}_{ma}^{\text{cr}}, \boldsymbol{r}') \\ & - \gamma_{ma}\gamma_{nb} l_m l_n \frac{i}{k_u} \frac{1}{A_{nb}} \int_{S_{nb}} d\boldsymbol{r}' g_u(\boldsymbol{r}_{ma}^{\text{cr}}, \boldsymbol{r}') \end{aligned} \tag{27}$$

$$\overline{\boldsymbol{K}}^{T}_{\mathrm{PV},\,u}[m,n,a,b] = \frac{\gamma_{ma}\gamma_{nb}l_m l_n}{4}(\boldsymbol{r}^{\mathrm{cr}}_{ma}-\boldsymbol{r}_{ma})\cdot(\boldsymbol{r}^{\mathrm{cr}}_{ma}-\boldsymbol{r}_{nb})\times\frac{1}{A_{nb}}\int_{\mathrm{PV},\,S_{nb}} d\boldsymbol{r}'\nabla' g_u(\boldsymbol{r}^{\mathrm{cr}}_{ma},\boldsymbol{r}'), \tag{28}$$

where $\{\boldsymbol{\rho}', \boldsymbol{\rho}_{ma}, \boldsymbol{\rho}_{nb}, \boldsymbol{\rho}^{\mathrm{cr}}_{ma}\}$ represent the projections of $\{\boldsymbol{r}', \boldsymbol{r}_{ma}, \boldsymbol{r}_{nb}, \boldsymbol{r}^{\mathrm{cr}}_{ma}\}$ onto the basis plane.

It is generally more efficient to compute the interactions via triangle by triangle (rather than RWG by RWG) since common integrals related to a basis triangle can be evaluated once and used in multiple interactions related to the triangle. For MCTF, interactions are calculated (for $a = 1, 2$ and $b = 1, 2$, and $u = o, p$) as

$$\overline{\boldsymbol{Z}}^{\mathrm{MCTF}}_{11}[m,n] \leftarrow \frac{\gamma_{ma}\gamma_{nb}l_m l_n}{4} ik_u\eta_u\left\{(\boldsymbol{\rho}^{\mathrm{cr}}_{ma}-\boldsymbol{\rho}_{ma})\cdot[\boldsymbol{I}^{B}_{ma,\,nb,\,u}+(\boldsymbol{\rho}^{\mathrm{cr}}_{ma}-\boldsymbol{\rho}_{nb})I^{A}_{ma,\,nb,\,u}]-\frac{4}{k_u^2}I^{A}_{ma,\,nb,\,u}\right\} \tag{29}$$

$$\overline{\boldsymbol{Z}}^{\mathrm{MCTF}}_{22}[m,n] \leftarrow \frac{\gamma_{ma}\gamma_{nb}l_m l_n}{4} ik_u\frac{\eta_o\eta_p}{\eta_u}\left\{(\boldsymbol{\rho}^{\mathrm{cr}}_{ma}-\boldsymbol{\rho}_{ma})\cdot[\boldsymbol{I}^{B}_{ma,\,nb,\,u}+(\boldsymbol{\rho}^{\mathrm{cr}}_{ma}-\boldsymbol{\rho}_{nb})I^{A}_{ma,\,nb,\,u}]-\frac{4}{k_u^2}I^{A}_{ma,\,nb,\,u}\right\} \tag{30}$$

$$\overline{\boldsymbol{Z}}^{\mathrm{MCTF}}_{12}[m,n] \leftarrow -\frac{\gamma_{ma}\gamma_{nb}l_m l_n}{4}(\boldsymbol{r}^{\mathrm{cr}}_{ma}-\boldsymbol{r}_{ma})\cdot[(\boldsymbol{r}^{\mathrm{cr}}_{ma}-\boldsymbol{r}_{nb})\times\boldsymbol{I}^{C,\,\mathrm{PV}}_{ma,\,nb,\,u}] \tag{31}$$

$$\overline{\boldsymbol{Z}}^{\mathrm{MCTF}}_{21}[m,n] \leftarrow \frac{\gamma_{ma}\gamma_{nb}l_m l_n}{4}\eta_o\eta_p(\boldsymbol{r}^{\mathrm{cr}}_{ma}-\boldsymbol{r}_{ma})\cdot[(\boldsymbol{r}^{\mathrm{cr}}_{ma}-\boldsymbol{r}_{nb})\times\boldsymbol{I}^{C,\,\mathrm{PV}}_{ma,\,nb,\,u}], \tag{32}$$

where $\leftarrow$ indicates the update operation. Each matrix element is obtained by combining the contributions of four triangle-triangle interactions. By using triangle-triangle interactions, a basis integral ($I^{A}_{ma,\,nb,\,u}$, $\boldsymbol{I}^{B}_{ma,\,nb,\,u}$, or $\boldsymbol{I}^{C,\,\mathrm{PV}}_{ma,\,nb,\,u}$) are used in nine different RWG-RWG interactions. These common integrals (with singularity extractions) can be listed as

$$\begin{aligned} I^{A}_{ma,\,nb,\,u} &= \frac{1}{A_{nb}}\int_{S_{nb}} d\boldsymbol{r}' g_u(\boldsymbol{r}^{\mathrm{cr}}_{ma},\boldsymbol{r}') = \frac{1}{A_{nb}}\int_{S_{nb}} d\boldsymbol{r}'\left[g_u(\boldsymbol{r}^{\mathrm{cr}}_{ma},\boldsymbol{r}')-\frac{1}{4\pi|\boldsymbol{r}^{\mathrm{cr}}_{ma}-\boldsymbol{r}'|}\right] \\ &\quad+\frac{1}{A_{nb}}\int_{S_{nb}} d\boldsymbol{r}'\frac{1}{4\pi|\boldsymbol{r}^{\mathrm{cr}}_{ma}-\boldsymbol{r}'|} \end{aligned} \tag{33}$$

$$\begin{aligned} \boldsymbol{I}^{B}_{ma,\,nb,\,u} &= \frac{1}{A_{nb}}\int_{S_{nb}} d\boldsymbol{r}'(\boldsymbol{\rho}'-\boldsymbol{\rho}^{\mathrm{cr}}_{ma})g_u(\boldsymbol{r}^{\mathrm{cr}}_{ma},\boldsymbol{r}') = \frac{1}{A_{nb}}\int_{S_{nb}} d\boldsymbol{r}'(\boldsymbol{\rho}'-\boldsymbol{\rho}^{\mathrm{cr}}_{ma})\left[g_u(\boldsymbol{r}^{\mathrm{cr}}_{ma},\boldsymbol{r}')-\frac{1}{4\pi|\boldsymbol{r}^{\mathrm{cr}}_{ma}-\boldsymbol{r}'|}\right] \\ &\quad+\frac{1}{A_{nb}}\int_{S_{nb}} d\boldsymbol{r}'(\boldsymbol{\rho}'-\boldsymbol{\rho}^{\mathrm{cr}}_{ma})\frac{1}{4\pi|\boldsymbol{r}^{\mathrm{cr}}_{ma}-\boldsymbol{r}'|} \end{aligned} \tag{34}$$

$$\begin{aligned} \boldsymbol{I}^{C,\,\mathrm{PV}}_{ma,\,nb,\,u} &= \frac{1}{A_{nb}}\int_{\mathrm{PV},\,S_{nb}} d\boldsymbol{r}'\nabla' g_u(\boldsymbol{r}^{\mathrm{cr}}_{ma},\boldsymbol{r}') = \frac{1}{A_{nb}}\int_{\mathrm{PV},\,S_{nb}} d\boldsymbol{r}'\nabla'\left[g_u(\boldsymbol{r}^{\mathrm{cr}}_{ma},\boldsymbol{r}')-\frac{1}{4\pi|\boldsymbol{r}^{\mathrm{cr}}_{ma}-\boldsymbol{r}'|}\right] \\ &\quad+\frac{1}{A_{nb}}\int_{\mathrm{PV},\,S_{nb}} d\boldsymbol{r}'\nabla'\left(\frac{1}{4\pi|\boldsymbol{r}^{\mathrm{cr}}_{ma}-\boldsymbol{r}'|}\right). \end{aligned} \tag{35}$$

Using the same convention and single-point testing, the elements of the right-hand-side vectors are evaluated as

$$\boldsymbol{w}_1^{\text{MCTF}}[m] \leftarrow -\frac{\gamma_{ma} l_m}{2}(\boldsymbol{r}_{ma}^{\text{cr}} - \boldsymbol{r}_{ma}) \cdot \boldsymbol{E}^{\text{inc}}(\boldsymbol{r}_{ma}^{\text{cr}}) \tag{36}$$

$$\boldsymbol{w}_2^{\text{MCTF}}[m] \leftarrow -\frac{\gamma_{ma} l_m}{2}\eta_o \eta_p(\boldsymbol{r}_{ma}^{\text{cr}} - \boldsymbol{r}_{ma}) \cdot \boldsymbol{H}^{\text{inc}}(\boldsymbol{r}_{ma}^{\text{cr}}) \tag{37}$$

for $a = \{1, 2\}$.

Matrix equations obtained as summarized in this section can be solved in different ways, particularly via iterative techniques accelerated via fast algorithms. Once the current coefficients $\boldsymbol{a}_J$ and $\boldsymbol{a}_M$ are found, electric and magnetic fields can be obtained at any location inside or outside the object. Using the RWG functions, secondary fields can be written as

$$\boldsymbol{E}^{\text{sec}}(\boldsymbol{r}) = \sum_{n=1}^{N}\sum_{b=1}^{2} \boldsymbol{a}_J[n]\frac{\gamma_{nb} l_n}{2} ik_u \eta_u \left\{ (\rho - \rho_{nb}) I_{nb,u}^{A}(\boldsymbol{r}) + \boldsymbol{I}_{nb,u}^{B}(\boldsymbol{r}) - \frac{2}{k_u^2}\boldsymbol{I}_{nb,u}^{C}(\boldsymbol{r}) \right\}$$

$$-\sum_{n=1}^{N}\sum_{b=1}^{2} \boldsymbol{a}_M[n]\frac{\gamma_{nb} l_n}{2}(\boldsymbol{r} - \boldsymbol{r}_{nb}) \times \boldsymbol{I}_{nb,u}^{C}(\boldsymbol{r}) \tag{38}$$

$$\boldsymbol{H}^{\text{sec}}(\boldsymbol{r}) = \sum_{n=1}^{N}\sum_{b=1}^{2} \boldsymbol{a}_M[n]\frac{\gamma_{nb} l_n}{2}\frac{ik_u}{\eta_u} \left\{ (\rho - \rho_{nb}) I_{nb,u}^{A}(\boldsymbol{r}) + \boldsymbol{I}_{nb,u}^{B}(\boldsymbol{r}) - \frac{2}{k_u^2}\boldsymbol{I}_{nb,u}^{C}(\boldsymbol{r}) \right\}$$

$$+\sum_{n=1}^{N}\sum_{b=1}^{2} \boldsymbol{a}_J[n]\frac{\gamma_{nb} l_n}{2}(\boldsymbol{r} - \boldsymbol{r}_{nb}) \times \boldsymbol{I}_{nb,u}^{C}(\boldsymbol{r}), \tag{39}$$

where

$$I_{nb,u}^{A}(\boldsymbol{r}) = \frac{1}{A_{nb}}\int_{S_{nb}} d\boldsymbol{r}' g_u(\boldsymbol{r}, \boldsymbol{r}') \tag{40}$$

$$\boldsymbol{I}_{nb,u}^{B}(\boldsymbol{r}) = \frac{1}{A_{nb}}\int_{S_{nb}} d\boldsymbol{r}'(\rho' - \rho) g_u(\boldsymbol{r}, \boldsymbol{r}') \tag{41}$$

$$\boldsymbol{I}_{nb,u}^{C}(\boldsymbol{r}) = \frac{1}{A_{nb}}\int_{S_{nb}} d\boldsymbol{r}' \nabla' g_u(\boldsymbol{r}, \boldsymbol{r}'). \tag{42}$$

Similar to the matrix elements, a triangle loop (rather than an RWG loop) can be used to efficiently perform the near-field computations. If the observation point **r** is close to the surface of the object, singularity extractions must be used for accurate integrations. If the medium parameters are set to ε_p and μ_p, the computations above lead to inner electromagnetic fields, while the fields outside the surface vanish due to the equivalence theorem. In fact, this can be used to assess the accuracy of numerical solutions, since any nonzero field outside corresponds to a numerical error. Similarly, using ε_o and μ_o, inner fields must be zero, while secondary fields are obtained outside. Then, the total fields outside the object can be obtained as

$$E(r) = E^{\text{inc}}(r) + E^{\text{sec}}(r) \tag{43}$$

$$H(r) = H^{\text{inc}}(r) + H^{\text{sec}}(r). \tag{44}$$

4. Matrix-vector multiplications with MLFMA

Plasmonic problems often involve large structures in terms of wavelength. In addition, typical $\lambda_o/10$ triangulations may not be sufficient to obtain accurate results, and dense discretizations are usually needed, leading to a large number of unknowns. Since direct solutions (e.g., Gaussian elimination) of the resulting matrix equations may not be feasible, fast iterative solvers are required for efficient analysis of plasmonic structures in reasonable processing times and using available memory. MLFMA is an efficient algorithm that can be used to perform fast matrix-vector multiplications with $\mathcal{O}(N\log N)$ complexity for an $N\times N$ dense matrix equation derived from an electrodynamic problem [25, 26]. Hence, MLFMA can be used within a Krylov subspace algorithm, such as the generalized minimal residual (GMRES) method, for efficient iterative solutions.

MLFMA is well known in the literature as a method with controllable accuracy. In practice, however, its accuracy heavily depends on the expansion method. In the most standard form, plane waves are used to diagonalize the addition theorem for Green's function. Then, the interaction distances, hence, the recursive clustering of the electrodynamic interactions, are limited by a low-frequency breakdown. For example, two to three digits of accuracy (1% and 0.1% maximum relative error) using a one-box-buffer scheme need a minimum box size of around λ_u. It is possible to use smaller boxes and/or to achieve higher accuracy, if alternative expansion tools [38, 39], such as a direct application of multipoles [40] or evanescent waves [41], are employed. In this chapter, where numerical solutions are performed with maximum 1% error, we restrict ourselves to the plane-wave expansion.

Using plane waves, Green's function is decomposed as

$$g_u(r, r') = \frac{\exp(ik_u|r-r'|)}{4\pi|r-r'|} = \frac{\exp(ik_u|w+v|)}{4\pi|w+v|} \approx \frac{ik_u}{(4\pi)^2}\int d^2\hat{k}\beta(k_u, v)\alpha_\tau(k_u, w) \tag{45}$$

for $w = |w| > v = |v|$, where $k_u = \hat{k}k_u$, $d^2\hat{k} = d\theta d\phi \sin\theta$, and

$$\beta(k_u, v) = \exp(ik_u \cdot v) \tag{46}$$

$$\alpha_\tau(k_u, w) = \sum_{t=0}^{\tau}(i)^t(2t+1)h_t^{(1)}(k_u w)P_t(\hat{k}\cdot\hat{w}) \tag{47}$$

are diagonal shift and translation operators, respectively. It is remarkable that, as a result of the factorization, the shift vector v and the translation vector w, which satisfy $w + v = r-r'$, are separated. In addition, with the help of the diagonalization, sampling of the shift and translation operators leads to diagonal matrices, as the shift or translation of a plane wave in a given direction does not contribute to plane waves in other directions. In the above, the translation operator is written in terms of the Legendre polynomials P_t and the spherical Hankel function of the first kind $h_t^{(1)}$, while τ is the truncation number that can be found via excess-bandwidth formulas [42]. We note that the derivatives of Green's function can also be obtained as

$$\nabla g_u(\boldsymbol{r},\boldsymbol{r}') \approx \frac{ik_u}{(4\pi)^2}\int d^2\hat{\boldsymbol{k}}(i\boldsymbol{k}_u)\beta(\boldsymbol{k}_u,\boldsymbol{v})\alpha_\tau(\boldsymbol{k}_u,\boldsymbol{w}) \tag{48}$$

$$\nabla\nabla' g_u(\boldsymbol{r},\boldsymbol{r}') \approx \frac{ik_u}{(4\pi)^2}\int d^2\hat{\boldsymbol{k}}(\boldsymbol{k}_u\boldsymbol{k}_u)\beta(\boldsymbol{k}_u,\boldsymbol{v})\alpha_\tau(\boldsymbol{k}_u,\boldsymbol{w}). \tag{49}$$

These expressions can directly be used to factorize the discretized operators by replacing Green's function with the diagonalized forms. In the context of MCTF, we have

$$\overline{\boldsymbol{T}}_u^T[m,n,a,b] = \left(\frac{ik_u}{4\pi}\right)^2\int d^2\hat{\boldsymbol{k}}\boldsymbol{R}_{ma}^{\mathcal{T}}(\boldsymbol{k}_u,\boldsymbol{r}_C)\cdot\alpha_\tau(\boldsymbol{k}_u,\boldsymbol{r}_C-\boldsymbol{r}_{C'})\boldsymbol{S}_{nb}(\boldsymbol{k}_u,\boldsymbol{r}_{C'}) \tag{50}$$

$$\overline{\boldsymbol{K}}_{\mathrm{PV},u}^T[m,n,a,b] = \left(\frac{ik_u}{4\pi}\right)^2\int d^2\hat{\boldsymbol{k}}\boldsymbol{R}_{ma}^{\mathcal{K}}(\boldsymbol{k}_u,\boldsymbol{r}_C)\cdot\alpha_\tau(\boldsymbol{k}_u,\boldsymbol{r}_C-\boldsymbol{r}_{C'})\boldsymbol{S}_{nb}(\boldsymbol{k}_u,\boldsymbol{r}_{C'}), \tag{51}$$

where $\boldsymbol{r}_C$ and $\boldsymbol{r}_{C'}$ are testing and basis centers, respectively. Using the RWG functions, the radiation and receiving patterns of the half basis and testing functions are derived as

$$\boldsymbol{S}_{nb}(\boldsymbol{k}_u,\boldsymbol{r}_{C'}) = \frac{\gamma_{nb}l_n}{2}(\overline{\boldsymbol{I}}_{3\times 3}-\hat{\boldsymbol{k}}\hat{\boldsymbol{k}})\cdot\frac{1}{A_{nb}}\int_{S_{nb}} d\boldsymbol{r}'\beta(\boldsymbol{k}_u,\boldsymbol{r}_{C'}-\boldsymbol{r}')(\boldsymbol{r}'-\boldsymbol{r}_{nb}) \tag{52}$$

$$\boldsymbol{R}_{ma}^{\mathcal{T}}(\boldsymbol{k}_u,\boldsymbol{r}_C) = \frac{\gamma_{ma}l_m}{2}(\overline{\boldsymbol{I}}_{3\times 3}-\hat{\boldsymbol{k}}\hat{\boldsymbol{k}})\cdot\frac{1}{A_{ma}}\int_{S_{ma}} d\boldsymbol{r}\beta(\boldsymbol{k}_u,\boldsymbol{r}-\boldsymbol{r}_C)(\boldsymbol{r}-\boldsymbol{r}_{ma}) \tag{53}$$

$$\boldsymbol{R}_{ma}^{\mathcal{K}}(\boldsymbol{k}_u,\boldsymbol{r}_C) = -\frac{\gamma_{ma}l_m}{2}\hat{\boldsymbol{k}}\times\frac{1}{A_{ma}}\int_{S_{ma}} d\boldsymbol{r}\beta(\boldsymbol{k}_u,\boldsymbol{r}-\boldsymbol{r}_C)(\boldsymbol{r}-\boldsymbol{r}_{ma}) = -\hat{\boldsymbol{k}}\times\boldsymbol{R}_{ma}^{\mathcal{T}}(\boldsymbol{k}_u,\boldsymbol{r}_C), \tag{54}$$

where $\overline{\boldsymbol{I}}_{3\times 3} = \hat{\boldsymbol{k}}\hat{\boldsymbol{k}} + \hat{\boldsymbol{\theta}}\hat{\boldsymbol{\theta}} + \hat{\boldsymbol{\phi}}\hat{\boldsymbol{\phi}}$. Using a single-point integration, the patterns can be calculated as

$$\boldsymbol{S}_{nb}(\boldsymbol{k}_u,\boldsymbol{r}_{C'}) = \frac{\gamma_{nb}l_n}{2}\beta(\boldsymbol{k}_u,\boldsymbol{r}_{C'}-\boldsymbol{r}_{nb}^{\mathrm{cr}})(\overline{\boldsymbol{I}}^{3\times 3}-\hat{\boldsymbol{k}}\hat{\boldsymbol{k}})\cdot(\boldsymbol{r}_{nb}^{\mathrm{cr}}-\boldsymbol{r}_{nb}) \tag{55}$$

$$\boldsymbol{R}_{ma}^{\mathcal{T}}(\boldsymbol{k}_u,\boldsymbol{r}_C) = \frac{\gamma_{ma}l_m}{2}\beta(\boldsymbol{k}_u,\boldsymbol{r}_{ma}^{\mathrm{cr}}-\boldsymbol{r}_C)(\overline{\boldsymbol{I}}^{3\times 3}-\hat{\boldsymbol{k}}\hat{\boldsymbol{k}})\cdot(\boldsymbol{r}_{ma}^{\mathrm{cr}}-\boldsymbol{r}_{ma}) \tag{56}$$

$$\boldsymbol{R}_{ma}^{\mathcal{K}}(\boldsymbol{k}_u,\boldsymbol{r}_C) = -\hat{\boldsymbol{k}}\times\boldsymbol{R}_{ma}^{\mathcal{T}}(\boldsymbol{k}_u,\boldsymbol{r}_C), \tag{57}$$

where $\boldsymbol{r}_{ma}^{\mathrm{cr}}$ and $\boldsymbol{r}_{nb}^{\mathrm{cr}}$ represent the centers of the associated testing and basis triangles, respectively. Then, the radiation/receiving patterns of the full RWG functions can be obtained as $\boldsymbol{S}_n = \boldsymbol{S}_{n1} + \boldsymbol{S}_{n2}$ and $\boldsymbol{R}_m^{\mathcal{K},\mathcal{T}} = \boldsymbol{R}_{m1}^{\mathcal{K},\mathcal{T}} + \boldsymbol{R}_{m2}^{\mathcal{K},\mathcal{T}}$ by combining the contributions of the half functions. These patterns, as well as the truncation operator, are sampled on the unit sphere, where the sampling scheme is a matter choice depending on the implementation.

In a standard implementation of MLFMA, the object is placed inside a computational cubic box, which is divided into sub-boxes until the smallest possible box size determined by the desired accuracy. Empty boxes that do not contain a part of the object (discretized surface) are omitted directly and are not divided further. This way, it is possible to construct a tree structure (consisting of $L = \mathcal{O}(\log N)$ levels) involving nonempty boxes at different levels with $\mathcal{O}(N)$ complexity. Using the child/parent relationship between the boxes, the stages of a matrix-vector multiplication, namely aggregation, translation, and disaggregation, are as follows.

In an aggregation stage, radiated fields of boxes are computed from bottom to top. At the lowest level, we have

$$\boldsymbol{a}[n]\boldsymbol{S}_n(\boldsymbol{k}_u, \boldsymbol{r}_{C'}) \rightarrow \boldsymbol{S}_{C'}(\boldsymbol{k}_u, \boldsymbol{r}_{C'}), \qquad (\boldsymbol{b}_n \in C'), \tag{58}$$

where the coefficients provided by the iterative solver are used to weight the contributions of the basis functions to the overall radiation patterns of the boxes C' at the lowest level. At higher levels ($l = 2, 3, \ldots, L$), aggregation is performed recursively as

$$\beta(\boldsymbol{k}_u, \boldsymbol{r}_{P\{C'\}} - \boldsymbol{r}_{C'})\boldsymbol{S}_{C'}(\boldsymbol{k}_u, \boldsymbol{r}_{C'}) \rightarrow \boldsymbol{S}_{P\{C'\}}(\boldsymbol{k}_u, \boldsymbol{r}_{P\{C'\}}), \tag{59}$$

where $P\{C'\}$ represents the parent of C'. Due to the exponential shifts from different locations within a box, the radiated fields become more oscillatory as the box size gets larger. Hence, the sampling rate must be increased, generally with $\mathcal{O}(k_u^2 D^2)$ where D is the box size.

After completing an aggregation stage, the radiated fields are translated between the boxes at the same level. For $l = 1, 2, \ldots, L$, this can be written as

$$\alpha_\tau(\boldsymbol{k}_u, \boldsymbol{r}_C - \boldsymbol{r}_{C'})\boldsymbol{S}_{C'}(\boldsymbol{k}_u, \boldsymbol{r}_{C'}) \rightarrow \boldsymbol{G}_C(\boldsymbol{k}_u, \boldsymbol{r}_C), \qquad (C' \in F\{C\}), \tag{60}$$

where $F\{C\}$ represents the far-zone boxes for a given box C. It is remarkable that $F\{C\}$ contains $\mathcal{O}(1)$ elements since interactions between too far boxes, for example, C and C' at level l, are made at a higher level ($l' > l$). Using a one-box-buffer scheme, the condition for translation is that the boxes should not intersect at a surface, line, or corner, while their parents must intersect at a surface, line, or corner.

In a translation stage, incoming fields are collected at the box centers, but they are only partial data, since the total incoming fields at the center of a box contain contribution from its parent (if exists) due to the translations at higher levels. Therefore, a disaggregation stage is performed recursively for $l = L-1, L-2, \ldots, 1$ as

$$\boldsymbol{G}_C(\boldsymbol{k}_u, \boldsymbol{r}_C) + \beta(\boldsymbol{k}_u, \boldsymbol{r}_C - \boldsymbol{r}_{P\{C\}})\boldsymbol{G}^{+}_{P\{C\}}(\boldsymbol{k}_u, \boldsymbol{r}_{P\{C\}}) \rightarrow \boldsymbol{G}^{+}_C(\boldsymbol{k}_u, \boldsymbol{r}_C). \tag{61}$$

At the lowest level, the testing functions receive the incoming fields as

$$\left(\frac{ik_u}{4\pi}\right)^2 \int d^2\hat{\boldsymbol{k}}\, \boldsymbol{R}_m^{\mathcal{K},\mathcal{T}}(\boldsymbol{k}_u, \boldsymbol{r}_C) \cdot \boldsymbol{G}^{+}_C(\boldsymbol{k}_u, \boldsymbol{r}_C) \rightarrow \boldsymbol{y}^{\mathrm{FF}}[m], \quad (\boldsymbol{t}_m \in C), \tag{62}$$

where

$$y^{FF}[m] = \sum_{n=1}^{N} \overline{Z}^{FF}[m, n]a[n] \tag{63}$$

for $m = 1, 2, \ldots, N$. The overall matrix-vector multiplication is completed by also considering the near-field interactions (that cannot be calculated via aggregation-translation-disaggregation stages) as

$$y[m] = y^{FF}[m] + \overline{Z}^{NF}[m, n]a[n]. \tag{64}$$

Using MLFMA, each matrix-vector multiplication can be performed in $\mathcal{O}(N\log N)$ time and using $\mathcal{O}(N\log N)$ memory.

For plasmonic objects with high negative permittivity values, electromagnetic interactions decay quickly with respect to the distance between the observation and source points. For a given accuracy, interactions at long distances can be omitted since the inner and outer interactions are combined in the surface formulations and outer interactions (related to the free space) dominate the related matrix elements [30]. The threshold distance for this purpose can also be found by considering the exponential behavior of the decay for large imaginary values of the wavenumber. This way, the processing time for the matrix-vector multiplication can significantly be reduced. As the negative permittivity increases, far-zone interactions related to the inner medium may completely vanish, leaving only near-zone interactions. In the limit, near-zone interactions further reduce into self interactions of basis/testing functions, leading to the Gram matrix to represent the inner medium.

5. Case example: numerical simulations of nanowires

Using surface integral equations and MLFMA, electrical properties of a plasmonic object are simply parameters, which can be used as variables in the implementations. For the electrical properties, that is, permittivity and permeability, alternative choices, including measurement data and those based on certain models for the materials, can be used. As an example, **Figure** presents the relative permittivity of silver (Ag) with respect to frequency from 200 to 1600 THz. In addition to measurement data [10], Drude (D) and Lorentz-Drude (LD) models are used to predict the real and imaginary parts of the relative permittivity. It can be observed that the real part of the permittivity has large negative values at the lower (infrared) frequencies and it increases smoothly toward unity as the frequency increases to the visible range and beyond. For imaginary values, which represent ohmic losses, we observe varying values between 0.01 and 10, while large discrepancies exist between measurement and D/LD models (especially considering the logarithmic scale of the y-axis). These discrepancies are responsible for different results in the simulations of plasmonic problems presented below.

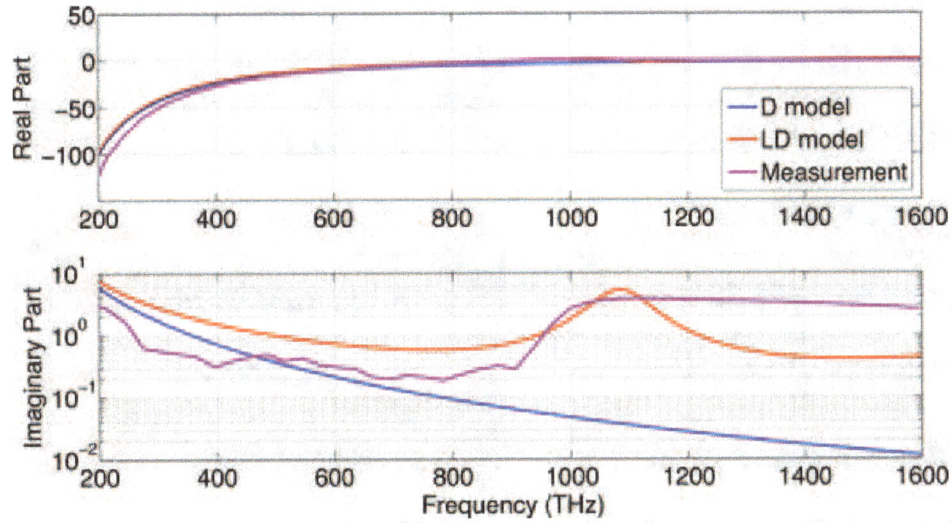

Figure. Real and imaginary parts of the relative permittivity of Ag with respect to frequency. In addition to measurement data [10], values based on Drude (D) and Lorentz-Drude (LD) models are depicted.

As a case study, we consider transmission though a pair of Ag nanowires described in **Figure**. The length of the nanowires is 5μm and each nanowire has 0.1 × 0.1μm (square) cross section. The distance between the nanowires is also 0.1μm. The transmission line is excited by a pair of Hertzian dipoles oriented in the opposite directions and located at 0.2μm distance from the nanowires. **Figure** presents the electromagnetic response of the transmission line in the infrared frequencies from 250 to 430 THz. The power density in dB scale (dBW/m^2) in the vicinity of the nanowires is depicted (normalized to 0 dB and using 40-dB dynamic range), when LD model and measurement data are used for the permittivity values. It can be observed that the electromagnetic power is effectively transmitted from the source region (right) to the transmission region (left). Coupling to the free space at the end of the line leads to two beams with decaying amplitudes due to propagation. Comparing the results, we observe very good agreement between the power density values when LD and measurement permittivity values are used. Considering **Figure**, the negative real permittivity dominates the response of the nanowires at these frequencies.

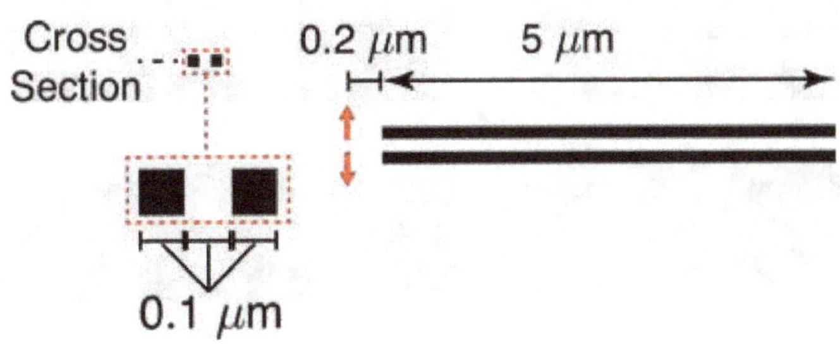

Figure. A transmission line involving two Ag nanowires of length 5μm. The nanowires are excited by a pair of dipoles located at 0.2μm distance.

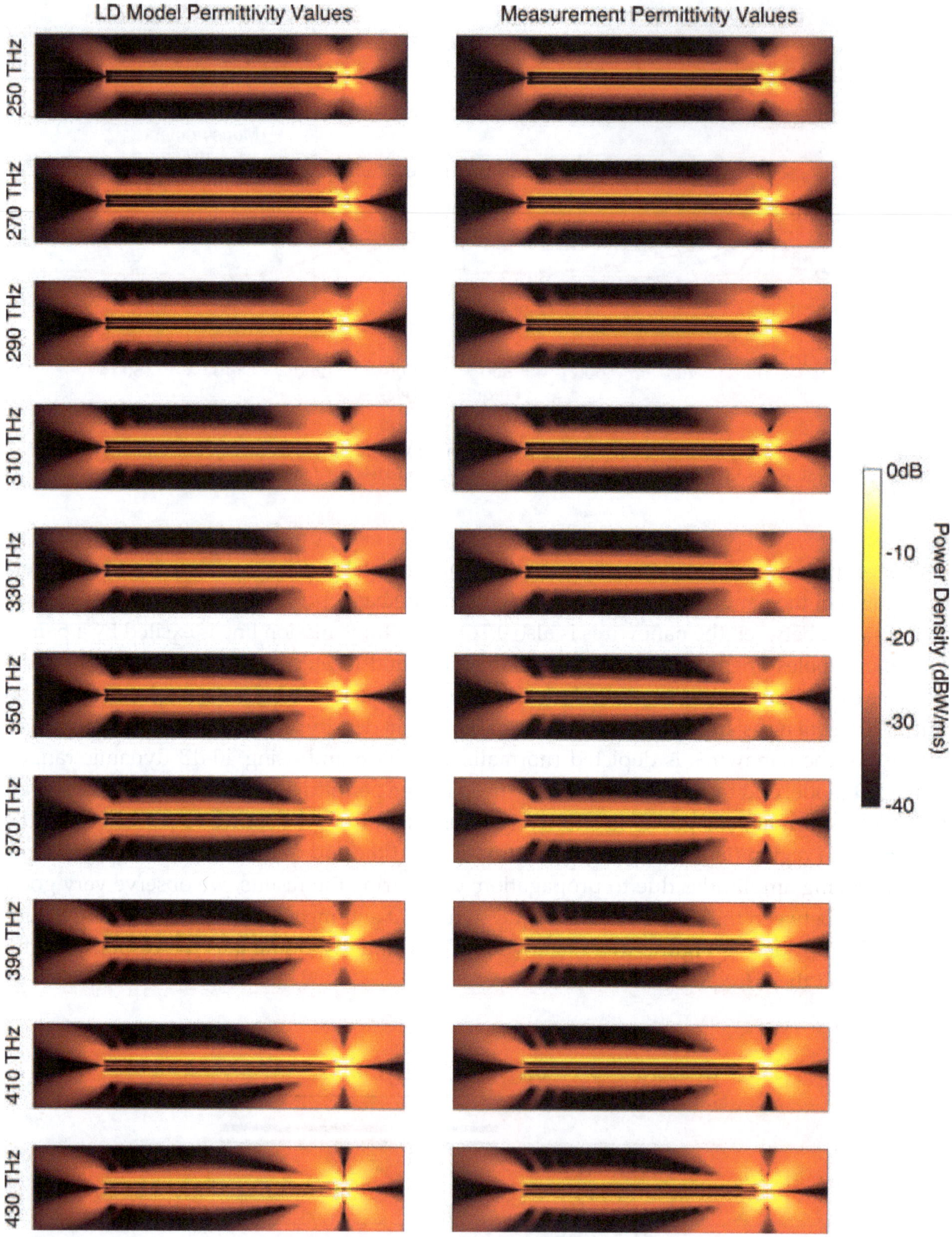

Figure. Power density in the vicinity of the nanowire system (**Figure**) from 250 to 430 THz. Numerical results obtained by using permittivity values derived from the LD model (left column) and those based on the measurement data (right column) are compared.

Figure presents similar results when the frequency is in the visible range. In this case, there are significant discrepancies between the power density values when the LD model and the measurement data for the permittivity are used. This is mainly due to the higher values for the imaginary permittivity predicted by the LD model. As the frequency increases, using the LD model, the transmission ability of the nanowire system deteriorates significantly. Specifically, the power density on the surfaces of the nanowires decreases and the transmitted power toward the left-hand side of the nanowires diminishes, leading to progressively weaker beams. It is remarkable that, using the measurement data that may be more accurate description of Ag, the transmission ability of the nanowire system is still at high levels, indicating that the transmission line operates as desired. These results may explain some of the contradictory results (especially simulations vs. measurements) for the nanowire and similar plasmonic systems investigated in the visible spectrum.

As depicted in **Figure**, nanowires cannot maintain a good transmission ability as the frequency increases. Using the measurement data, the transmission of the nanowire system deteriorates significantly at the higher frequencies of the visible spectrum (e.g., at 770 THz). At 750 THz, the power density drops to less than −40 dB after a few μm along the nanowires. We note that the effective length of the nanowires increases with the frequency. For example, at 250 THz, the length of the nanowires is approximately $4.17\lambda_0$, while it is around $12.5\lambda_0$ at 750 THz. In addition, the effective distance between the sources and the nanowires increases. However, investigating the power values on the nanowire surfaces close to the source, it is obvious that the poor power transmission cannot be explained only with the increasing effective lengths at the higher frequencies. Since the power cannot be coupled to the free space, the power density along the nanowires possesses an oscillatory behavior. At the end of the visible spectrum, the discrepancy between the results obtained by using the LD and measurement values decreases, both predicting reduced interaction between the sources and nanowires.

Figure presents the results even at higher (lower-ultraviolet) frequencies. In this range, the nanowires are not expected to demonstrate transmission abilities, as predicted by both LD model and measurement data for the permittivity values. At lower frequencies of the range, the nanowires are more visible close to the source region, while, as the frequency increases, their effects diminish and the power distribution becomes close to that of two dipoles in free space. **Figures** present the summary of input/output of the transmission line, for the LD model and measurement data, respectively, from 450 to 750 THz. For the input, the power density is sampled at 30 nm distance from the nanowires on a horizontal line from −1to1μm. The double-peak pattern due to two dipoles in opposite directions is clearly visible, with some variations due to reflections from the nanowires. For the output, samples are selected again on a horizontal line from −1t o 1μm in the transmission side at 40 μm distance from the nanowires. Using the LD model, the output pattern deteriorates significantly as the frequency increases. Using measured permittivity values, however, the double-peak pattern is effectively maintained for most frequencies until 750 THz, at which the transmission fails. **Figure** pre-sents the average input/output graphics, confirming consistency between the LD model and measurement data at lower and higher frequencies. On the other hand, at some frequencies in the visible range, there is more than 30 dB difference between the predicted output levels.

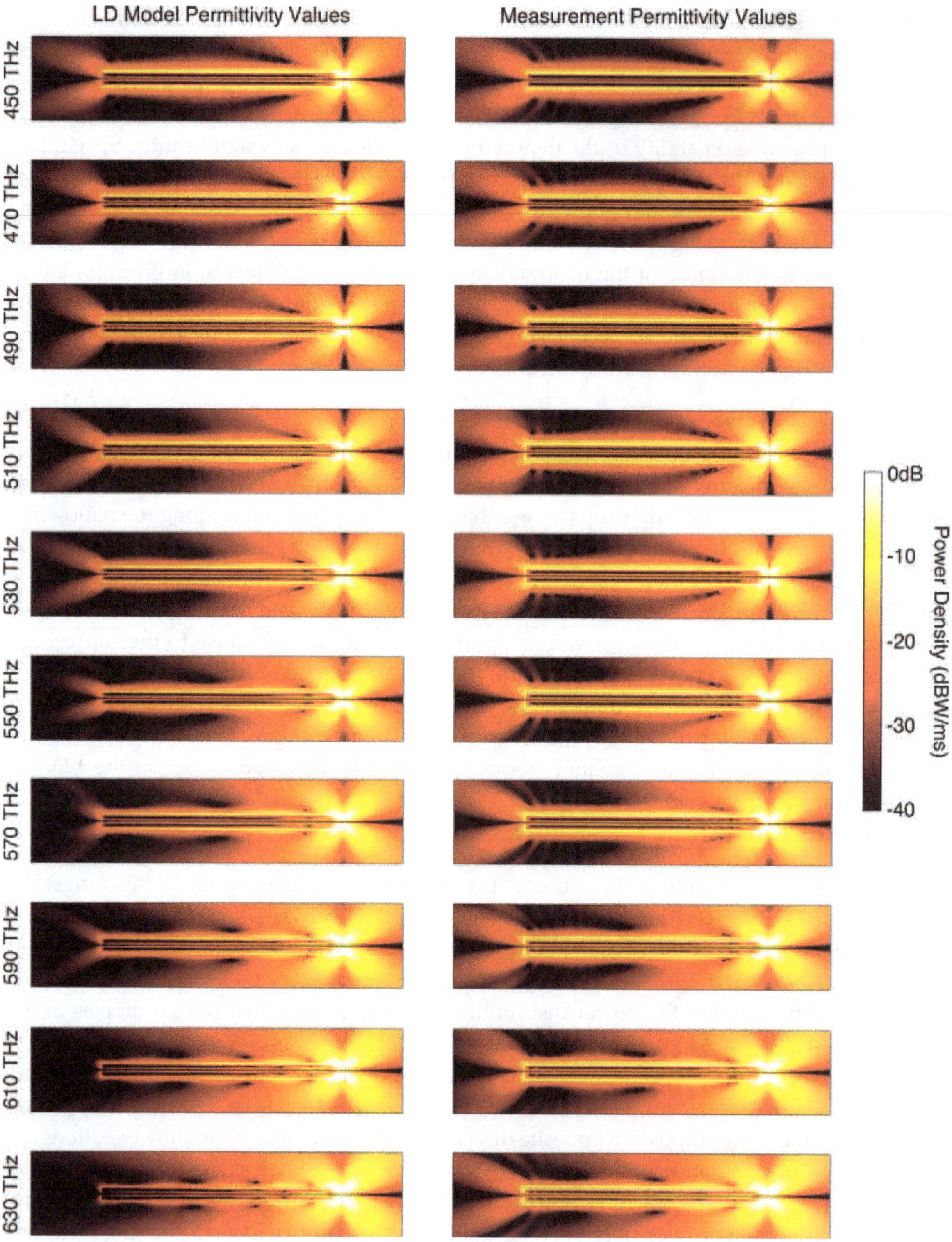

Figure. Power density in the vicinity of the nanowire system (**Figure**) from 450 to 630 THz. Numerical results obtained by using permittivity values derived from the LD model (left column) and those based on the measurement data (right column) are compared.

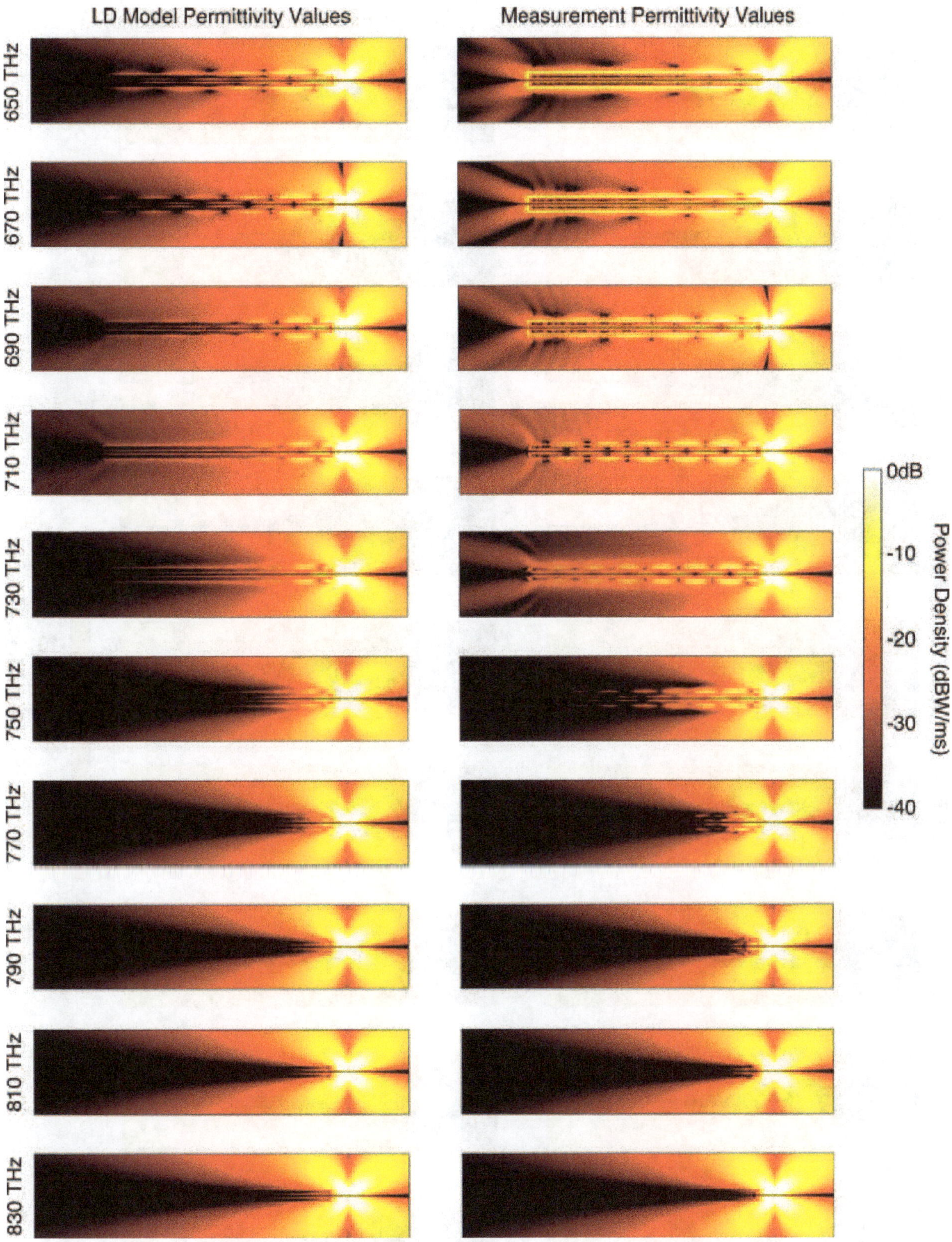

Figure. Power density in the vicinity of the nanowire system (**Figure**) from 650 to 830 THz. Numerical results obtained by using permittivity values derived from the LD model (left column) and those based on the measurement data (right column) are compared.

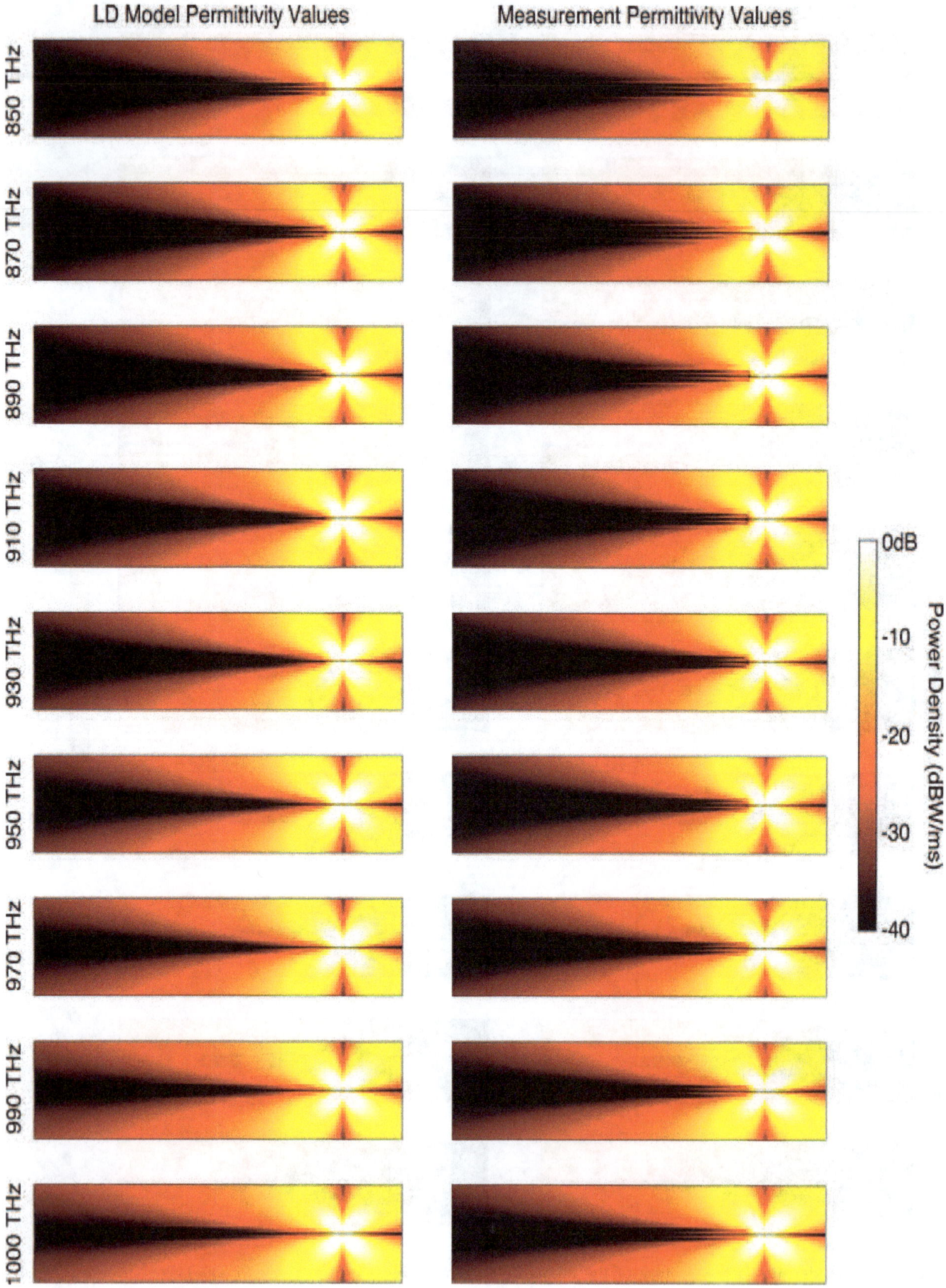

Figure. Power density in the vicinity of the nanowire system (**Figure**) from 850 to 1000 THz. Numerical results obtained by using permittivity values derived from the LD model (left column) and those based on the measurement data (right column) are compared.

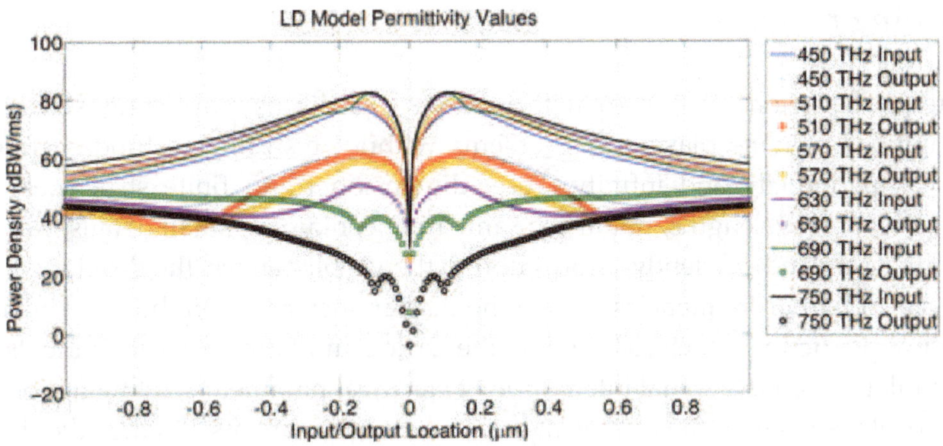

Figure. Power density on the input and output sides of the nanowire system (**Figure**) from 450 to 750 THz. For the input and output, the samples are selected on 2μm lines at 30 and 40 nm distances from the nanowires. LD model is used for the permittivity values.

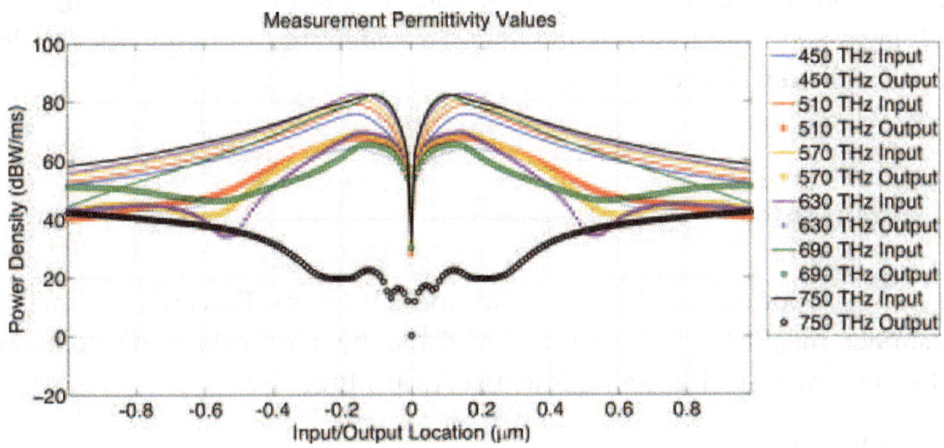

Figure. Power density on the input and output sides of the nanowire system (**Figure**) from 450 to 750 THz. For the input and output, the samples are selected on 2μm lines at 30 and 40 nm distances from the nanowires. Measurement data are used for the permittivity values.

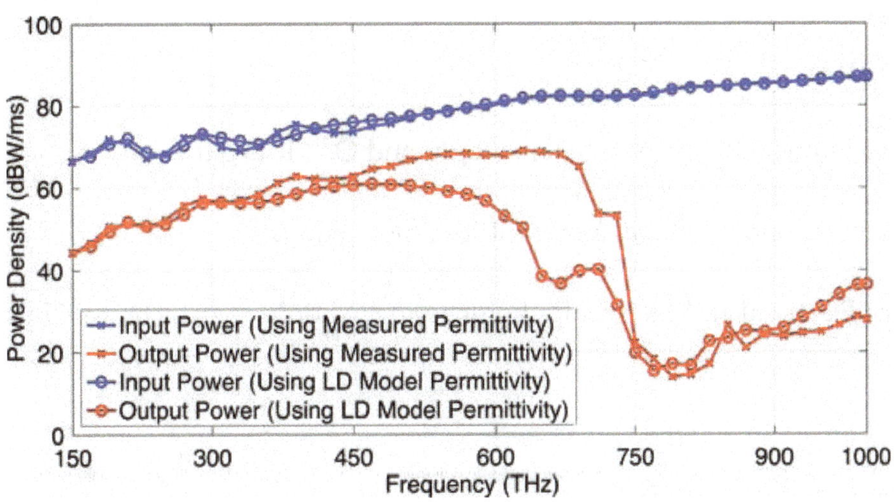

Figure. Average input and output power density values for the nanowire system (**Figure**) from 150 to 1000 THz. Despite the consistency of the inputs, significant discrepancies in the output values obtained when LD model and measurement data for the permittivity values are used from 450 to 750 THz.

6. Concluding remarks

Surface integral equations combined with iterative algorithms employing MLFMA provide accurate solutions of nano-plasmonic problems without resorting to fundamental assumptions, such as periodicity and infinity. Three-dimensional and finite structures, which are typically of tens of wavelengths, but at the same time containing small details, can be investigated both precisely and efficiently. In addition to the visible ranges, the developed solvers are very beneficial at higher frequencies, where the discrepancy between the experimental results and theoretical predictions, such as based on the Drude and Lorentz-Drude models, increases. Surface formulations enable trivial integration of electrical parameters, allowing for fast tuning of the numerical results with the increasingly precise measurements. On the other hand, such a reliable simulation environment can be constructed only with appropriate combinations of surface integral equations, discretizations, numerical integrations, fast algorithms, and iterative techniques, as shown in this chapter. We present how to construct such an implementation with all details from formulations to iterative solutions using MLFMA, along with a set of results involving a nanowire transmission line in a wide range of frequencies to demonstrate the capabilities of the developed solvers.

Acknowledgements

This work was supported by the Scientific and Technical Research Council of Turkey (TUBITAK) under the Research Grant 113E129 and by the Turkish Academy of Sciences (TUBA) in the framework of the Young Scientist Award Program.

Authors contributions

Ö.E. developed the theory and formulations, B.K. and Ö.E. implemented the solvers. A.Ç. conducted the numerical experiments.

Author details

Abdulkerim Çekinmez, Barışcan Karaosmanoğlu and Özgür Ergül*

*Address all correspondence to: ozgur.ergul@eee.metu.edu.tr

Department of Electrical and Electronics Engineering, Middle East Technical University, Ankara, Turkey

References

[1] Yao J, Liu Z, Liu Y, Wang Y, Sun C, Bartal G, Stacy AM, Zhang X. Optical negative refraction in bulk metamaterials of nanowires. Science. 2008;**321**:930.

[2] Liu Y, Bartal G, Zhang X. All-angle negative refraction and imaging in a bulk medium made of metallic nanowires in the visible region. Opt. Exp. 2008;**16**:15439–15448.

[3] Schuck PJ, Fromm DP, Sundaramurthy A, Kino GS, Moerney WE. Improving the mismatch between light and nanoscale objects with gold bowtie nanoantennas. Phys. Rev. Lett. 2005;**94**:017402–1-4.

[4] Alda J, Rico-Garcia JM, Lopez-Alonso JM, Boreman G. Optical antennas for nano-photonic applications. Nanotechnology. 2005;**16**:230–234.

[5] Muhlschlegel P, Eisler HJ, Martin OJF, Hecht B, Pohl DW. Resonant optical antennas. Science. 2005;**308**:1607–1609.

[6] Kinkhabwala A, Yu Z, Fan S, Avlasevich Y, Mullen K, Moerney WE. Large single-molecule fluorescence enhancements produced by a bowtie nanoantenna. Nat. Photonics. 2009;**3**:654–657.

[7] Kosako T, Kadoya Y, Hofmann HF. Directional control of light by a nano-optical Yagi-Uda antenna. Nat. Photonics. 2010;**4**:312–315.

[8] Solis DM, Taboada JM, Obelleiro F, Landesa L. Optimization of an optical wireless nanolink using directive nanoantennas. Opt. Exp. 2013;**21**:2369–2377.

[9] Karaosmanoğlu B, Gür UM, Ergül Ö. Investigation of nanoantennas using surface integral equations and the multilevel fast multipole algorithm. In Proceedings of the Progress in Electromagnetics Research Symposium (PIERS); 2015. p. 2026–2030.

[10] Johnson PB, Christy RW. Optical constants of the noble metals. Phys. Rev. B. 1972;**6**:4370–4379.

[11] Poggio AJ, Miller EK. Integral equation solutions of three-dimensional scattering problems. In Mittra R, editor. Computer Techniques for Electromagnetics. Oxford: Pergamon Press; 1973. Chap. 4.

[12] Ylä-Oijala P, Taskinen M, Järvenpää S. Surface integral equation formulations for solving electromagnetic scattering problems with iterative methods. Radio Sci. 2005;**40**:6002–1-19.

[13] Hohenester U, Krenn J. Surface plasmon resonances of single and coupled metallic nanoparticles: a boundary integral method approach. Phys. Rev. B. 2005;**72**:195429–1-9.

[14] Sondergaard T. Modeling of plasmonic nanostructures: Green's function integral equation methods. Phys. Status Solidi B. 2007;**244**:3448–3462.

[15] Kern AM, Martin OFJ. Surface integral formulation for 3D simulations of plasmonic and high permittivity nanostructures. J. Opt. Soc. Am. A. 2009;**26**:732–740.

[16] Gallinet B, Martin OFJ. Scattering on plasmonic nanostructures arrays modeled with a surface integral formulation. Photon. Nanostruct. Fund. Appl. 2010;**8**:278–284.

[17] Rodriguez-Oliveros R, Sanchez-Gil JA. Localized surface-plasmon resonances on single and coupled nanoparticles through surface integral equations for flexible surfaces. Opt. Exp. 2011;**16**:12208–12219.

[18] Araujo MG, Taboada JM, Solis DM, Rivero J, Landesa L, Obelleiro F. Comparison of surface integral equation formulations for electromagnetic analysis of plasmonic nanoscatterers. Opt. Exp. 2012;**20**:9161–9171.

[19] Ergül Ö. Analysis of composite nanoparticles with surface integral equations and the multilevel fast multipole algorithm. J. Opt. 2012;**14**:062701–1-4.

[20] Landesa L, Araujo MG, Taboada JM, Bote L, Obelleiro F. Improving condition number and convergence of the surface integral-equation method of moments for penetrable bodies. Opt. Exp. 2012;**20**:17237–17249.

[21] Araujo MG, Taboada JM, Rivero J, Solis DM, Obelleiro F. Solution of large-scale plasmonic problems with the multilevel fast multipole algorithm. Opt. Lett. 2012;**37**:416–418.

[22] Mäkitalo J, Kauranen M, Suuriniemi S. Modes and resonances of plasmonic scatterers. Phys. Rev. B. 2014;**89**:165429–1-11.

[23] Solis DM, Taboada JM, Rubinos-Lopez O, Obelleiro F. Improved combined tangential formulation for electromagnetic analysis of penetrable bodies. J. Opt. Soc. Am. B. 2015;**32**:1780–1787.

[24] Solis DM, Taboada JM, Obelleiro F. Surface integral equation method of moments with multiregion basis functions applied to plasmonics. IEEE Trans. Antennas Propag. 2015;**63**:2141–2152.

[25] Chew WC, Jin JM, Michielssen E, Song J. Fast and Efficient Algorithms in Computational Electromagnetics. Boston: Artech House; 2001.

[26] Ergül Ö, Gürel L. The Multilevel Fast Multipole Algorithm (MLFMA) for Solving Large-Scale Computational Electromagnetics Problems. Wiley-IEEE; Chichester, West Sussex, UK 2014.

[27] Ergül Ö, Gürel L. Comparison of integral-equation formulations for the fast and accurate solution of scattering problems involving dielectric objects with the multilevel fast multipole algorithm. IEEE Trans. Antennas Propag. 2009;**57**:176–187.

[28] Ergül Ö. Solutions of large-scale electromagnetics problems involving dielectric objects with the parallel multilevel fast multipole algorithm. J. Opt. Soc. Am. A. 2011;**28**:2261–2268.

[29] Yılmaz A, Karaosmanoğlu B, Ergül Ö. Computational electromagnetic analysis of deformed nanowires using the multilevel fast multipole algorithm. Sci. Rep. 2015;**5**: 8469–1-9.

[30] Karaosmanoğlu B, Yılmaz A, Gür UM, Ergül Ö. Solutions of plasmonic structures using the multilevel fast multipole algorithm. Int. J. RF Microwave Comput. Aided. Eng. 2016;**26**:335–341.

[31] Gomez-Sousa H, Rubinos-Lopez O, Martinez-Lorenzo JA. Comparison of iterative solvers for electromagnetic analysis of plasmonic nanostructures using multiple surface integral equation formulations. J. Electromagn. Waves Appl. 2016;**30**:456–472.

[32] Karaosmanoğlu B, Yılmaz A, Ergül Ö. On the accuracy and efficiency of surface formulations in fast analysis of plasmonic structures via MLFMA. In Proceedings of the Progress in Electromagnetics Research Symposium (PIERS); 2016. p. 2629–2633.

[33] Rao SM, Wilton DR, Glisson AW. Electromagnetic scattering by surfaces of arbitrary shape. IEEE Trans. Antennas Propag. 1982;**30**:409–418.

[34] Medgyesi-Mitschang LN, Putnam JM, Gedera MB. Generalized method of moments for three-dimensional penetrable scatterers. J. Opt. Soc. Am. A. 1994;**11**:1383–1398.

[35] Ylä-Oijala P, Taskinen M. Application of combined field integral equation for electromagnetic scattering by dielectric and composite objects. IEEE Trans. Antennas Propag. 2005;**53**:1168–1173.

[36] Ergül Ö, Gürel L. Discretization error due to the identity operator in surface integral equations. Comput. Phys. Comm. 2009;**180**:1746–1752.

[37] Cools K, Bogaert I, Peeters J, Vande Ginste D, Rogier H, De Zutter D. Accurate and efficient algorithms for boundary element methods in electromagnetic scattering: a tribute to the work of F. Olyslager. Radio Sci. 2011;**46**:RS0E21-1-10.

[38] Ergül Ö, Karaosmanoğlu B. Approximate stable diagonalization of the Green's function for low frequencies. IEEE Antennas Wireless Propag. Lett. 2014;**13**:1054–1056.

[39] Ergül Ö, Karaosmanoğlu B. Broadband multilevel fast multipole algorithm based on an approximate diagonalization of the Green's function. IEEE Trans. Antennas Propag. 2015;**63**:3035–3041.

[40] Jiang LJ, Chew WC. A mixed-form fast multipole algorithm. IEEE Trans. Antennas Propag. 2005;**53**:4145–4156.

[41] Bogaert I, Peeters J, Olyslager F. A nondirective plane wave MLFMA stable at low frequencies. IEEE Trans. Antennas Propag. 2008;**56**:3752–3767.

[42] Ohnuki S, Chew WC. Truncation error analysis of multipole expansions. SIAM J. Sci. Comput. 2003;**25**:1293–1306.

Permissions

All chapters in this book were first published in DSACT, by InTech Open; hereby published with permission under the Creative Commons Attribution License or equivalent. Every chapter published in this book has been scrutinized by our experts. Their significance has been extensively debated. The topics covered herein carry significant findings which will fuel the growth of the discipline. They may even be implemented as practical applications or may be referred to as a beginning point for another development.

The contributors of this book come from diverse backgrounds, making this book a truly international effort. This book will bring forth new frontiers with its revolutionizing research information and detailed analysis of the nascent developments around the world.

We would like to thank all the contributing authors for lending their expertise to make the book truly unique. They have played a crucial role in the development of this book. Without their invaluable contributions this book wouldn't have been possible. They have made vital efforts to compile up to date information on the varied aspects of this subject to make this book a valuable addition to the collection of many professionals and students.

This book was conceptualized with the vision of imparting up-to-date information and advanced data in this field. To ensure the same, a matchless editorial board was set up. Every individual on the board went through rigorous rounds of assessment to prove their worth. After which they invested a large part of their time researching and compiling the most relevant data for our readers.

The editorial board has been involved in producing this book since its inception. They have spent rigorous hours researching and exploring the diverse topics which have resulted in the successful publishing of this book. They have passed on their knowledge of decades through this book. To expedite this challenging task, the publisher supported the team at every step. A small team of assistant editors was also appointed to further simplify the editing procedure and attain best results for the readers.

Apart from the editorial board, the designing team has also invested a significant amount of their time in understanding the subject and creating the most relevant covers. They scrutinized every image to scout for the most suitable representation of the subject and create an appropriate cover for the book.

The publishing team has been an ardent support to the editorial, designing and production team. Their endless efforts to recruit the best for this project, has resulted in the accomplishment of this book. They are a veteran in the field of academics and their pool of knowledge is as vast as their experience in printing. Their expertise and guidance has proved useful at every step. Their uncompromising quality standards have made this book an exceptional effort. Their encouragement from time to time has been an inspiration for everyone.

The publisher and the editorial board hope that this book will prove to be a valuable piece of knowledge for researchers, students, practitioners and scholars across the globe.

List of Contributors

Sergei Soldatenko and Rafael Yusupov
St. Petersburg Institute for Informatics and Automation of the Russian Academy of Sciences, St. Petersburg, Russia

Guillermo Fernadez-Anaya, Luis Alberto Quezada-Téllez, Jorge Antonio López-Rentería and Guillermo Manuel Mallen-Fullerton
Departamento de Física y Matemáticas, Universidad Iberoamericana, Ciudad de México, México

Oscar A. Rosas-Jaimes
Facultad de Ingeniería, Universidad Autónoma del Estado de México, Toluca, Estado de, México

Rodrigo Muñoz-Vega
Universidad Autónoma de la Ciudad de México, Ciudad de México, México

José Job Flores-Godoy
Departamento de Matemática, Facultad de Ingeniería y Tecnologías, Universidad Católica del Uruguay, Uruguay

Gabino Torres-Vega
Physics Department, Cinvestav, México

Said R. Grace
Department of Engineering Mathematics, Faculty of Engineering, Cairo University, Giza, Egypt

Irena Jadlovská
Department of Mathematics and Theoretical Informatics, Faculty of Electrical Engineering and Informatics, Technical University of Košice, Košice, Slovakia

Francesco Aldo Costabile, Maria Italia Gualtieri and Anna Napoli
Department of Mathematics and Informatics, University of Calabria, Rende (Cs), Italy

Elvan Akın and Özkan Öztürk
Missouri University of Science and Technology, Missouri, USA

Dušan Krokavec and Anna Filasová
Department of Cybernetics Artificial Intelligence, Faculty of Electrical Engineering Informatics, Technical University of Košice, Košice, Slovakia

Cheng Yung Ming
Department of Civil and Environmental Engineering, Hong Kong Polytechnic University, Hong Kong SAR, China

Abdulkerim Çekinmez, Barişcan Karaosmanoğlu and Özgür Ergül
Department of Electrical and Electronics Engineering, Middle East Technical University, Ankara, Turkey

Index

www.ingramcontent.com/pod-product-compliance
Lightning Source LLC
LaVergne TN
LVHW060604210926
848485LV00005B/1365